CLASSICS OF
INDIAN MATHEMATICS

Algebra, with Arithmetic and Mensuration, From the Sanskrit of Brahmagupta and Bhāṣkara

Translated by

Henry Thomas Colebrooke Esq.

F.R.S.; M.LINN. AND GEOL. SOC. AND
R. INST. LONDON; AS. SOC. BENGAL;
AC.SC. MUNICH

Foreword

S.R. Sarma

Sharada Publishing House
Delhi-110035
2005

First Published : 1817
Reprinted in 2005

ISBN : 81-88934-24-0

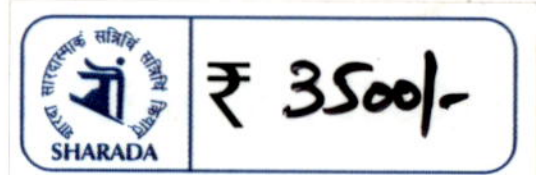

Published by :
B.L. Bansal
SHARADA PUBLISHING HOUSE
(Publishers on India's Past and Present)
2094/165, Ganeshpura, Tri Nagar
Delhi-110035, (India)
Ph. : 011-30921560, 27552102
Email : bansal_sph @ indiatimes.com

Printed at :
Salasar Imaging Systems, Delhi-110035

Foreword

It is indeed gratifying that Colebrooke's celebrated work, *Algebra, with Arithmetic and Mensuration, from the Sanskrit of Brahmagupta and Bhāskara,* is being reprinted. It was one of the earliest fruits of the European encounter with the scientific heritage of India. While many of the other books of this period became dated and are now merely historical curiosities, Colebrooke's work remains as useful today as when it appeared first in 1817.

Henry Thomas Colebrooke was born in London on 15 June 1765, as the third son of Sir George Colebrooke who was a member of the British Parliament and a director of the East India Company. Because of his father's association with the East India Company, a career in India was envisaged for the young Colebrooke. He was educated privately at home in classical languages, French, German and mathematics and was packed off, when he was barely seventeen, to India to join the service of the Company as "writer" at Calcutta in 1782. Four years later, he was appointed Assistant Collector at Tirhut. He began his study of oriental languages, especially Sanskrit. In 1789 he was sent to Purnia. Just about this time, in 1784, William Jones founded the Asiatic Society at Calcutta for "inquiring into the history and antiquities, the arts, sciences, and literature of Asia." Colebrooke was attracted by the activities of this newly founded Society, and decided to pursue the study of Sanskrit legal texts in order to gain a better knowledge of Hindu law—a knowledge that was essential for his professional career as well. His posting in 1795 as magistrate at Mirzapur, nor far from Benares, gave him the welcome opportunity to widen his knowledge of Sanskrit literature through the acquaintance of the pandits in the Benares Sanskrit College, which was founded by Jonathan Duncan in the year 1791. After an infructuous mission to the court of Nagpur from 1799 to 1801, he was appointed judge in the Calcutta Court of Appeal in 1801. The next fourteen thirteen years, which he spent at Calcutta, were his most productive period, both professionally and academically. In 1805 he was elevated to the position of the President of the Bench in the Court of Appeal and was concurrently appointed Professor of Hindu Law and Sanskrit at the College of Fort William. In 1807 he was made a member of the Governor-General's Council. He also took active part in the transactions of the Asiatic Society, first as a member and subsequently as its president, and contributed well-researched articles regularly to the Society's journal *Asiatick Researches.* He returned to England in 1814.

During his 32 years' stay in India, Colebrooke collected a large number of Sanskrit manuscripts. Several of these were copied especially for him. After his return to England, he donated this valuable collection of Indian manuscripts to the India Office. In 1823, he founded the Royal Asiatic Society in London, on the lines of the Asiatic Society of Calcutta. In 1824 he was elected President of the Astronomical Society. In subsequent years he was elected foreign member of the French Institute of Paris and of the Imperial Academy of St. Petersburg. He died on 10 March 1837. The articles he contributed to the various volumes of the *Asiatick Researches* and the prefaces he wrote to some of his publications were collected in two volumes and published under the title *Miscellaneous Essays* in the same year.

William Jones and Henry Thomas Colebrooke, mutually complementary personalities, can be said to be the pioneers of Indology. While Jones introduced Europe to the aesthetic qualities of Sanskrit *kāvya* through his English translation of Kālidāsa's *śakuntalā,* Colebrooke drew attention to the wealth of systematized knowledge in the *Śāstra* literature in Sanskrit. Colebrooke was the first European to master several branches of Sanskrit learning—both original texts and commentaries—and his writings were pioneering efforts in every one of these branches. His essay "On the Vedas or Sacred Writings of Hindus" (*Asiatick Researches,* vol. 8, 1805, pp. 369-476) was the first authentic account in a European language of the vast Vedic literature, and can be read even today with profit. Likewise his "Observations on the Sect of Jains," (*Asiatick Researches,* vol. 9, 1806, pp. 287-322; reprinted in *Miscellaneous Essays,* vol. 2, pp. 191-224) provides detailed information about the Jaina religious principles and practices. His *Grammar of Sanskrit Language* (Calcutta 1805) foreshadowed comparative linguistics. He was also the first to give a systematic account of the Sanskrit and Prakrit metres, with copious illustrations, and a list of metres arranged with mathematical precision ("On Sanskrit and Prakrit Poetry," *Asiatick Researches,* vol. 10, 1807, pp. 389-474; reprinted in *Miscellaneous Essays,* vol. 2, pp. 62-165). He edited original texts like the *Hitopadeśa* (Calcutta 1804) and *Amarakośa* (Serampore 1808) and several Sanskrit inscriptions.

Two areas received his special attention : Hindu law and exact sciences. His *Digest of Hindu Laws* (Calcutta 1798) was highly acclaimed; so also his *Two Treatises of Hindu Law of Inheritance* (Calcutta 1810), which contains his translation of Vijñāneśvara's *Mitākṣarā* and Jīmūtavāhana's *Dāyabhāga*. These two works were used extensively in the law courts.

Colebrooke's various writings on Indian mathematics and astronomy laid the foundations for the study of the history of science in India. Some of his notable contributions are "On Indian Weights and Measures," (*Asiatick Researches,* vol. 5, 1797, pp. 91-109); "On the Indian and Arabian Division of the Zodiac," (*Asiatick Researches,* vol. 9, 1807, pp. 323-376; reprinted in *Miscellaneous Essays,* vol. 2, pp. 321-373); and "On the Notion of the Hindu Astronomers Concerning the Precession of the Equinoxes and Motions of the Planets," (*Asiatick Researches,* vol. 12, 1816, pp. 209-250; reprinted in *Miscellaneous Essays,* vol. 2, pp. 374-416).

The most important of these writings is, without any doubt, the *Algebra, with Arithmetic and Mensuration, from the Sanscrit of Brahmegupta and Bhāscara,* which was published in 1817 by John Murray at London. This work contains English translations of two classics of Indian mathematics, namely, Bhāskara's *Līlāvāti* and *Bījagaṇita*; and of the twelfth and eighteenth chapters of Brahmagupta's *Brāhmasphuṭasiddhānta.* Colebrooke's main purpose was to elucidate the state of algebra as reflected in the writings of Bhāskara. But since Bhāskara himself mentions Brahmagupta as one of his sources, Colebrooke thought it necessary to append the translation of two relevant chapters from Brahmagupta's *Brāhmasphuṭasiddhānta*; the twelfth and eighteenth which deal with arithmetic and algebra respectively. These translations are enriched by copious extracts from various commentaries: by Gaṅgādhara, Śūryadāsa, Gaṇeśa and Rāmakṛṣṇa on the *Līlāvatī;* by Kṛṣṇa Daivajña and Rāmakṛṣṇa on the *Bījagaṇita* Moreover, he also made use of the Persian translations of these mathematical treatises, viz. Abū al-Faiḍ Faiḍī's translation of the *Līlāvatī,* which he wrote at the instance of Akbar in 1587, and 'Ata' Allah Rushdī's translation of the *Bījagaṇita,* which was prepared at the court of Shah Jahan in 1634.

These four translations are preceded by a preface entitled "Dissertation (on the Algebra of the Hindus)," where Colebrooke seeks to situate Indian Algebra in the context of developments in other parts of the world. This is followed by fourteen highly informative appendices dealing, among others, with the dates of various Sanskrit mathematical texts and their authors, knowledge of Algebra among the Arabs and Italians, exchanges between India and the west in astrology and astronomy, and so on.

It should be remembered that none of these texts or commentaries were available in print at the beginning of the nineteenth century, nor were there any manuscript libraries with catalogued collections. Colebrooke's entire work is based on manuscripts which had to be located in the private collections of individual pandits and which had to be copied afresh for Colebrooke's use.

Colebrooke's "Dissertation" commences with the following words : "The history of sciences, if it want the prepossessing attractions of political history and narration of events, is nevertheless not wholly devoid of interest and instruction. A laudable curiosity prompts to inquire the sources of knowledge; and a review of its progress furnishes suggestions tending to promote the same or some kindred study. We would know the people and the names at least of individuals, to whom we owe particular discoveries and successive steps in the advancement of knowledge. If no more be obtained by the research, still the inquiry has not been wasted, which points aright the gratitude of mankind." Our gratitude is due to Colebrooke for illuminating an important part of India's scientific heritage.

Sharada Publishing House deserves our felicitations for bringing out the first Indian reprint of this unique document. Colebrooke's orthography of Sanskrit words is, of course, outdated but not his translation of the mathematical texts. It is hoped that this facsimile edition will arouse fresh interest in the history of mathematics in India.

S.R. Sarma

(Formerly Professor of Sanskrit,
Aligarh Muslim University)

Contents

ALGEBRA (*Vīja-gaṇita*)

BRAHMEGUPTA

DISSERTATION.

THE history of sciences, if it want the prepossessing attractions of political history and narration of events, is nevertheless not wholly devoid of interest and instruction. A laudable curiosity prompts to inquire the sources of knowledge; and a review of its progress furnishes suggestions tending to promote the same or some kindred study. We would know the people and the names at least of the individuals, to whom we owe particular discoveries and successive steps in the advancement of knowledge. If no more be obtained by the research, still the inquiry has not been wasted, which points aright the gratitude of mankind.

In the history of mathematical science, it has long been a question to whom the invention of Algebraic analysis is due? among what people, in what region, was it devised? by whom was it cultivated and promoted? or by whose labours was it reduced to form and system? and finally from what quarter did the diffusion of its knowledge proceed? No doubt indeed is entertained of the source from which it was received immediately by modern Europe; though the channel have been a matter of question. We are well assured, that the Arabs were mediately or immediately our instructors in this study. But the Arabs themselves scarcely pretend to the discovery of Algebra. They were not in general inventors but scholars, during the short period of their successful culture of the sciences: and the germ at least of the Algebraic analysis is to be found among the Greeks in an age not precisely determined, but more than probably anterior to the earliest dawn of civilization among the Arabs: and this science in a more advanced state subsisted among the Hindus prior to the earliest disclosure of it by the Arabians to modern Europe.

The object of the present publication is to exhibit the science in the state in which the Hindus possessed it, by an exact version of the most approved treatise on it in the ancient language of India, with one of the earlier treatises (the only extant one) from which it was compiled. The design of this preliminary dissertation is to deduce from these and from the evidence which will be here offered, the degree of advancement to which the science had arrived in a remote age. Observations will be added, tending to a comparison of the Indian, with the Arabian, the Grecian, and the modern Algebra: and the subject will be left to the consideration of the learned, for a conclusion to be drawn by them from the internal, no less than the external proof, on the question who can best vindicate a claim to the merit of having originally invented or first improved the methods of computation and analysis, which are the groundwork of both the simple and abstruser parts of Mathematics; that is, Arithmetic and Algebra: so far at least as the ancient inventions are affected; and also in particular points, where recent discoveries are concerned.

In the actual advanced condition of the analytic art, it is not hoped, that this version of ancient *Sanscrit* treatises on Algebra, Arithmetic, and Mensuration, will add to the resources of the art, and throw new light on mathematical science, in any other respect, than as concerns its history. Yet the remark may not seem inapposite, that had an earlier version of these treatises been completed, had they been translated and given to the public, when the notice of mathematicians was first drawn to the attainments of the Hindus in astronomy and in sciences connected with it, some addition would have been then made to the means and resources of Algebra for the general solution of problems by methods which have been re-invented, or have been perfected, in the last age.

The treatises in question, which occupy the present volume, are the *Víja-gańita* and *Lílávatí* of BHÁSCARA ÁCHÁRYA and the *Gańitád'haya* and *Cuttacád'hyaya* of BRAHMEGUPTA. The two first mentioned constitute the preliminary portion of BHÁSCARA's Course of Astronomy, entitled *Sidd'hánta-śirómańi*. The two last are the twelfth and eighteenth chapters of a similar course of astronomy, by BRAHMEGUPTA, entitled *Brahma-sidd'hánta*.

The questions to be first examined in relation to these works are their authenticity and their age. To the consideration of those points we now proceed.

The period when BHÁSCARA, the latest of the authors now named, flourished, and the time when he wrote, are ascertained with unusual precision.

He completed his great work, the *Sidd'hánta-sirómańi*, as he himself informs us in a passage of it,[1] in the year 1072 *Saca*. This information receives corroboration, if any be wanted, from the date of another of his works, the *Carańa-cutuhala*, a practical astronomical treatise, the epoch of which is 1105 *Saca*;[2] 33 years subsequent to the completion of the systematic treatise. The date of the *Sidd'hánta-śirómańi*, of which the *Víja-gańita* and *Lílávatí* are parts, is fixt then with the utmost exactness, on the most satisfactory grounds, at the middle of the twelfth century of the Christian era, A.D. 1150.[3]

The genuineness of the text is established with no less certainty by numerous commentaries in *Sanscrit*, besides a Persian version of it. Those commentaries comprise a perpetual gloss, in which every passage of the original is noticed and interpreted: and every word of it is repeated and explained. A comparison of them authenticates the text where they agree; and would serve, where they did not, to detect any alterations of it that might have taken place, or variations, if any had crept in, subsequent to the composition of the earliest of them. A careful collation of several commentaries,[4] and of three copies of the original work, has been made; and it will be seen in the notes to the translation how unimportant are the discrepancies.

From comparison and collation, it appears then, that the work of BHÁSCARA, exhibiting the same uniform text, which the modern transcripts of it do, was in the hands of both Mahommedans and Hindus between two and three centuries ago: and, numerous copies of it having been diffused throughout India, at an earlier period, as of a performance held in high estimation, it was the subject of study and habitual reference in countries and places so remote from each other as the north and west of India and the southern peninsula: or, to speak with the utmost precision, *Jambusara* in the west, *Agra* in North *Hindustan*, and *Párthapúra*, *Gólagráma*, *Amarávatí*, and *Nandigráma*, in the south.

[1] *Góládhyáya*; or lecture on the sphere. c. 11. § 56. As. Res. vol. 12. p. 214.

[2] As. Res. *ibid.*

[3] Though the matter be introductory, the preliminary treatises on arithmetic and algebra may have been added subsequently, as is hinted by one of the commentators of the astronomical part. (*Vártic.*) The order there intimated places them after the computation of planets, but before the treatise on spherics; which contains the date.

[4] Note A.

This, though not marking any extraordinary antiquity, nor approaching to that of the author himself, was a material point to be determined: as there will be in the sequel occasion to show, that modes of analysis, and, in particular, general methods for the solution of indeterminate problems both of the first and second degrees, are taught in the *Vija-gańita,* and those for the first degree repeated in the *Lílávatí,* which were unknown to the mathematicians of the west until invented anew in the last two centuries by algebraists of France and England. It will be also shown, that BHÁSCARA, who himself flourished more than six hundred and fifty years ago, was in this respect a compiler, and took those methods from Indian authors as much more ancient than himself.

That BHÁSCARA's text (meaning the metrical rules and examples, apart from the interspersed gloss;) had continued unaltered from the period of the compilation of his work until the age of the commentaries now current, is apparent from the care with which they have noticed its various readings, and the little actual importance of these variations; joined to the consideration, that earlier commentaries, including the author's own explanatory annotations of his text, were extant, and lay before them for consultation and reference. Those earlier commentaries are occasionally cited by name: particularly the *Gańita-caumudí,* which is repeatedly quoted by more than one of the scholiasts.[1]

No doubt then can be reasonably entertained, that we now possess the arithmetic and algebra of BHÁSCARA, as composed and published by him in the middle of the twelfth century of the Christian era. The age of his precursors cannot be determined with equal precision. Let us proceed, however, to examine the evidence, such as we can at present collect, of their antiquity.

Towards the close of his treatise on Algebra,[2] BHÁSCARA informs us, that it is compiled and abridged from the more diffuse works on the same subject, bearing the names of BRAHME, (meaning no doubt BRAHMEGUPTA,) ŚRÍD'HARA and PADMANÁBHA; and in the body of his treatise, he has cited a passage of ŚRÍD'HARA's algebra,[3] and another of PADMANÁBHA's.[4] He repeatedly adverts to preceding writers, and refers to them in general terms,

[1] For example, by SÚRYADÁSA, under *Lílávatí,* § 74; and still more frequently by RANGANÁT'HA.

[2] *Vija-gańita,* § 218.

[3] Ibid. § 131.

[4] Ibid. § 142.

where his commentators understand him to allude to ÁRYA-BHAṬṬA, to BRAHMEGUPTA, to the latter's scholiast CHATURVEDA PRIT'HÚDACA SWÁMÍ,[1] and to the other writers above mentioned.

Most, if not all, of the treatises, to which he thus alludes, must have been extant, and in the hands of his commentators, when they wrote; as appears from their quotations of them; more especially those of BRAHMEGUPTA and ARYA-BHAṬṬA, who are cited, and particularly the first mentioned, in several instances.[2] A long and diligent research in various parts of India, has, however, failed of recovering any part of the *Padmanábha víja*, (or Algebra of PADMANÁBHA,) and of the Algebraic and other works of ÁRYA-BHAṬṬA.[3] But the translator has been more fortunate in regard to the works of ŚRÍD'HARA and BRAHMEGUPTA, having in his collection ŚRÍD'HARA's compendium of arithmetic, and a copy, incomplete however, of the text and scholia of BRAHMEGUPTA's *Brahma-sidd'hánta*, comprising among other no less interesting matter, a chapter treating of arithmetic and mensuration; and another, the subject of which is algebra: both of them fortunately complete.[4]

The commentary is a perpetual one; successively quoting at length each verse of the text; proceeding to the interpretation of it, word by word; and subjoining elucidations and remarks: and its colophon, at the close of each chapter, gives the title of the work and name of the author.[5] Now the name, which is there given, CHATURVÉDA PRIT'HÚDACA SWÁMÍ, is that of a celebrated scholiast of BRAHMEGUPTA, frequently cited as such by the commentators of BHÁSCARA and by other astronomical writers: and the title of the work, *Bráhma-sidd'hánta*, or sometimes *Bráhma sphuṭa-sidd'hánta*, corresponds, in the shorter form, to the known title of BRAHMEGUPTA's treatise in the usual references to it by BHÁSCARA's commentators;[6] and answers, in the longer form, to the designation of it, as indicated in an introductory couplet which is quoted from BRAHMEGUPTA by LACSHMÍDÁSA, a scholiast of BHÁSCARA.[7]

Remarking this coincidence, the translator proceeded to collate, with the

[1] *Vij.-gan.* Ch. 5. note of SÚRYADASA. Also *Vij.-gań.* § 174; and *Líl.* § 246 ad finem.

[2] For example, under *Líl.* Ch. 11.

[3] Note G.

[4] Note B.

[5] *Vásaná-bháshya* by CHATURVÉDA PRIT'HÚDACA SWÁMÍ, son of MAD'HUSÚDANA, on the *Brahma-sidd'hánta*; (or sometimes *Brahma-sphuṭa-sidd'hánta*.)

[6] They often quote from the *Bráhma-sidd'hánta* after premising a reference to BRAHMEGUPTA.

[7] Note C.

text and commentary, numerous quotations from both, which he found in BHÁSCARA's writings or in those of his expositors. The result confirmed the indication, and established the identity of both text and scholia as BRAHMEGUPTA's treatise, and the gloss of PRIT'HÚDACA. The authenticity of this *Brahma-sidd'hánta* is further confirmed by numerous quotations in the commentary of BHAṬṬÓTPALA on the *sanhitá* of VARÁHA MIHIRA: as the quotations from the *Brahma-sidd'hánta* in that commentary, (which is the work of an author who flourished eight hundred and fifty years ago,) are verified in the copy under consideration. A few instances of both will suffice; and cannot fail to produce conviction.[1]

It is confidently concluded, that the Chapters on Arithmetic and Algebra, fortunately entire in a copy, in many parts imperfect, of BRAHMEGUPTA's celebrated work, as here described, are genuine and authentic. It remains to investigate the age of the author.

Mr. DAVIS, who first opened to the public a correct view of the astronomical computations of the Hindus,[2] is of opinion, that BRAHMEGUPTA lived in the 7th century of the Christian era.[3] Dr. WILLIAM HUNTER, who resided for some time with a British Embassy at *Ujjayaní*, and made diligent researches into the remains of Indian science, at that ancient seat of Hindu astronomical knowledge, was there furnished by the learned astronomers whom he consulted, with the ages of the principal ancient authorities. They assigned to BRAHMEGUPTA the date of 550 *Śaca;* which answers to A. D. 628. The grounds, on which they proceeded, are unfortunately not specified: but, as they gave BHÁSCARA's age correctly, as well as several other dates right, which admit of being verified; it is presumed, that they had grounds, though unexplained, for the information which they communicated.[4]

Mr. BENTLEY, who is little disposed to favour the antiquity of an Indian astronomer, has given his reasons for considering the astronomical system which BRAHMEGUPTA teaches, to be between twelve and thirteen hundred years old ($1263\frac{2}{3}$ years in A. D. 1799).[5] Now, as the system taught by this author is professedly one corrected and adapted by him to conform with the observed positions of the celestial objects when he wrote,[6] the age, when their positions would be conformable with the results of computations made as by him directed, is precisely the age of the author himself: and so far as

[1] Note D.
[2] As. Res. 2. 225.
[3] Ibid. 9. 242.
[4] Note E.
[5] As. Res. 6. 586.
[6] Supra.

Mr. BENTLEY's calculations may be considered to approximate to the truth, the date of BRAHMEGUPTA's performance is determined with like approach to exactness, within a certain latitude however of uncertainty for allowance to be made on account of the inaccuracy of Hindu observations.

The translator has assigned on former occasions[1] the grounds upon which he sees reason to place the author's age, soon after the period, when the vernal equinox coincided with the beginning of the lunar mansion and zodiacal asterism *Aświní*, where the Hindu ecliptic now commences. He is supported in it by the sentiments of BHÁSCARA and other Indian astronomers, who infer from BRAHMEGUPTA's doctrine concerning the solstitial points, of which he does not admit a periodical motion, that he lived when the equinoxes did not, sensibly to him, deviate from the beginning of *Aświní* and middle of *Chitrá* on the Hindu sphere.[2] On these grounds it is maintained, that BRAHMEGUPTA is rightly placed in the sixth or beginning of the seventh century of the Christian era; as the subjoined calculations will more particularly show.[3] The age when BRAHMEGUPTA flourished, seems then, from the concurrence of all these arguments, to be satisfactorily settled as antecedent to the earliest dawn of the culture of sciences among the Arabs; and consequently establishes the fact, that the Hindus were in possession of algebra before it was known to the Arabians.

BRAHMEGUPTA's treatise, however, is not the earliest work known to have been written on the same subject by an Indian author. The most eminent scholiast of BHÁSCARA[4] quotes a passage of ÁRYA-BHAŤŤA specifying algebra under the designation of *Víja*, and making separate mention of *Cuttaca*, which more particularly intends a problem subservient to the general method of resolution of indeterminate problems of the first degree: he is understood by another of BHÁSCARA's commentators[5] to be at the head of the elder writers, to whom the text then under consideration adverts, as having designated by the name of *Mad'hyamáharańa* the resolution of affected quadratic equations by means of the completion of the square. It is to be presumed, therefore, that the treatise of ARYA-BHAŤŤA then extant, did extend to quadratic equations in the determinate analysis; and to indeterminate problems of the first degree; if not to those of the second likewise, as most probably it did.

This ancient astronomer and algebraist was anterior to both VARÁHA-MIHIRA

[1] As. Res. 9. 329. [2] Ibid. 12. p. 215. [3] Note F.

[4] GAŃÉŚA, a distiguished mathematician and astronomer. [5] SÚR. on *Víj.-gań.* § 128.

and BRAHMEGUPTA; being repeatedly named by the latter; and the determination of the age when he flourished is particularly interesting, as his astronomical system, though on some points agreeing, essentially disagreed on others, with that which those authors have followed, and which the Hindu astronomers still maintain.[1]

He is considered by the commentators of the *Súrya sidd'hánta* and *Śirómańi,*[2] as the earliest of uninspired and mere human writers on the science of astronomy; as having introduced requisite corrections into the system of PARÁŚARA, from whom he took the numbers for the planetary mean motions; as having been followed in the tract of emendation, after a sufficient interval to make further correction requisite, by DURGASINHA and MIHIRA; who were again succeeded after a further interval by BRAHMEGUPTA son of JISHNU.[3]

In short, ÁRYA-BHAŤŤA was founder of one of the sects of Indian astronomers, as PULÍŚA, an author likewise anterior to both VARÁHAMIHIRA and BRAHMEGUPTA, was of another: which were distinguished by names derived from the discriminative tenets respecting the commencement of planetary motions at sun-rise according to the first, but at midnight according to the latter;[4] on the meridian of *Lancá*, at the beginning of the great astronomical cycle. A third sect began the astronomical day, as well as the great period, at noon.

His name accompanied the intimation which the Arab astronomers (under the Abbasside Khalifs, as it would appear,) received, that three distinct astronomical systems were current among the Hindus of those days: and it is but slightly corrupted, certainly not at all disguised, in the Arabic representation of it *Arjabahar*, or rather *Arjabhar.*[5] The two other systems were, first, BRAHMEGUPTA's *Sidd'hánta*, which was the one they became best acquainted with, and to which they apply the denomination of the *sind-hind;* and second, that

[1] Note G. [2] *Nrisinha* on *Súr.* GAŃÉŚA pref. to *Grah. lágh.*

[3] As. Res. 2. 235, 242, and 244; and Note H.

[4] BRAHMEGUPTA, ch. 11. The names are *Audayaca* from *Udaya* rising; and *Árdharátrica* from *Ardharátri*, midnight. The third school is noticed by BHAŤŤÓTPALA the scholiast of VARÁHA MIHIRA, under the denomination of *Mádhyandinas,* as alleging the commencement of the astronomical period at noon: (from *Madhyandina*, mid-day.)

[5] The *Sanscrit ť*, it is to be remembered, is the character of a peculiar sound often mistaken for *r*, and which the Arabs were likely so to write, rather than with a *te* or with a *tau.* The *Hindi ť* is generally written by the English in India with an *r.* Example: *Ber (vaťa)*, the Indian fig. vulg. Banian tree.

of *Árca* the sun, which they write *Árcand,* a corruption still prevalent in the vulgar *Hindi.*[1]

Áryabhat́t́a appears to have had more correct notions of the true explanation of celestial phenomena than Brahmegupta himself; who, in a few instances, correcting errors of his predecessor, but oftener deviating from that predecessor's juster views, has been followed by the herd of modern Hindu astronomers, in a system not improved, but deteriorated, since the time of the more ancient author.

Considering the proficiency of Áryabhat́t́a in astronomical science, and adverting to the fact of his having written upon Algebra, as well as to the circumstance of his being named by numerous writers as the founder of a sect, or author of a system in astronomy, and being quoted at the head of algebraists, when the commentators of extant treatises have occasion to mention early and original[2] writers on this branch of science, it is not necessary to seek further for a mathematician qualified to have been the great improver of the analytic art, and likely to have been the person, by whom it was carried to the pitch to which it is found to have attained among the Hindus, and at which it is observed to be nearly stationary through the long lapse of ages which have since passed: the later additions being few and unessential in the writings of Brahmegupta, of Bháscara, and of Jnyána rája, though they lived at intervals of centuries from each other.

Áryabhat́t́a then being the earliest author known to have treated of Algebra among the Hindus, and being likely to be, if not the inventor, the improver, of that analysis, by whom too it was pushed nearly to the whole degree of excellence which it is found to have attained among them; it becomes in an especial manner interesting to investigate any discoverable trace in the absence of better and more direct evidence, which may tend to fix the date of his labours, or to indicate the time which elapsed between him and Brahmegupta, whose age is more accurately determined.[3]

Taking Áryabhat́t́a, for reasons given in the notes,[3] to have preceded Brahmegupta and Varáhamihira by several centuries; and Brahmegupta to have flourished about twelve hundred years ago;[4] and Varáha mihira, concerning whose works and age some further notices will be found in a sub-

[1] See notes I, K, and N.

[2] Súrya-dása on *Víja-gańita,* ch. 5.

[3] Note I.

[4] See before and note F.

joined note,[1] to have lived at the beginning of the sixth century after Christ,[2] it appears probable that this earliest of known Hindu algebraists wrote as far back as the fifth century of the Christian era; and, perhaps, in an earlier age. Hence it is concluded, that he is nearly as ancient as the Grecian algebraist DIOPHANTUS, supposed, on the authority of ABULFARAJ,[3] to have flourished in the time of the Emperor JULIAN, or about A. D. 360.

Admitting the Hindu and Alexandrian authors to be nearly equally ancient, it must be conceded in favour of the Indian algebraist, that he was more advanced in the science; since he appears to have been in possession of the resolution of equations involving several unknown, which it is not clear, nor fairly presumable, that DIOPHANTUS knew; and a general method for indeterminate problems of at least the first degree, to a knowledge of which the Grecian algebraist had certainly not attained; though he displays infinite sagacity and ingenuity in particular solutions; and though a certain routine is discernible in them.

A comparison of the Grecian, Hindu, and Arabian algebras, will more distinctly show, which of them had made the greatest progress at the earliest age of each, that can be now traced.

The notation or algorithm of Algebra is so essential to this art, as to deserve the first notice in a review of the Indian method of analysis, and a comparison of it with the Grecian and Arabian algebras. The Hindu algebraists use abbreviations and initials for symbols: they distinguish negative quantities by a dot;[4] but have not any mark, besides the absence of the negative sign, to discriminate a positive quantity. No marks or symbols indicating operations of addition, or multiplication, &c. are employed by them: nor any announcing equality[5] or relative magnitude (greater or less).[6] But a factum is denoted by the initial syllable of a word of that import,[7] subjoined to the terms which compose it, between which a dot is sometimes interposed. A fraction is indicated by placing the divisor under the dividend,[8] but without a line of separation. The two sides of an equation are ordered in the same

[1] Note K. [2] See before and note E. [3] Pococke's edition and translation, p. 39.

[4] *Víj.-gan.* § 4.

[5] The sign of equality was first used by Robert Recorde, because, as he says, no two things can be more equal than a pair of parallels, or *gemowe* lines of one length. *Hutton.*

[6] The signs of relative magnitude were first introduced into European algebra by Harriot.

[7] *Víj.-gan.* § 21. [8] *Líl.* § 33.

manner, one under the other:[1] and, this method of placing terms under each other being likewise practised upon other occasions,[2] the intent is in the instance to be collected from the recital of the steps of the process in words at length, which always accompanies the algebraic process. That recital is also requisite to ascertain the precise intent of vertical lines interposed between the terms of a geometric progression, but used also upon other occasions to separate and discriminate quantities. The symbols of unknown quantity are not confined to a single one: but extend to ever so great a variety of denominations: and the characters used are initial syllables of the names of colours,[3] excepting the first, which is the initial of *yávat-távat*, as much as; words of the same import with BOMBELLI's *tanto;* used by him for the same purpose. Colour therefore means unknown quantity, or the symbol of it: and the same *Sanscrit* word, *varńa*, also signifying a literal character, letters are accordingly employed likewise as symbols; either taken from the alphabet;[4] or else initial syllables of words signifying the subjects of the problem; whether of a general nature,[5] or specially the names of geometric lines in algebraic demonstrations of geometric propositions or solution of geometric problems.[6] Symbols too are employed, not only for unknown quantities, of which the value is sought; but for variable quantities of which the value may be arbitrarily put, (*Vij.* Ch. 6, note on commencement of § 153—156,) and, especially in demonstrations, for both given and sought quantities. Initials of the terms for square and solid respectively denote those powers; and combined they indicate the higher. These are reckoned not by the sums of the powers; but by their products.[7] An initial syllable is in like manner used to mark a surd root.[8] The terms of a compound quantity are ordered according to the powers; and the absolute number invariably comes last. It also is distinguished by an initial syllable, as a discriminative token of known quantity.[9] Numeral coefficients are employed, inclusive of unity which is always noted, and comprehending fractions;[10] for the numeral divisor is generally so placed, rather than under the symbol of the unknown: and in like manner the negative dot is set over the numeral coefficient: and not over the literal character. The coefficients are placed after the symbol of the

[1] *Vij.-gań.* and *Brahm.* 18, passim.
[2] *Vij.-gań.* § 55.
[3] *Vij.-gań.* § 17. *Brahm* c. 18, § 2.
[4] *Vij.-gań.* ch. 6.
[5] *Vij.-gań.* § 111.
[6] *Vij.-gań.* § 146.
[7] *Líl.* § 26.
[8] *Vij.-gań.* § 29.
[9] *Vij.-gań.* § 17.
[10] Stevinus in like manner included fractions in coefficients.

unknown quantity.[1] Equations are not ordered so as to put all the quantities positive; nor to give precedence to a positive term in a compound quantity: for the negative terms are retained, and even preferably put in the first place. In stating the two sides of an equation, the general, though not invariable, practice is, at least in the first instance, to repeat every term, which occurs in the one side, on the other: annexing nought for the coefficient, if a term of that particular denomination be there wanting.

If reference be made to the writings of DIOPHANTUS, and of the Arabian algebraists, and their early disciples in Europe, it will be found, that the notation, which has been here described, is essentially different from all theirs; much as they vary. DIOPHANTUS employs the inverted medial of ἔλλειψις, defect or want (opposed to υπαρξις, substance or abundance[2]) to indicate a negative quantity. He prefixes that mark ⋔ to the quantity in question. He calls the unknown, αριθμος; representing it by the final ς, which he doubles for the plural; while the Arabian algebraists apply the equivalent word for number to the constant or known term; and the *Hindus*, on the other hand, refer the word for numerical character to the coefficient. He denotes the monad, or unit absolute, by μ°; and the linear quantity is called by him *arithmos;* and designated, like the unknown, by the final *sigma*. He marks the further powers by initials of words signifying them: δ^υ, κ^υ, δδ^υ, δκ^υ, κκ^υ, &c. for *dynamis*, power (meaning the square); *cubos*, cube; *dynamo-dynamis*, biquadrate, &c. But he reckons the higher by the sums, not the products, of the lower. Thus the sixth power is with him the *cubo-cubos*, which the *Hindus* designate as the quadrate-cube, (cube of the square, or square of the cube).

The Arabian Algebraists are still more sparing of symbols, or rather entirely destitute of them.[4] They have none, whether arbitrary or abbreviated, either for quantities known or unknown, positive or negative, or for the steps and operations of an algebraic process: but express every thing by words, and phrases, at full length. Their European scholars introduced a few and very few abbreviations of names: c°, c^e, c^u, for the three first powers; c°, q^a, for the first and second unknown quantities; p, m, for plus and minus; and

[1] VIETA did so likewise.

[2] A word of nearly the same import with the *Sanscrit d'hana*, wealth, used by *Hindu* algebraists for the same signification.

[3] Def. 9.   [4] As. Res. 12. 183.

℞ for the note of radicality; occur in the first printed work which is that of PACIOLO.[1] LEONARDO BONACCI of *Pisa*, the earliest scholar of the Arabians,[2] is said by TARGIONI TOZZETTI to have used the small letters of the alphabet to denote quantities.[3] But LEONARDO only does so because he represents quantities by straight lines, and designates those lines by letters, in elucidation of his Algebraic solutions of problems.[4]

The Arabians termed the unknown (and they wrought but on one) *shai* thing. It is translated by LEONARDO of *Pisa* and his disciples, by the correspondent Latin word *res* and Italian *cosa;* whence *Regola de la Cosa*, and Rule of *Coss*, with *Cossike* practise and *Cossike* number of our older authors,[5] for Algebra or Speculative practice, as PACIOLO[6] denominates the analytic art; and *Cossic* number, in writers of a somewhat later date, for the root of an equation.

The *Arabs* termed the square of the unknown *mál*, possession or wealth; translated by the Latin *census* and Italian *censo;* as terms of the same import: for it is in the acceptation of amount of property or estate[7] that *census* was here used by LEONARDO.

The cube was by the Arabs termed *Cáb*, a die or cube; and they combined these terms *mál* and *cáb* for compound names of the more elevated powers; in the manner of DIOPHANTUS by the sums of the powers; and not like the *Hindus* by their products. Such indeed, is their method in the modern elementary works: but it is not clear, that the same mode was observed by their earlier writers; for their Italian scholars denominated the biquadrate and higher powers Relato primo, secundo, tertio, &c.

Positive they call *záïd* additional; and negative *nákis* deficient: and, as before observed, they have no discriminative marks for either of them.

The operation of *restoring* negative quantities, if any there be, to the positive form, which is an essential step with them, is termed *jebr*, or with the article *Aljebr*, the mending or restoration. That of *comparing* the terms and taking like from like, which is the next material step in the process of resolu-

[1] Or PACIOLI, PACIUOLO,—LI, &c. For the name is variously written by Italian authors.

[2] See Note L.

[3] *Viaggi*, 2d Edit. vol. 2, p. 62.

[4] COSSALI, Origine dell'Algebra, i.

[5] Robert Recorde's Whetstone of Witte.

[6] Secondo noi detta Pratica Speculativa. *Summa* 8. 1.

[7] *Census*, quicquid fortunarum quis habet. *Steph. Thes.*

tion, is called by them *mukábalah* comparison. Hence the name of *Tarík aljebr wa almukabala*, 'the method of restoration and comparison,' which obtained among the Arabs for this branch of the Analytic art; and hence our name of Algebra, from LEONARDO of *Pisa's* exact version of the Arabic title. *Fi istakhráju'l majhulát ba tarik aljebr wa almukábalah*,[1] De solutione quarundam quæstionum secundum modum *Algebræ* et *Almuchabalæ*.[2]

The two steps or operations, which have thus given name to the method of analysis, are precisely what is enjoined without distinctive appellations of them, in the introduction of the arithmetics of DIOPHANTUS, where he directs, that, if the quantities be positive on both sides, like are to be taken from like until one species be equal to one species; but, if on either side or on both, any species be negative, the negative species must be added to both sides, so that they become positive on both sides of the equation: after which like are again to be taken from like, until one species remain on each side.[3]

The *Hindu* Algebra, not requiring the terms of the equation to be all exhibited in the form of positive quantity, does not direct the preliminary step of *restoring* negative quantity to the affirmative state: but proceeds at once to the operation of equal subtraction *(samasódhana)* for the difference of like terms which is the process denominated by the Arabian Algebraists comparison *(mukábalah)*. On that point, therefore, the Arabian Algebra has more affinity to the Grecian than to the Indian analysis.

As to the progress which the Hindus had made in the analytic art, it will be seen, that they possessed well the arithmetic of surd roots;[4] that they were aware of the infinite quotient resulting from the division of finite quantity by cipher;[5] that they knew the general resolution of equations of the second degree; and had touched upon those of higher denomination; resolving them in the simplest cases, and in those in which the solution happens to be practicable by the method which serves for quadratics:[6] that they had attained a general solution of indeterminate problems of the first degree:[7] that they had arrived at a method for deriving a multitude of solutions of answers to pro-

[1] *Khulásatúl hisáb*. c. 8. *Calcutta*.

[2] *Liber abbaci*, 9. 15. 3. M.S. in Magliab. Libr.

[3] Def. 11.

[4] *Brahm*. 18. § 27—29. *Víj.-gań*. § 29—52.

[5] *Líl*. § 45. *Víj.-gań*. § 15—16 and § 135.

[6] *Víj.-gań*. § 129. and § 137—138.

[7] *Brahm*. 18. § 3—18. *Víj.-gań*. 53—73. *Líl*. § 248—265.

blems of the second degree from a single answer found tentatively;[1] which is as near an approach to a general solution of such problems, as was made until the days of LAGRANGE, who first demonstrated, that the problem, on which the solutions of all questions of this nature depend, is always resolvable in whole numbers.[2] The *Hindus* had likewise attempted problems of this higher order by the application of the method which suffices for those of the first degree;[3] with indeed very scanty success, as might be expected.

They not only applied algebra both to astronomy[4] and to geometry;[5] but conversely applied geometry likewise to the demonstration of Algebraic rules.[6] In short, they cultivated Algebra much more, and with greater success, than geometry; as is evident from the comparatively low state of their knowledge in the one,[7] and the high pitch of their attainments in the other: and they cultivated it for the sake of astronomy, as they did this chiefly for astrological purposes. The examples in the earliest algebraic treatise extant (BRAHMEGUPTA'S) are mostly astronomical: and here the solution of indeterminate problems is sometimes of real and practical use. The instances in the later treatise of Algebra by BHA'SCARA are more various: many of them geometric; but one astronomical; the rest numeral: among which a great number of indeterminate; and of these some, though not the greatest part, resembling the questions which chiefly engage the attention of DIOPHANTUS. But the general character of the Diophantine problems and of the Hindu unlimited ones is by no means alike: and several in the style of Diophantine are noticed by BHASCARA in his arithmetical, instead of his algebraic, treatise.[8]

To pursue this summary comparison further, DIOPHANTUS appears to have been acquainted with the direct resolution of affected quadratic equations; but less familiar with the management of them, he seldom touches on it. Chiefly busied with indeterminate problems of the first degree, he yet seems to have possessed no general rule for their solution. His elementary instructions for the preparation of equations are succinct.[9] His notation, as

[1] *Brahm.* 18. § 29—49. *Vij.-gan.* § 75—99.

[2] Mem. of Acad. of Turin: and of Berlin.

[3] *Vij.-gan.* § 206—207.

[4] *Brahm.* 18. passim. *Vij.-gan.*

[5] *Vij.-gan.* § 117—127. § 146—152.

[6] *Vij.-gan.* § 212—214.

[7] *Brahm.* 12. § 21; corrected however in *Líl.* § 169—170.

[8] *Líl.* § 59—61, where it appears, however, that preceding writers had treated the question algebraically. See likewise § 139—146.

[9] Def. 11.

before observed, scanty and inconvenient. In the whole science, he is very far behind the Hindu writers: notwithstanding the infinite ingenuity, by which he makes up for the want of rule: and although presented to us under the disadvantage of mutilation; if it be, indeed, certain that the text of only six, or at most seven, of thirteen books which his introduction announces, has been preserved.[1] It is sufficiently clear from what does remain, that the lost part could not have exhibited a much higher degree of attainment in the art. It is presumable, that so much as we possess of his work, is a fair specimen of the progress which he and the Greeks before him (for he is hardly to be considered as the inventor, since he seems to treat the art as already known;) had made in his time.

The points, in which the Hindu Algebra appears particularly distinguished from the Greek, are, besides a better and more comprehensive algorithm,—1st, The management of equations involving more than one unknown term. (This adds to the two classes noticed by the Arabs, namely simple and compound, two, or rather three, other classes of equation.) 2d, The resolution of equations of a higher order, in which, if they achieved little, they had, at least, the merit of the attempt, and anticipated a modern discovery in the solution of biquadratics. 3d, General methods for the solution of indeterminate problems of 1st and 2d degrees, in which they went far, indeed, beyond DIOPHANTUS, and anticipated discoveries of modern Algebraists. 4th, Application of Algebra to astronomical investigation and geometrical demonstration: in which also they hit upon some matters which have been reinvented in later times.

This brings us to the examination of some of their anticipations of modern discoveries. The reader's notice will be here drawn to three instances in particular.

The first is the demonstration of the noted proposition of PYTHAGORAS, concerning the square of the base of a rectangular triangle, equal to the squares of the two legs, containing a right angle. The demonstration is given two ways in BHÁSCARA's Algebra, (*Vij.-gaṅ.* § 146.) The first of them is the same which is delivered by WALLIS in his treatise on angular sections, (Ch. 6.) and, as far as appears, then given for the first time.[2]

[1] Note M.

[2] He designates the sides C. D. Base B. Segments $\varkappa$, δ. Then

$$\left.\begin{matrix} B : C :: C : \varkappa \\ B : D :: D : \delta \end{matrix}\right\} \text{ and therefore } \left\{\begin{matrix} C^2 = B\varkappa \\ D^2 = B\delta \end{matrix}\right.$$

On the subject of demonstrations, it is to be remarked that the Hindu mathematicians proved propositions both algebraically and geometrically: as is particularly noticed by BHÁSCARA himself, towards the close of his Algebra, where he gives both modes of proof of a remarkable method for the solution of indeterminate problems, which involve a factum of two unknown quantities. The rule, which he demonstrates, is of great antiquity in Hindu Algebra: being found in the works of his predecessor BRAHMEGUPTA, and being there a quotation from a more ancient treatise; for it is injudiciously censured, and a less satisfactory method by unrestricted arbitrary assumption given in its place. BHÁSCARA has retained both.

The next instance, which will be here noticed, is the general solution of indeterminate problems of the first degree. It was first given among moderns by BACHET *de* MEZIRIAC in 1624.[1] Having shown how the solution of equations of the form $ax-by=c$ is reduced to $ax-by=\pm1$, he proceeds to resolve this equation: and prescribes the same operation on a and b as to find the greatest common divisor. He names the residues c, d, e, f, &c. and the last remainder is necessarily unity: a and b being prime to each other. By retracing the steps from $e\mp1$ or $f\pm1$ (according as the number of remainders is even or odd) $e\mp1=\varepsilon$, $\frac{\varepsilon d\pm1}{e}=\delta$, $\frac{\delta c\mp1}{d}=\gamma$, $\frac{\gamma b\pm1}{c}=\beta$, $\frac{\beta a\mp1}{b}=\alpha$

or $f\pm1=\zeta$, $\frac{\zeta e\mp1}{f}=\varepsilon$, $\frac{\varepsilon d\pm1}{e}=\delta$, &c.

The last numbers β and α will be the smallest values of x and y. It is observed, that, if a and b be not prime to each other, the equation cannot subsist in whole numbers unless c be divisible by the greatest common measure of a and b.

Here we have precisely the method of the Hindu algebraists, who have not failed, likewise, to make the last cited observation. See *Brahm.* Algebra,

Therefore $C^2+D^2=(B\varkappa+B\delta=B$ into $\varkappa+\delta=)\ B^2$.

The Indian demonstration, with the same symbols, is

$$\left.\begin{array}{l} B:C::C:\varkappa \\ B:D::D:\delta \end{array}\right\}\text{ Therefore }\left\{\begin{array}{l} \varkappa=\frac{C^2}{B} \\ \delta=\frac{D^2}{B} \end{array}\right.$$

Therefore $B=\varkappa+\delta=\frac{C^2}{B}+\frac{D^2}{B}$ and $B^2=C^2+D^2$.

[1] Problèmes plaisans et délectables qui se font par les nombres. 2d Edit. (1624). LAGRANGE's additions to EULER's Algebra, ij. 382. (Edit. 1807.)

section 1. and *Bhásc. Lil.* ch. 12. *Vij.* ch. 2. It is so prominent in the Indian Algebra as to give name to the oldest treatise on it extant; and to constitute a distinct head in the enumeration of the different branches of mathematical knowledge in a passage cited from a still more ancient author. See *Lil.* § 248.

Confining the comparison of Hindu and modern Algebras to conspicuous instances, the next for notice is that of the solution of indeterminate problems of the 2d degree: for which a general method is given by BRAHMEGUPTA, besides rules for subordinate cases: and two general methods (one of them the same with BRAHMEGUPTA's) besides special cases subservient however to the universal solution of problems of this nature; and, to obtain whole numbers in all circumstances, a combination of the method for problems of the first degree with that for those of the second, employing them alternately, or, as the Hindu algebraist terms it, proceeding in a circle.

BHÁSCARA's second method (*Vij.* § 80—81) for a solution of the problem on which all indeterminate ones of this degree depend, is exactly the same, which Lord BROUNCKER devised to answer a question proposed by way of challenge by FERMAT in 1657. The thing required was a general rule for finding the innumerable square numbers, which multiplied by a proposed (non-quadrate) number, and then assuming an unit, will make a square. Lord BROUNCKER's rule, putting n for any given number, r^2 for any square taken at pleasure, and d for difference between n and r^2 $(r^2 \backsim n)$ was $\frac{4r^2}{d^2}\left(=\frac{2r}{d}\times\frac{2r}{d}\right)$ the square required. In the Hindu rule, using the same symbols, $\frac{2r}{d}$ is the square root required.[1] But neither BROUNCKER, nor WALLIS, who himself contrived another method, nor FERMAT, by whom the question was proposed, but whose mode of solution was never made known by him, (probably because he had not found anything better than WALLIS and BROUNCKER discovered,[2]) nor FRENICLE, who treated the subject without, however, adding to what had been done by WALLIS and BROUNCKER,[3] appear to have been aware of the importance of the problem and its universal use: a discovery, which, among the moderns, was reserved for EULER in the middle of the last century. To him, among the moderns, we owe the remark, which the Hindus had made more than a thousand years before,[4] that the problem was requisite to find all the

[1] *Vij.-gań.* § 80—81. [2] Wallis, Alg. c. 98. [3] Ibid.

[4] *Bháscara Vij.* § 173, and § 207. See likewise *Brahm.* Alg. sect. 7.

possible solutions of equations of this sort. LAGRANGE takes credit for having further advanced the progress of this branch of the indeterminate analysis, so lately as 1767;[1] and his complete solution of equations of the 2d degree appeared no earlier than 1769.[2]

It has been pretended, that traces of the art are to be dicovered in the writings of the Grecian geometers, and particularly in the five first propositions of EUCLID's thirteenth book; whether, as WALLIS conjectures, what we there have be the work of THEON or some other antient scholiast, rather than of EUCLID himself:[3] Also examples of analytic investigation in PAPPUS;[4] and indications of a method somewhat of a like nature with algebra, or at least the effects of it, in the works of ARCHIMEDES and APOLLONIUS; though they are supposed to have very studiously concealed this their art of invention.[5]

This proceeds on the ground of considering Analysis and Algebra, as interchangeable terms; and applying to Algebra EUCLID's or THEON's definition of Analysis, 'a taking of that as granted, which is sought; and thence by consequences arriving at what is confessedly true.'[6]

Undoubtedly they possessed a geometrical analysis; hints or traces of which exist in the writings of more than one Greek mathematician, and especially in those of ARCHIMEDES. But this is very different from the Algebraic Calculus. The resemblance extends, at most, to the method of inversion; which both Hindus and Arabians consider to be entirely distinct from their respective Algebras; and which the former, therefore, join with their arithmetic and mensuration.[7]

In a very general sense, the analytic art, as Hindu writers observe, is merely sagacity exercised; and is independent of symbols, which do not constitute the art. In a more restricted sense, according to them, it is calculation attended with the manifestation of its principles: and, as they further intimate, a method aided by devices, among which symbols and literal signs are conspicuous.[8] Defined, as analysis is by an illustrious modern

[1] Mem. de l'Acad. de Berlin, vol. 24.

[2] See French translation of Euler's Algebra, Additions, p. 286. And Legendre Theorie des Nombres 1. § 6. No. 36.

[3] WALLIS, Algebra, c. 2.

[4] Ibid. and Preface.

[5] Ibid. and Nunez Algebra 114.

[6] WALLIS, following VIETA's version, Alg. c. 1.

[7] *Líl.* 3. 1. § 47. *Khulásat. Hisáb.* c. 5.

[8] *Víj.-gań.* § 110, 174, 215, 224.

mathematician,[1] 'a method of resolving mathematical problems by reducing them to equations,' it assuredly is not to be found in the works of any Grecian writer extant, besides DIOPHANTUS.

In his treatise the rudiments of Algebra are clearly contained. He delivers in a succinct manner the Algorithm of affirmative and negative quantities; teaches to form an equation; to transpose the negative terms; and to bring out a final simple equation comprising a single term of each species known and unknown.

Admitting on the ground of the mention of a mathematician of his name, whose works were commented by HYPATIA about the beginning of the fifth century;[2] and on the authority of the Arabic annals of an Armenian Christian;[3] which make him contemporary with JULIAN; that he lived towards the middle of the fourth century of the Christian era; or, to speak with precision, about the year 360;[4] the Greeks will appear to have possessed in the fourth century so much of Algebra, as is to be effected by dexterous application of the resolution of equations of the first degree, and even the second, to limited problems; and to indeterminate also, without, however, having attained a general solution of problems of this latter class.

The Arabs acquired Algebra extending to simple and compound (meaning quadratic) equations; but it was confined, so far as appears, to limited problems of those degrees: and they possessed it so early as the close of the eighth century, or commencement of the ninth. Treatises were at that period written in the Arabic language on the Algebraic Analysis, by two distinguished mathematicians who flourished under the Abbasside ALMÁMÚN: and the more ancient of the two, MUHAMMED BEN MUSA *Al Khuwárezmí*, is recognised among the Arabians as the first who made Algebra known to them. He is the same, who abridged, for the gratification of ALMÁMÚN, an astronomical work taken from the Indian system in the preceding age, under ALMANSÚR. He framed tables likewise, grounded on those of the Hindus; which he professed to correct. And he studied and communicated to his

[1] D'ALEMBERT.

[2] SUIDAS, in voce *Hypatia*.

[3] GREGORY ABULFARAJ. Ex iis etiam [nempe philosophis qui prope tempora Juliani floruerunt] Diophantus, cujus liber, quem Algebram vocant, celebris est, in quem si immiserit se Lector, oceanum hoc in genere reperiet.—*Pococke*.

[4] JULIAN was emperor from 360 to 363. See note M.

countrymen the Indian compendious method of computation; that is, their arithmetic, and, as is to be inferred, their analytic calculus also.[1]

The Hindus in the fifth century, perhaps earlier,[2] were in possession of Algebra extending to the general solution of both determinate and indeterminate problems of the 1st and 2d degrees: and subsequently advanced to the special solution of biquadratics wanting the second term; and of cubics in very restricted and easy cases.

Priority seems then decisive in favour of both Greeks and Hindus against any pretensions on the part of the Arabians, who in fact, however, prefer none, as inventors of Algebra. They were avowed borrowers in science: and, by their own unvaried acknowledgment, from the Hindus they learnt the science of numbers. That they also received the Hindu Algebra, is much more probable, than that the same mathematician who studied the Indian arithmetic and taught it to his Arabian brethren, should have hit upon Algebra unaided by any hint or suggestion of the Indian analysis.

The Arabs became acquainted with the Indian astronomy and numerical science, before they had any knowledge of the writings of the Grecian astronomers and mathematicians: and it was not until after more than one century, and nearly two, that they had the benefit of an interpretation of DIOPHANTUS, whether version or paraphrase, executed by MUHAMMED ABULWAFÁ *Al Buzjání*; who added, in a separate form, demonstrations of the propositions contained in DIOPHANTUS; and who was likewise author of Commentaries on the Algebraic treatises of the *Khuwarezmite* MUHAMMED BEN MUSA, and of another Algebraist of less note and later date, ABI YAHYA, whose lectures he had personally attended.[3] Any inference to be drawn from their knowledge and study of the *Arithmetics* of DIOPHANTUS and their seeming adoption of his preparation of equations in their own Algebra, or at least the close resemblance of both on this point, is of no avail against the direct evidence, with which we are furnished by them, of previous instruction in Algebra and the publication of a treatise on the art, by an author conversant with the Indian science of computation in all its branches.

But the age of the earliest known Hindu writer on Algebra, not being with certainty carried to a period anterior, or even quite equal to that in which DIOPHANTUS is on probable grounds placed, the argument of priority, so far as investigation has yet proceeded, is in favour of Grecian invention.

[1] Note N. [2] See note I. [3] See note N.

The Hindus, however, had certainly made distinguished progress in the science, so early as the century immediately following that in which the Grecian taught the rudiments of it. The Hindus had the benefit of a good arithmetical notation: the Greeks, the disadvantage of a bad one. Nearly allied as algebra is to arithmetic, the invention of the algebraic calculus was more easy and natural where arithmetic was best handled. No such marked identity of the Hindu and Diophantine systems is observed, as to demonstrate communication. They are sufficiently distinct to justify the presumption, that both might be invented independently of each other.

If, however, it be insisted, that a hint or suggestion, the seed of their knowledge, may have reached the Hindu mathematicians immediately from the Greeks of Alexandria, or mediately through those of Bactria, it must at the same time be confessed, that a slender germ grew and fructified rapidly, and soon attained an approved state of maturity in Indian soil.

More will not be here contended for: since it is not impossible, that the hint of the one analysis may have been actually received by the mathematicians of the other nation; nor unlikely, considering the arguments which may be brought for a probable communication on the subject of astrology; and adverting to the intimate connexion between this and the pure mathematics, through the medium of astronomy.

The Hindus had undoubtedly made some progress at an early period in the astronomy cultivated by them for the regulation of time. Their calendar, both civil and religious, was governed chiefly, not exclusively, by the moon and sun: and the motions of these luminaries were carefully observed by them: and with such success, that their determination of the moon's synodical revolution, which was what they were principally concerned with, is a much more correct one than the Greeks ever achieved.[1] They had a division of the ecliptic into twenty-seven and twenty-eight parts, suggested evidently by the moon's period in days; and seemingly their own: it was certainly borrowed by the Arabians.[2] Being led to the observation of the fixed stars, they obtained a knowledge of the positions of the most remarkable; and noticed, for religious purposes, and from superstitious notions, the heliacal rising, with other phœnomena of a few. The adoration of the sun, of the planets, and of the stars, in common with the worship of the elements,

[1] As. Res. 2 and 12.

[2] As. Res. 9, Essay vi.

held a principal place in their religious observances, enjoined by the *Vèdas:*[1] and they were led consequently by piety to watch the heavenly bodies. They were particularly conversant with the most splendid of the primary planets; the period of Jupiter being introduced by them, in conjunction with those of the sun and moon, into the regulation of their calendar, sacred and civil, in the form of the celebrated cycle of sixty years, common to them and to the Chaldeans, and still retained by them. From that cycle they advanced by progressive stages, as the Chaldeans likewise did, to larger periods; at first by combining that with a number specifically suggested by other, or more correctly determined, revolutions of the heavenly bodies; and afterwards, by merely augmenting the places of figures for greater scope, (preferring this to the more exact method of combining periods of the planets by an algebraic process; which they likewise investigated[2]): until they arrived finally at the unwieldy cycles named *Mahâyugas* and *Calpas.* But it was for the sake of astrology, that they pushed their cultivation of astronomy, especially that of the minor planets, to the length alluded to. Now divination, by the relative position of the planets, seems to have been, in part at least, of a foreign growth, and comparatively recent introduction, among the Hindus. The belief in the influence of the planets and stars, upon human affairs, is with them, indeed, remotely antient; and was a natural consequence of their creed, which made the sun a divine being, and the planets gods. But the notion, that the tendency of that supposed influence, or the manner in which it will be exerted, may be foreseen by man, and the effect to be produced by it foretold, through a knowledge of the position of the planets at a particular moment, is no necessary result of that creed: for it takes from beings believed divine, free-agency in other respects, as in their visible movements.

Whatever may have been the period when the notion first obtained, that foreknowledge of events on earth might be gained by observations of planets and stars, and by astronomical computation; or wherever that fancy took its rise; certain it is, that the Hindus have received and welcomed communications from other nations on topics of astrology: and although they had astrological divinations of their own as early as the days of PARÁŚARA and GARGA, centuries before the Christian era, there are yet grounds to presume

[1] As. Res. 8.

[2] BRAHMEGUPTA, Algebra.

that communications subsequently passed to them on the like subject, either from the Greeks, or from the same common source (perhaps that of the Chaldeans) whence the Greeks derived the grosser superstitions engrafted on their own genuine and antient astrology, which was meteorological.

This opinion is not now suggested for the first time. Former occasions have been taken of intimating the same sentiment on this point:[1] and it has been strengthened by further consideration of the subject. As the question is closely connected with the topics of this dissertation, reasons for this opinion will be stated in the subjoined note.[2]

Joining this indication to that of the division of the zodiac into twelve signs, represented by the same figures of animals, and named by words of the same import with the zodiacal signs of the Greeks; and taking into consideration the analogy, though not identity, of the Ptolemaic system, or rather that of HIPPARCHUS, and the Indian one of excentric deferents and epicycles, which in both serve to account for the irregularities of the planets, or at least to compute them, no doubt can be entertained that the Hindus received hints from the astronomical schools of the Greeks.

It must then be admitted to be at least possible, if not probable, in the absence of direct evidence and positive proof, that the imperfect algebra of the Greeks, which had advanced in their hands no further than the solution of equations, involving one unknown term, as it is taught by DIOPHANTUS, was made known to the Hindus by their Grecian instructors in improved astronomy. But, by the ingenuity of the Hindu scholars, the hint was rendered fruitful, and the algebraic method was soon ripened from that slender beginning to the advanced state of a well arranged science, as it was taught by ÁRYABHAṬṬA, and as it is found in treatises compiled by BRAHMEGUPTA and BHÁSCARA, of both which versions are here presented to the public.

[1] As. Res. 12. [2] Note O.

NOTES AND ILLUSTRATIONS.

A.

SCHOLIASTS OF BHÁSCARA.

THE oldest commentary of ascertained date, which has come into the translator's hands, and has been accordingly employed by him for the purpose of collation, as well as in the progress of translation, is one composed by GANGÁD'HARA son of GÓBARD'HANA and grandson of DIVÁCARA, inhabitant of *Jambusara.*[1] It appears from an example of an astronomical computation, which it exhibits,[2] to have been written about the year 1342 *Śaca* (A.D. 1420). Though confined to the *Lílávatí*, it expounds and consequently authenticates a most material chapter of the *Víja-gańita*, which recurs nearly verbatim in both treatises; but is so essential a part of the one, as to have given name to the algebraic analysis in the works of the early writers.[3] His elder brother VISHŃU PAŃDITA was author of a treatise of arithmetic, &c. named *Gańita-sára*, a title borrowed from the compendium of ŚRÍD'HARA. It is frequently quoted by him.

The next commentary in age, and consequent importance for the objects now under consideration, is that of SÚRYA SURI also named SÚRYADÁSA, native of *Párthapura*, near the confluence of the *Gódá* and *Vidarbhá* rivers.[4] He was author of a complete commentary on the *Sidd'hánta-śirómańi*; and of a distinct work on calculation, under the title of *Gańita-málatí*; and of a compilation of astronomical and astrological doctrines, Hindu and Muhammedan, under the name of *Sidd'hánta-sánhitá-sára-samuchchaya*; in which he makes mention of his commentary on the *Śirómańi*. The gloss on the *Lilávatí*, en-

[1] A town situated in Gujra. (*Gurjara*), twenty-eight miles north of the town of *Broach.*

[2] *Líl.* § 264.

[3] *Cuttacád'hyáya*, the title of BRAHMEGUPTA's chapter on Algebra, and of a chapter in ÁRYABHATTA's work.

[4] *Gódávarí* and *Werdá.*

titled *Gańitámrĭta*, and that on the *Vija-gańitc*, named *Súrya-pracáśa*, both excellent works, containing a clear interpretation of the text, with a concise explanation of the principles of the rules, are dated the one in 1460, the other in 1463 *Śaca;* or A. D. 1538 and 1541. His father Jnyánarája, son of Nágganát'ha, a Brahmen and astronomer, was author, among other works, of an astronomical course, under the title of *Sidd'hánta-sundara*, still extant,[1] which, like the *Sidd'hánta-śirómańi*, comprises a treatise on algebra. It is repeatedly cited by his son.

Gańéśa, son of Césava, a distinguished astronomer, native of *Nandigráma*, near *Dévagiri*, (better known by the Muhammedan name of *Dauletábád)*,[2] was author of a commentary on the *Sidd'hánta-śirómańi*, which is mentioned by his nephew and scholiast Nrĭsinha; in an enumeration of his works, contained in a passage quoted by Viśwanát'ha on the *Grahalághava.* His commentary on the *Lílávatí* bears the title of *Budd'hivilásiní*, and date of 1467 *Śaca*, or A. D. 1545. It comprises a copious exposition of the text, with demonstrations of the rules: and has been used throughout the translation as the best interpreter of it. He, and his father Céśava, and nephew Nrĭsinha, as well as his cousin Lacshmídása, were authors of numerous works both on astronomy and divination. The most celebrated of his own performances, the *Grahalághava*, bears date 1442 *Śaca*, answering to A. D. 1520.

The want of a commentary by Gańéśa on the *Vija-gańita*, is supplied by that of Crĭshńa, son of Ballála, and pupil of Vishńu, the disciple of Gańéśa's nephew Nrĭsinha. It contains a clear and copious exposition of the sense, with ample demonstrations of the rules, much in the manner of Gańéśa, on the *Lilávatí;* whom also he imitated in composing a commentary on that treatise, and occasionally refers to it. His work is entitled *Calpalatávatára*. Its date is determined, at the close of the sixteenth century of the Christian era, by the notice of it and of the author in a work of his brother Ranganát'ha, dated 1524 *Śaca* (A. D. 1602), as well as in one by his nephew Muníśwara. He appears to have been astrologer in the service of the Emperor Jehángír, who reigned at the beginning of the seventeenth century.

The gloss of Ranganát'ha on the *Vásaná*, or demonstratory annotations

[1] The astronomical part is in the library of the East India Company.

[2] *Nandigrám* retains its ancient name; and is situated west of *Dauletabad*, about sixty-five miles.

of BHA'SCARA, which is entitled *Mita-bháshiṅí*, contains no specification of date; but is determined, with sufficient certainty, towards the middle of the sixteenth century of the *Śaca* era, by the writer's relation of son to NRĬSINHA, the author of a commentary on the *Súrya-sidd'hánta*, dated 1542 *Śaca*, and of the *Vásaná-vártica* (or gloss on BHA'SCARA's annotations of the *Śirómaṅi)*, which bears date in 1543 *Śaca*, or A. D. 1621; and his relation of brother, as well as pupil, to CAMALA'CARA, author of the *Sidd'hánta-tatwa-vivéca*, also composed towards the middle of the same century of the *Śaca* era. NRĬSINHA, and his uncle VIŚWANA'T'HA, (author of astrological commentaries,) describe their common ancestor DIVA'CARA, and his grandfather RA'MA, as *Maháráshtra* Brahmens, living at *Gólagráma*,[1] on the northern bank of the *Gódávarí*, and do not hint a migration of the family. NRĬSINHA's own father, CRĬSHṄA, was author of a treatise on algebra in compendious rules *(sutra)*, as his son affirms.

The *Víja-prabód'ha*, a commentary on the *Víja-gaṅita*, by RA'MA CRĬSHṄA, son of LACSHMAṄA, and grandson of NRĬSINHA, inhabitant of *Amarávatí*,[2] is without date or express indication of its period; unless his grandfather NRĬSINHA be the same with the nephew of VIŚWANA'T'HA just now mentioned: or else identified with the nephew of GAṄÉŚA and preceptor of VISHṄU, the instructor of CRĬSHṄA, author of the *Calpalatávatára*. The presumption is on either part consistent with proximity of country: *Amarávatí* not being more than 150 miles distant from *Nandigráma*, nor more than 200 from *Gólagráma*. It is on one side made probable by the author's frequent reference to a commentary of his preceptor CRĬSHṄA, which in substance corresponds to the *Calpalatávatára*; but the title differs, for he cites the *Naváncura*. On the other side it is to be remarked, that CRĬSHṄA, father of the NRĬSINHA, who wrote the *Vasaná-vártica*, was author of a a treatise on Algebra, which is mentioned by his son, as before observed.

The *Manóranjana*, another commentary on the *Lilávati*, which has been used in the progress of the translation, bears no date, nor any indication whatsoever of the period when the author RA'MA-CRĬSHṄA DÉVA, son of SADA'DÉVA, surnamed APADÉVA, wrote.

The *Gaṅita-caumudí*, on the *Lílávati*, is frequently cited by the modern

[1] Gólgám of the maps, in lat. 18° N. long. 78° E.

[2] A great commercial town in Berár.

commentators, and in particular by SÚRYA-SURI and RANGANÁT'HA: but has not been recovered, and is only known from their quotations.

Of the numerous commentaries on the astronomical portion of BHÁSCARA's *Sidd'hánta-śirómańi*, little use having been here made, either for settling the text of the algebraic and arithmetical treatises of the author, or for interpreting particular passages of them, a reference to two commentaries of this class, besides those of SÚRYA-SURI and GAŃÉŚA, (which have not been recovered,) and the author's own annotations and the interpretation of them by NRĬSINHA above noticed, may suffice: viz. the *Gańita-tatwa-chintámańi*, by LACSHMÍDÁSA, grandson of CÉSAVA, (probably the same with the father of GAŃÉŚA before mentioned,) and son of VÁCHESPATI, dated 1423 *Śaca*, (A.D. 1501); and the *Márícha*, by MUNÍŚWARA, surnamed VIŚWARUPA, grandson of BALLÁLA, and son of RANGANÁT'HA, who was compiler of a work dated 1524 *Śaca* (A.D. 1602), as before mentioned. MUNÍŚWARA himself is the author of a distinct treatise of astronomy entitled *Sidd'hánta-sárvabhauma*.

Persian versions of both the *Lílávatí* and *Víja-gańita* have been already noticed, as also contributing to the authentication of the text. The first by FAIZÍ, undertaken by the command of the Emperor ACBER, was executed in the 32d year of his reign; A. H. 995, A.D. 1587. The translation of the *Víja-gańita* is later by half a century, having been completed by ATA ULLAH RASHÍDÍ, in the 8th year of the reign of SHÁH JEHÁN; A. H. 1044, A.D. 1634.

B.

ASTRONOMY OF BRAHMEGUPTA.

BRAHMEGUPTA's entire work comprises twenty-one lectures or chapters; of which the ten first contain an astronomical system, consisting (1st and 2d) in the computation of mean motions and true places of the planets; 3d, solution of problems concerning time, the points of the horizon, and the position of places; 4th and 5th, calculation of lunar and solar eclipses; 6th, rising and setting of the planets; 7th, position of the moon's cusps; 8th, observation of altitudes by the gnomon; 9th, conjunctions of the planets; and, 10th, their conjunction with stars. The next ten are supplementary, including five chapters of problems with their solutions: and the twenty-first explains the principles of the astronomical system in a compendious treatise on spherics, treating of the astronomical sphere and its circles, the construction of sines, the rectification of the apparent planet from mean motions, the cause of lunar and solar eclipses, and the construction of the armillary sphere.

The copy of the scholia and text, in the translator's possession, wants the whole of the 6th, 7th, and 8th chapters, and exhibits gaps of more or less extent in the preceding five; and appears to have been transcribed from an exemplar equally defective. From the middle of the 9th, to near the close of the 15th chapters, is an uninterrupted and regular series, comprehending a very curious chapter, the 11th, which contains a revision and censure of earlier writers: and next to it the chapter on arithmetic and mensuration, which is the 12th of the work. It is followed in the 13th, and four succeeding chapters, by solutions of problems concerning mean and true motions of planets, finding of time, place, and points in the horizon; and relative to other matters, which the defect of the two last of five chapters renders it impracticable to specify. Next comes, (but in a separate form, being transcribed from a different exemplar,) the 18th chapter on Algebra. The two, which should succeed, (and one of which, as appears from a reference to a chapter on this subject, treats of the various measures of time under the several denominations of solar, siderial, lunar, &c.; and the other, from like references to it, is known to treat of the delineation of celestial phœnomena by diagram,) are entirely wanting, the remainder of the copy being defective. The twenty-first chapter, however, which is last in the author's arrangement, (as the corresponding book on spherics of BHÁSCARA's *Sidd'hánta-sirómaṅi* is in his,) has been transposed and first expounded by the scholiast: and very properly so, since its subject is naturally preliminary, being explanatory of the principles of astronomy. It stands first in the copy under consideration; and is complete, except one or two initial couplets.

C.

BRAHMA-SIDD'HÁNTA, TITLE OF BRAHMEGUPTA'S ASTRONOMY.

THE passage is this: " BRAHMÓ*cta-graha-gańitam mahatá cálèna yat c'hilí-bhútam, abhid'híyatè* sphutan *tat* JISHŃU-*suta* BRAHMEGUPTÉNA."

' The computation of planets, taught by BRAHMA, which had become imperfect by great length of time, is propounded correct by BRAHMEGUPTA son of JISHŃU.'

The beginning of PRĬT'HÚDACA's commentary on the *Brahma-sidd'hánta*, where the three initial couplets of the text are expounded, being deficient, the quotation cannot at present be brought to the test of collation. But the title is still more expressly given near the close of the eleventh chapter, (§ 59) " *Bráhmè sphuta-sidd'hántè* ravíndu-bhú yógam, &c."

And again, (§ 61) " *Chandra-ravigrahańèndu-ch'háyádishu sarvadá yató Bráhmè, drĭg-gańitaicyam bhavati,* sphutá-sidd'hántas *tató* Bráhmah." ' As observation and computation always agree in respect of lunar and solar eclipses, moon's shadow (i. e. altitude) and other particulars, according to the *Bráhma*, therefore is the *Bráhma* a correct system, *(sphuta-sidd'hánta)*.'

It appears from the purport of these several passages compared, that BRAHMEGUPTA's treatise is an emendation of an earlier system, (bearing the same name of *Brahma-sidd'hánta*, or an equivalent title, as *Pitámaha-sidd'hánta*, or adjectively *Paitámaha,)* which had ceased to agree with the phœnomena, and into which requisite corrections were therefore introduced by him to reconcile computation and observation; and he entitled his amended treatise ' Correct *Brahma-sidd'hánta*.' That earlier treatise is considered to be the identical one which is introduced into the *Vishńu-d'hermóttara puráńa*, and from which parallel passages are accordingly cited by the scholiasts of BHÁSCARA. (See following note.) It is no doubt the same which is noticed by VARÁHAMIHIRA under the title of *Paitámaha* and *Bráhma siddhanta*. Couplets, which are cited by his commentator BHATTÓTPALA from the *Brahma-sidd'hánta*, are found in BRAHMEGUPTA's work. But whether the original or the amended treatise be the one to which the scholiast referred, is nevertheless a disputable point, as the couplets in question may be among passages which BRAHMEGUPTA retained unaltered.

D.

VERIFICATION OF THE TEXT OF BRAHMEGUPTA'S TREATISE OF ASTRONOMY.

A PASSAGE, referring the commencement of astronomical periods and of planetary revolutions, to the supposed instant of the creation, is quoted from BRAHMEGUPTA, with a parallel passage of another *Brahma sidd'hánta* (comprehended in the *Vishnú-dhermóttara-purána*) in a compilation by MUNÍSWARA one of BHÁSCARA's glossators.[1] It is verified as the 4th couplet of BRAHMEGUPTA's first chapter (upon mean motions) in the translator's copy.

Seven couplets, specifying the mean motions of the planets' nodes and apogees, are quoted after the parallel passage of the other *Brahma sidd'hánta*, by the same scholiast of BHÁSCARA, as the text of BRAHMEGUPTA: and they are found in the same order from the 15th to the 21st in the first chapter of his work in the copy above mentioned.

This commentator, among many other corresponding passages noticed by him on various occasions, has quoted one from the same *Brahma sidd'hánta* of the *Vishńu-dharmóttara* concerning the orbits of the planets deduced from the magnitude of the sky computed there, as it also is by BRAHMEGUPTA (ch. 21, § 9), but in other words, at a circumference of 18712069200000000 *yójanas:* he goes on to quote the subsequent couplet of BRAHMEGUPTA declaring that planets travel an equal measured distance in their orbits in equal times: and then cites his scholiast (*tícacára*) CHATURVÉDÁCHÁRYA.

The text of BRAHMEGUPTA (ch. 1, § 21) specifying the diurnal revolutions of the siderial sphere, or number of siderial days in a *calpa*, with the correspondent one of the *Paitámaha sidd'hánta* in the *Vishńu-dhermóttara*, is another of the quotations of the same writer in his commentary on BHÁSCARA.

A passage relating to oval epicycles,[2] cited by the same author in another place, is also verified in the 2d chapter (in the rectification of a planet's place).

A number of couplets on the subject of eclipses[3] is cited by LACSHMÍDÁSÁ, a commentator of BHÁSCARA. They are found in the 5th chapter (on eclipses) § 10 and 24; and in a section of the 21st (on the cause of eclipses) § 37 to 46, in the copy in question.

Several couplets, relating to the positions of the constellations and to the

[1] As. Res. 12, p. 232. [2] Ibid. 12, p. 236. [3] Ibid. 12, p. 241.

longitudes and latitudes of principal fixt stars, are cited from BRAHMEGUPTA in numerous compilations, and specifically in the commentaries on the *Súrya-sidd'hánta* and *Sidd'hánta-śirómańi.*[1] They are all found correct in the 10th chapter, on the conjunctions of planets with fixt stars.

A quotation by GAŃÉŚA on the *Lílávatí* (A. D. 1545) describing the attainments of a true mathematician,[2] occurs with exactness as the first couplet of the 12th chapter, on arithmetic; and one adduced by BHÁSCARA himself, in his arithmetical treatise (§ 190), giving a rule for finding the diagonal of a trapezium,[3] is precisely the 28th of the same chapter.

A very important passage, noticed by BHÁSCARA in his notes on his *Sidd'hánta-śirómańi,* and alluded to in his text, and fully quoted by his commentator in the *Máricha,* relative to the rectification of a planet's true place from the mean motions,[4] is found in the 21st chapter, § 27. BHÁSCARA has, on that occasion, alluded to the scholiast, who is accordingly quoted by name in the commentary of LACSHMÍDÁSA (A. D. 1501): and here again the correspondence is exact.

The identity of the text as BRAHMEGUPTA'S, and of the gloss as his scholiast's, being (by these and many other instances, which have been collated,) satisfactorily established; as the genuineness of the text is by numerous quotations from the *Brahma-sidd'hánta* (without the author's name) in the more ancient commentary of BHAŤŤÓTPALA (A. D. 968) on the works of VARÁHAMIHIRA, which also have been verified in the mutilated copy of the *Brahma-sidd'hánta* under consideration; the next step was the examination of the detached copy of a commentary on the 18th chapter, upon Algebra, which is terminated by a colophon so describing it, and specifying the title of the entire book *Brahma-sidd'hánta,* and the name of its author BRAHMEGUPTA.

For this purpose materials are happily presented in the scholiast's enumeration, at the close of the chapter on arithmetic, of the topics treated by his author in the chapter on Algebra, entitled *Cuťťaca*:[5] in a general reference to the author's algorithm of unknown quantities, affirmative and negative terms, cipher and surd roots, in the same chapter;[6] and the same scholiast's quotations of the initial words of four rules; one of them relative to surd roots;[7] the other three regarding the resolution of quadratic equations:[8] as

1 As. Res. 9. Essay 6.
2 *Líl.* ch. 11.
3 *Líl.* § 190.
4 As. Res. 12, 239.
5 Arithm. of *Brahm.* § 66.
6 Ibid. § 13.
7 Ibid. § 39.
8 Ibid. § 15 & 18.

also in the references of the scholiast of the Algebraic treatise to passages in the astronomical part of his author's work.[1]

The quotations have been verified; and they exactly agree with the rule concerning surds (§ 26) and the three rules which compose the section relating to quadratic equations (§ 32—34); and with the rule in the chapter on the solution of astronomical problems concerning mean motions (ch. 13, § 22): and this verification and the agreement of the more general references demonstrate the identity of this treatise of Algebra, consonantly to its colophon, as BRAHMEGUPTA's Algebra entitled *Cuttaca* and a part of his *Brahma-sidd'hánta.*

E.

CHRONOLOGY OF ASTRONOMICAL AUTHORITIES ACCORDING TO ASTRONOMERS OF UJJAYANI.

THE names of astronomical writers with their dates, as furnished by the astronomers of *Ujjayani* who were consulted by Dr. WILLIAM HUNTER sojourning there with a British embassy, are the following:

	VARÁHA-MIHIRA	122 Śaca	[A. D. 200-1]
Another	VARÁHA-MIHIRA	427	[A. D. 505-6]
	BRAHMEGUPTA	550	[A. D. 628-9]
	MUNJÁLA	854	[A. D. 932-3]
	BHATTÓTPALA	890	[A. D. 1068-9]
	ŚWÉTÓTPALA	939	[A. D. 1017-8]
	VARUŃA-BHATTA	962	[A. D. 1040-1]
	BHÓJA-RÁJA	964	[A. D. 1042-3]
	BHÁSCARA	1072	[A. D. 1150-1]
	CALYÁNA-CHANDRA	1101	[A.D. 1179-80]

The grounds, on which this chronology proceeds, are unexplained in the note which Dr. HUNTER preserved of the communication. But means exist for verifying two of the dates specified and corroborating others.

The date, assigned to BHÁSCARA, is precisely that of his *Sidd'hánta-śirómańi,* plainly concluded from a passage of it, in which he declares, that it was

[1] Alg. of Brahm. § 96 (Rule 55).

completed by him, being thirty-six years of age; and that his birth was in 1036 *Śaca.*

Rájá BHÓJA-DÉVA, or BHÓJA-RÁJA, is placed in this list of Hindu astronomers apparently on account of his name being affixed as that of the author, to an astrological treatise on the calendar, which bears the title of *Rája-mártańda,* and which was composed probably at his court and by astrologers in his service. It contains no date; or at least none is found in the copy which has been inspected. But the age assigned to the prince is not inconsistent with Indian History: and is supported by the colophon of a poem entitled *Subháshita ratna-sańdóha,* composed by a *Jaina* sectary named AMITAGATI who has given the date of his poem in 1050 of *Vicramáditya,* in the reign of MUNJA. Now MUNJA was uncle and predecessor of BHÓJA-RÁJA, being regent, with the title of sovereign, during his nephew's minority: and this date, which answers to A. D. 993-4, is entirely consistent with that given by the astronomers of *Ujjayaní,* viz. 964 *Śaca* corresponding to A. D. 1042-3: for the reign of BHÓJA-DÉVA was long: extending, at the lowest computation, to half a century, and reaching, according to an extravagant reckoning, to the round number of an hundred years.

The historical notices of this King of *Dhárá*[1] are examined by Major WILFORD and Mr. BENTLEY in the 9th and 8th volumes of Asiatic Researches: and they refer him to the tenth century of the Christian era; the one making him ascend the throne in A. D. 982; the other, in A. D. 913. The former, which takes his reign at an entire century, including of course his minority, or the period of the administration, reign, or regency, of his uncle MUNJA, is compatible with the date of AMITAGATI's poem (A. D. 993) and with that of the *Rája-mártańda* or other astrological and astronomical works ascribed to him (A. D. 1042) according to the chronology of the astronomers of *Ujjayaní.*

The age, assigned to BRAHMEGUPTA, is corroborated by the arguments adduced in the text. That, given to MUNJÁLA, is consistent with the quotation of him as at the head of a tribe of authors, by BHÁSCARA at the distance of two centuries. The period allotted to VARÁHAMIHIRA, that is, to the second and most celebrated of the name, also admits corroboration. This point, however, being specially important, to the history of Indian astronomy, and collaterally to that of the Hindu Algebra, deserves and will receive a full and distinct consideration.

[1] The modern *Dhar.* WILFORD. As. Res.

F.

AGE OF BRAHMEGUPTA INFERRED FROM ASTRONOMICAL DATA.

The star *Chitrá*, which unquestionably is *Spica Virginis*,[1] was referred by Brahmegupta to the 103d degree counted from its origin to the intersection of the star's circle of declination;[2] whence the star's right ascension is deduced 182° 45′. Its actual right ascension in A. D. 1800 was 198° 40′ 2″.* The difference, 15° 55′ 2″, is the quantity, by which the beginning of the first zodiacal asterism and lunar mansion, *Aświní*, as inferible from the position of the star *Chitrá*, has receded from the equinox: and it indicates the lapse of 1216 years (to A. D. 1800,) since that point coincided with the equinox; the annual precession of the star being reckoned at 47″, 14.†

The star *Rêvatí*, which appears to be ζ Piscium,[5] had no longitude, according to the same author, being situated precisely at the close of the asterism and commencement of the following one, *Aświní*, without latitude or declination, exactly in the equinoctial point. Its actual right ascension in 1800 was 15° 49′ 15″.[6] This, which is the quantity by which the origin of the Indian ecliptic, as inferible from the position of the star *Rêvatí*, has receded from the equinox, indicates a period of 1221 years, elapsed to the end of the eighteenth century; the annual precession for that star being 46″, 63.[7]

The mean of the two is 1218½ years; which, taken from 1800, leave 581 or 582 of the Christian era. Brahmegupta then appears to have observed and written towards the close of the sixth, or the beginning of the following century; for, as the Hindu astronomers seem not to have been very accurate observers, the belief of his having lived and published in the seventh century, about A. D. 628, which answers to 550 *Śaca*, the date assigned to him by the astronomers of *Ujjayaní*, is not inconsistent with the position, that the vernal equinox did not sensibly to his view deviate from the begin-

[1] As. Res. vol. 9, p. 339. (8vo.)

[2] Ibid. 9, 327, (8vo.), and 12, p. 240.

* Zach's Tables for 1800 deduced from Maskelyne's Catalogue.

† Maskelyne's Catalogue: the mean precession of the equinoctial points being reckoned 50″, 3.

[5] As. Res. 9, p. 346. (8vo.)

[6] Zach's Tables.

[7] Zach's Tables.

ning of Aries or *Mésha*, as determined by him from the star *Rêvatí* (ζ Piscium) which he places at that point.

The same author assigns to *Agastya* or *Canopus* a distance of 87°, and to *Lubd'haca* or *Sirius* 86°, from the beginning of *Mésha*. From these positions a mean of 1280 years is deducible.

The passage in which this author denies the precession of the colures, as well as the comment of his scholiast on it, being material to the present argument, they are here subjoined in a literal version.

'The very fewest hours of night occur at the end of *Mit'huna*; and the seasons are governed by the sun's motion. Therefore the pair of solstices appears to be stationary, by the evidence of a pair of eyes.'[1]

Scholia: 'What is said by VISHŃU CHANDRA at the beginning of the chapter on the *yuga* of the solstice: ("Its revolutions through the asterisms are here [in the *calpa*] a hundred and eighty-nine thousand, four hundred and eleven. This is termed a *yuga* of the solstice, as of old admitted by BRAHMA, ARCA, and the rest.") is wrong: for the very fewest hours of night to us occur when the sun's place is at the end of *Mit'huna* [Gemini]; and of course the very utmost hours of day are at the same period. From that limitary point, the sun's progress regulates the seasons; namely, the cold season *(śiśira)* and the rest, comprising two months each, reckoned from *Macara* [Capricorn]. Therefore what has been said concerning the motion of the limitary point is wrong, being contradicted by actual observation of days and nights.

'The objection, however, is not valid: for now the greatest decrease and increase of night and day do not happen when the sun's place is at the end of *Mit'huna:* and passages are remembered, expressing "The southern road of the sun was from the middle of *Aślésha;* and the northern one at the beginning of *Dhanisht'ha;*"[2] and others [of like import]. But all this only proves, that there is a motion; not that the solstice has made many revolutions through the asterisms.'[3]

It was hinted at the beginning of this note, that BRAHMEGUPTA's longitude *(dhruvaca)* of a star is the arc of the ecliptic intercepted by the star's circle of declination, and counted from the origin of the ecliptic at the be-

[1] *Brahma-sidd'hánta*, 11, § 54.

[2] This quotation is from VARÁHA-MIHIRA's *sanhitá*, ch. 3, § 1 and 2.

[3] PRĬT'HÚDACA SWÁMÍ CHATURVÉDA on *Brahm.*

ginning of *Mésha;* as his latitude (*vicshépa*) of a star is the star's distance on a circle of declination from its point of intersection with the ecliptic. In short, he, like other Hindu astronomers, counts longitude and latitude of stars by the intersection of circles of declination with the ecliptic. The subject had been before noticed.[1] To make it more clear, an instance may be taken: and that of the scholiast's computation of the zenith distance and meridian altitude of Canopus for the latitude of *Canyacubja (Canouj)* may serve as an apposite example.

From the *vicshépa* of the star *Agastya*, 77°, he subtracts the declination of the intersected point of the ecliptic 23° 58′; to the remainder, which is the declination of the star, 53° 2′, he adds the latitude of the place 26° 35′; the sum, 79° 37′, is the zenith distance; and its complement to ninety degrees, 10° 23′, is the meridian altitude of the star.[2]

The annual variation of the star in declination, 1″, 7, is too small to draw any inference as to the age of the scholiast from the declination here stated. More especially as it is taken from data furnished by his author; and as he appears to have been, like most of the Hindu astronomers, no very accurate observer; the latitude assigned by him to the city, in which he dwelt, being no less than half a degree wrong: for the ruins of the city of *Canouj* are in 27° 5′ N.

G.

ÁRYABHAŤŤA'S DOCTRINE.

ÁRYABHAŤŤA was author of the *Áryáshťaśata* (800 couplets) and *Daśagíticá* (ten stanzas), known by the numerous quotations of BRAHMEGUPTA, BHAŤŤÓTPALA, and others, who cite both under these respective titles. The *laghu Árya-sidd'hánta*, as a work of the same author, and, perhaps, one of those above-mentioned, is several times quoted by BHÁSCARA's commentator MUNÍŚWARA. He likewise treated of Algebra, &c. under the distinct heads of *Cuťťaca*, a problem serving for the resolution of indeterminate ones, and *Víja* principle of computation, or analysis in general.—*Líl.* c. 11.

[1] As. Res. 9, p. 327. (8vo.), and 12, p. 240; (4to.)

[2] PRĪT'HÚDACA SWÁMÍ on *Brahm.* ch. 10, § 35.

From the quotations of writers on astronomy, and particularly of BRAHMEGUPTA, who in many instances cites ÁRYABHAṬṬA to controvert his positions, (and is in general contradicted in his censure by his own scholiast PRIT'HÚDACA, either correcting his quotations, or vindicating the doctrine of the earlier author), it appears, that ÁRYABHAṬṬA affirmed the diurnal revolution of the earth on its axis; and that he accounted for it by a wind or current of aerial fluid, the extent of which, according to the orbit assigned to it by him, corresponds to an elevation of little more than a hundred miles from the surface of the earth; that he possessed the true theory of the causes of lunar and solar eclipses, and disregarded the imaginary dark planets of the mythologists and astrologers; affirming the moon and primary planets (and even the stars) to be essentially dark, and only illumined by the sun: that he noticed the motion of the solstitial and equinoctial points, but restricted it to a regular oscillation, of which he assigned the limit and the period: that he ascribed to the epicycles, by which the motion of a planet is represented, a form varying from the circle and nearly elliptic: that he recognised a motion of the nodes and apsides of all the primary planets, as well as of the moon; though in this instance, as in some others, his censurer imputes to him variance of doctrine.

The magnitude of the earth, and extent of the encompassing wind, is among the instances wherein he is reproached by BRAHMEGUPTA with versatility, as not having adhered to the same position throughout his writings; but he is vindicated on this, as on most occasions, by the scholiast of his censurer. Particulars of this question, leading to rather curious matter, deserve notice.

ÁRYABHAṬṬA's text specifies the earth's diameter, 1050 *yójanas;* and the orbit or circumference of the earth's wind [spiritus vector] 3393 *yójanas;* which, as the scholiast rightly argues, is no discrepancy. The diameter of this orbit, according to the remark of BRAHMEGUPTA, is 1080.

On this, it is to be in the first place observed, that the proportion of the circumference to the diameter of a circle, here employed, is that of 22 to 7; which, not being the same which is given by BRAHMEGUPTA's rule, (Arithm. § 40,) must be presumed to be that, which ÁRYABHAṬṬA taught. Applying it to the earth's diameter as by him assigned, viz. 1050, the circumference of the earth is 3300; which evidently constitutes the dimensions by him intended: and that number is accordingly stated by a commentator of BHÁSCARA. See *Gań.* on *Líl.* § 4.

This approximation to the proportion of the diameter of a circle to its periphery, is nearer than that which both BRAHMEGUPTA and ŚRÍD'HARA, though later writers, teach in their mensuration, and which is employed in the *Súrya-sidd'hánta;* namely, one to the square-root of ten. It is adopted by BHÁSCARA, who adds, apparently from some other authority, the still nearer approximation of 1250 to 3927.—(*Líl.* § 201.)

ÁRYABHAṬṬA appears, however, to have also made use of the ratio which afterwards contented both BRAHMEGUPTA and ŚRÍD'HARA; for his rule adduced by GAŃÉŚA (*Líl.* § 207) for finding the arc from the chord and versed sine, is clearly founded on the proportion of the diameter to the periphery, as one to the square root of ten: as will be evident, if the semicircle be computed by that rule: for it comes out the square root of $\frac{10}{4}$, the diameter being 1.

A more favourable notion of his proficiency in geometry, a science, however, much less cultivated by the Hindus than Algebra, may be received from his acquaintance with the theorem containing the fundamental property of the circle, which is cited by PRĬT'HÚDACA.—(*Brahm.* 12, § 21.)

The number of 3300 *yójanas* for the circumference of the earth, or $9\frac{1}{6}$ *yójanas* for a degree of a great circle, is not very wide of the truth, and is, indeed, a very near approach, if the *yójana*, which contains four *cróśas*, be rightly inferred from the modern computed *cróśa* found to be 1, 9 B. M.[1] For, at that rate of 7, 6 miles to a *yójana*, the earth's circumference would be 25080 B. miles.

The difference between the diameter of the earth, and that of its air (*váyu*), by which term ÁRYABHAṬṬA seems to intend a current of wind whirling as a vortex, and causing the earth's revolution on its axis, leaves 15 *yójanas*, or 114 miles, for the limit of elevation of this atmospheric current.

[1] As. Res. 5, 105. (8vo.)

H.

SCANTINESS OF THE ADDITIONS BY LATER WRITERS ON ALGEBRA.

THE observation in the text on the scantiness of the improvements or additions made to the Algebra of the Hindus in a long period of years after ÁRYABHAṬṬA probably, and after BRAHMEGUPTA certainly, is extended to authors whose works are now lost, on the faith of quotations from them. ŚRÍD'HARA's rule, which is cited by BHÁSCARA (*Víj.-gań.* § 131) concerning quadratics, is the same in substance with one of BRAHMEGUPTA's (Ch. 18. § 32—33). PADMANÁBHA, indeed, appears from the quotation from his treatise (*Víj.-gań.* § 142.) to have been aware of quadratic equations affording two roots; which BRAHMEGUPTA has not noticed; and this is a material accession which the science received. There remains an uncertainty respecting the author, from whom BHÁSCARA has taken the resolution of equations of the third and fourth degrees in their simple and unaffected cases.

The only names of Algebraists, who preceded BHÁSCARA, to be added to those already mentioned, are 1st an earlier writer of the same name (BHÁSCARA) who was at the head of the commentators of ÁRYABHAṬṬA; and 2d, the elder scholiast of the *Brahma-sidd'hánta*, named BHAṬṬA BALABHADRA. Both are repeatedly cited by the successor of the latter in the same task of exposition, PRĬT'HÚDACA SWÁMÍ; who was himself anterior to the author of the *Śirómańi*; being more than once quoted by him. As neither of those earlier commentators is named by the younger BHÁSCARA; nor any intimation given of his having consulted and employed other treatises besides the three specified by him, in the compilation of the *Víja-gańita*, it is presumable, that the few additions, which a comparison with the *Cuṭṭaca* of BRAHMEGUPTA exhibits, are properly ascribable either to ŚRÍD'HARA or to PADMANÁBHA: most likely to the latter; as he is cited for one such addition;[1] and as ŚRÍD'HARA's treatise of arithmetic and mensuration, which is extant, is not seemingly the work of an author improving on the labours of those who went before him.[2] The corrections and improvements introduced by BHÁSCARA himself, and of which he carefully apprizes his readers,[3] are not very numerous, nor in general important.[4]

[1] *Víj.-gań.* § 142. [2] *Líl.* § 147. *Brahm.* 12, § 21 and 40. *Gań. Sár.* § 126.

[3] *Víj.-gań.* before § 44, and after § 57. also Ch. 1, towards the end; and Ch. 5. § 142.

[4] Unless *Líl.* § 170 and 190.

I.

AGE OF ÁRYABHAṬṬA.

UNDER the Abbasside Khalifs ALMANSÚR and ALMÁMÚN, in the middle of the eighth and beginning of the ninth centuries of the Christian era, the Arabs became conversant with the Indian astronomy. It was at that period, as may be presumed, that they obtained information of the existence and currency of three astronomical systems among the Indians;[1] one of which bore the name of ÁRYABHAṬṬA, or, as written in Arabic characters, ÁRJABAHAR,[2] (perhaps intended for ÁRJABHAR) which is as near an approximation as the difference of characters can be expected to exhibit. This then unquestionably was the system of the astronomer whose age is now to be investigated; and who is in a thousand places cited by Hindu writers on Astronomy, as author of a system and founder of a sect in this science. It is inferred from the acquaintance of the Arabs with the astronomical attainments of the Hindus, at that time, when the court of the Khalif drew the visit of a Hindu astrologer and mathematician, and when the Indian determination of the mean motions of the planets was made the basis of astronomical tables compiled by order of the Khalifs, 'for a guide in matters pertaining to the stars,' and when Indian treatises on the science of numbers were put in an Arabic dress; adverting also to the difficulty of obtaining further insight into the Indian sciences, which the author of the *Tárikhu'l hukmá* complains of, assigning for the cause the distance of countries, and the various impediments to intercourse: it is inferred, we say, from these, joined to other considerations, that the period in question was that in which the name of ÁRYABHAṬṬA was introduced to the knowledge of the Arabs. This, as a first step in inquiring the antiquity of this author, ascertains his celebrity as an astronomical authority above a thousand years ago.

He is repeatedly named by Hindu authors of a still earlier date: particularly by BRAHMEGUPTA, in the first part of the seventh century of the Christian era. He had been copied by writers whom BRAHMEGUPTA cites. VARÁHA-MIHIRA has allusions to him, or employs his astronomical determi-

[1] *Tárikhu'l hukmá*, or Bibl. Arab. Phil. quoted by CASIRI: Bibl. Arab. Hisp. 426. See below, Note M.

[2] COSSALI's *Argebakr* is a misprint (Orig. &c. dell' Alg. i. 207). CASIRI gives, as in the Arabic, *Argebahr:* which, in the orthography here followed, is *Arjabahr*.

nations in an astrological work at the beginning of the sixth century. These facts will be further weighed upon as we proceed.

For determining ÁRYABHAṬṬA's age with the greater precision of astronomical chronology, grounds are presented, at the first view promising, but on examination insufficient.

In the investigation of the question upon astronomical grounds, recourse was in the first place had to his doctrine concerning the precession of the equinoxes. As quoted by MUNÍŚWARA, a scholiast of BHÁSCARA, he maintained an oscillation of the equinoctial points to twenty-four degrees on either side; and he reckoned 578159 such librations in a *calpa*.[1] From another passage cited by BHAṬṬÓTPALA on VARAHA-MIHIRA,[2] his position of the mean equinoxes was the beginning of Aries and of Libra.[3] From one more passage quoted by the scholiast of BRAHMEGUPTA,[4] it further appears, that he reckoned 1986120000 years expired[5] before the war of the *Bhárata:* and the duration of the *Calpa*, if he be rightly quoted by BRAHMEGUPTA,[6] is 1008 quadruple *yugas* of 4320000 years each.

From these data it follows that according to him, the equinoctial point had completed 263699 oscillations at the epoch of the war of the *Bhárata*. But we are without any information as to the progress made in the current oscillation when he wrote; or the actual distance of the equinox from the beginning of *Mésha:* the position of which, also, as by him received, is uncertain.

His limit of the motion in trepidation, 24°, was evidently suggested to him by the former position of the colures declared by PÁRÁSÁRA; the exact difference being 23° 20′. But the commencement of PÁRÁSÁRA's *Áślésha*, in his sphere, or the origin of his siderial *Mésha*, are unascertained. Whether his notions of the duodecimal division of the Zodiac were taken from the Grecian or Egyptian spheres, or from what other immediate source, is but matter of conjecture.

Quotations of this author furnish the revolutions of Jupiter in a *yuga*,[7] and of Saturn's aphelion in a *Calpa*;[8] and those of the moon in the latter

[1] As. Res. 12. 213.

[2] Vrihat-sanhitá. 2.

[3] 'From the beginning of *Mésha* to the end of *Canyá* (Virgo), the half the ecliptic passes through the north. From the beginning of *Tula* to the end of (the fishes) *Mina*, the remaining half passes by the south.'

[4] PRIT'HÚDACA on Brahm. c. 1. § 10 and 30. c. 11. § 4.

[5] Six *menus*, twenty-seven *yugas* and three quarters.

[6] PRIT'HÚDACA on Brahm. c. 1. § 12.

[7] As. Res. 3. 215.

[8] *Mun.* on *Bhás.* c. 1. § 33.

period: but the same passage,[1] in which the number of lunar revolutions in that great period are given, supplies those of the sun; namely 4320000000; differing from the duration of the *Calpa* according to this author as cited by more ancient compilers. The truth is, as appears from another quotation,[2] that ÁRYABHATT'A, after delivering one complete astronomical system, proceeds in a second and distinct chapter to deliver another and different one as the doctrine of PA'RA'SARA; whose authority, he observes, prevails in the *Cali* age: and though he seems to indicate the *Calpa* as the same in both, he also hints that in one a deduction is made for the time employed in creation; and we have seen, that the duration of the *Calpa* differs in the quotations of compilers from this author.

The ground then being insufficient, until a more definitive knowledge of either system, as developed by him, be recovered, to support any positive conclusion, recourse must be had, on failure of precise proof, to more loose presumption. It is to be observed, that he does not use the *Śaca* or *Sambat* of VICRAMA'DITYA, nor the *Śaca* era of SA'LIVA'HANA: but exclusively employs the epoch of the war of the *Bhárata*, which is the era of YUDHISHT'HIRA and the same with the commencement of the *Cali yuga*. Hence it is to be argued, that he flourished before this era was superseded by the introduction of the modern epochas. VARA'HA-MIHIRA, on the other hand, does employ the *Śaca*, termed by him *Śaca-bhúpa-cála* and *Śacéndra-cála:* which the old scholiast interprets "the time when the barbarian kings called *Śaca* were discomfited by VICRAMA'DITYA:"[3] and BRAHMEGUPTA uses the modern *Śaca* era; which he expresses by *Śaca-nrǐpánté*, interpreted by the scholiast of BHA'SCARA "the end [of the life or reign] of VICRAMA'DITYA who slew a people of barbarians named *Śacas*." VARA'HA-MIHIRA's epoch of *Śaca* appears to have been understood by his scholiast BHATT'ÓTPALA to be the same with the era of VICRAMA'DITYA, which now is usually called *Sambat*; and which is reckoned to commence after 3044 years of the *Cali* age were expired: and BRAHMEGUPTA's epoch of *Śaca* is the era of SA'LIVA'HANA beginning at the expiration of 3179 years of the *Cali yuga:* and accordingly this number is specified in his *Brahma-sidd'hánta.* When those eras were first introduced is not at present with certainty known. If that of VICRAMA'DITYA, dating with a most memorable event of his reign, came into use

[1] *Mun.* on *Bhás.* c. 1. § 16—18. [2] *Várt.* and *Mun.* on *Bhás.* [3] *Vrǐhat-sanhitá.*

during its continuance, still its introduction could not be from the first so general as at once and universally to supersede the former era of YUDHISH-T'HIRA. But the argument drawn from ÁRYABHAT́T́A's use of the ancient epoch, and his silence respecting the modern, so far as it goes, favours the presumption that he lived before the origin of the modern eras. Certainly he is anterior to BRAHMEGUPTA, who cites him in more than a hundred places by name; and to VARÁHA-MIHIRA, whose compilation is founded, among other authorities, on the *Rómaca* of ŚRÍSHÉŃA, and *Vásisht'ha* of VISHNU-CHANDRA, which BRAHMEGUPTA affirms to be partly taken from ÁRYABHAT́T́A.[1] The priority of this author is explicitly asserted likewise by the celebrated astronomer GAŃÉŚA, who, in explanation of his own undertaking, says, "Rules framed by other holy sages were right in the *Trétá* "and *Dwápara;* but, in the present age, PÁRÁŚARA's. ÁRYABHAT́T́A, "however, finding his imperfect, after great lapse of time, reformed the sys-"tem. It grew inaccurate and was therefore amended by DURGASINHA, "MIHIRA, and others. This again became insufficient: and correct rules "were framed by the son of JISHNU [BRAHMEGUPTA] founded upon "BRAHMA's revelation. His sytem also, after a long time, came to exhibit "differences. CÉŚAVA rectified it. Now, finding this likewise, a little in-"correct after sixty years, his son GAŃÉŚA has perfected it, and reconciled "computation and experience."[2]

ÁRYABHAT́T́A then preceded BRAHMEGUPTA who lived towards the middle of the sixth century of the *Śaca* era; and VARÁHA-MIHIRA placed by the chronologers of *Ujjayaní* at the beginning of the fifth or of the second; (for they notice two astronomers of the name.) He is prior also to VISHŃU CHANDRA, ŚRÍSHÉŃA, and DURGASINHA; all of them anterior to the second VARÁHA-MIHIRA; and an interval of two or of three centuries is not more than adequate to a series of astronomers following each other in the task of emendation, which process of time rendered successively requisite.

On these considerations it is presumed, that ÁRYABHAT́T́A is unquestionably to be placed earlier than the fifth century of the *Śaca:* and probably so, by several (by more than two or three) centuries: and not unlikely before the commencement of either *Śaca* or *Sambat* eras. In other words, he flourished some ages before the sixth century of the Christian era: and perhaps lived before, or, at latest, soon after its commencement. Between these limits, either

[1] *Brahm. Sidd'h.* c. 11. § 48—51.

[2] Citation by NRÍSINHA on *Súr. Sidd'h.*

the third or the fourth century might be assumed as a middle term. We shall, however, take the fifth of CHRIST as the latest period to which ÁRYABHAṬṬA can, on the most moderate assumption, be referred.

K.

WRITINGS AND AGE OF VARÁHA-MIHIRA.

THIS distinguished astrological writer, a native of *Ujjayaní*, and son of ÁDITYADÁSA,[1] was author of a copious work on astrology, compiled, and, as he declares, abridged from earlier writers. It is comprised in three parts; the first on astronomy; the second and third, on divination: together constituting a complete course. Such a course, he observes in his preface to the third part, has been termed by ancient writers *Sanhitá*, and consists of three *Scand'has* or parts: the first, which teaches to find a planet's place by computation *(gańita)*, is called *tantra;* the second, which ascertains lucky and unlucky indications, is named *hórá;* it relates chiefly to nativities, journeys, and weddings; the third, on prognostics relative to various matters, is denominated *Sác'há*. The direct and retrograde motions of planets, with their rising and setting, and other particulars, he goes on to say, had been propounded by him in a treatise termed *Carańa*, meaning, as the scholiast remarks, his compilation entitled *Pancha-sidd'hánticá:* which constitutes the first and astronomical portion of his entire work. What relates to the first branch of astrology *(hórá)*, the author adds, had likewise been delivered by him, including nativities and prognostics concerning journeys and weddings. These astrological treatises of his author, the scholiast observes, are entitled *Vrĭhat-játaca*, *Vrĭhad-yátrá*, and *Vrĭhad-viváha-paṭala*. The author proceeds to deliver the third part of his course, or the second on divination, omitting, as he says, superfluous and pithless matter, which abounds in the writings of his predecessors: such as questions and replies in dialogue, legendary tales, and the mythological origin of the planets.

[1] *Vrĭhat-játaca*, c. 26 § 5; where the author so describes himself. His scholiast also calls him *Avantica* from his native city *Ujjayaní*, and terms him a *Magadha Brahmen*, and a compiler of astronomical science. BHAṬṬÓTPALA on *Vrĭ.-ját.* 1. The same scholiast similarly describes him in the introduction of a commentary on a work of his son PRĬTHUYASÁS.

The third part is extant, and entire; and is generally known and cited by the title of *Vrĭhat-sanhitá;* or great course of astrology: a denomination well deserved; for, notwithstanding the author's professions of conciseness, it contains about four thousand couplets distributed in more than a hundred chapters, or precisely (including the metrical table of contents) 106.

Of the second part, the first section, on casting of nativities, called *Vrĭhat-játaca*, is also extant, and comprises twenty-five chapters; or, with the metrical table of contents and peroration which concludes it, twenty-six. The other two sections of this part of the course have not been recovered, though probably extant in the hands of Hindu astrologers.

The scholia of the celebrated commentator of this author's works, who is usually called BHAṬṬÓTPALA, and who in several places of his commentary names himself UTPALA, (quibbling with simulated modesty on his appellation; for the word signifies stone:[1]) are preserved; and are complete for the third part of the author's course; and for the first section of the second: and the remainder of it likewise is probably extant; as the copy of the first section, in the possession of the author of this dissertation, terminates abruptly after the commencement of the second.

This commentator is noticed in the list of authorities furnished by the astronomers of *Ujjayaní;* and is there stated as of the year 890 of the *Śaca* era (A. D. 1068). Sir *William* JONES supposed him to be the son of the author, whose work is expounded by him. The grounds of this notion, which is not, however, very positively advanced by that learned orientalist,[2] are not set forth. No intimation of such relation of the scholiast to his author, appears in the preface or the conclusion, nor in the colophon, of the commentary which has been inspected: nor in the body of the work; where the author is of course repeatedly named or referred to, without however any addition indicative of filial respect, as Hindu writers usually do employ when speaking of a parent or ancestor. Neither is there any hint of relationship in the commentary of the same scholiast BHAṬṬÓTPALA on a brief treatise of divination, entitled *Praśna-cóshṭí*, comprizing fifty-six stanzas, by

[1] Preface to the Commentary on the *Vrĭhat-játaca.* Conclusion of the gloss on ch. 18 of *Vrĭhat-sanhitá*, &c. 'Stone *(utpala)* frames the raft of interpretation to cross the ocean composed by *Varáhamihira.*'

[2] The words are 'the comment written by BHAṬṬÓTPALA, who, it seems, was a son of the author.' *As. Res.* 2. 390.

PRĬT'HUYAŚAS son of VARÁHA-MIHIRA. The suggestion of the filial relation of the scholiast is probably therefore a mere error.

The *Pancha-sidd'hánticá* of VARÁHA-MIHIRA has not yet been recovered; and is only at present known from quotations of authors; and particularly a number of passages cited from it by his scholiast in course of interpreting his astrological writings. An important passage of it so quoted will be noticed forthwith.

It is a compilation, as its name implies, from five *sidd'hántas;* and they are specified in the second chapter of the *Vrĭhat-sanhitá,* where the author is enumerating the requisite qualifications of an astronomer competent to calculate a calendar: among other attainments he requires him to be conversant with time measured by *yugas,* &c. as taught in the five *sidd'hántas* upon astronomy named *Pauliśa, Rómaca, Vásisht'ha, Saura,* and *Paitámaha.*[1]

The title of VARÁHA-MIHIRA's compilation misled a writer on Hindu astronomy[2] into an unfounded supposition, that he was the acknowledged author of the five *sidd'hántas;* the names of two of which moreover are mistaken, *Sóma* and *Paulastya* being erroneously substituted for *Rómaca* and *Pauliśa.* These two, as well as the *Vásisht'ha,* are the works of known authors, namely, PULIŚA, ŚRÍSHÉNA, and VISHNU-CHANDRA; all three mentioned by BRAHMEGUPTA: by whom also the whole five *sidd'hántas* are noticed under the very same names and in the same order;[3] and who has specified the authors of the first three.[4] The *Vásisht'ha* of VISHŃU-CHANDRA was indeed preceded by an earlier work (so entitled) of an unknown author, from which that, as well as the *Rómaca,* is in part taken;[5] and it may be deemed an amended edition: but the *Rómaca* and *Pauliśa* are single of the names; and no Hindu astronomer, possessing any knowledge of the history of the science cultivated by him, ever could imagine, that VARÁHA-MIHIRA composed the work which takes its name from PULIŚA, the distinguished founder of a sect or school in astronomy opposed to that of ÁRYA-BHAŤŤA.

The passage of the *Pancha-sidd'hánticá* cited by the scholiast,[6] and promised to be here noticed, has been quoted in an essay inserted in the researches of the Asiatic Society,[7] as well as a parallel passage of the *Vrĭhat-*

[1] *Vrĭhat-sanhitá,* c. 2. § 7.
[2] As. Res. 8. 196.
[3] *Brahm. Sidd'h.* c. 14.
[4] Ibid. c. 11.
[5] Ibid.
[6] On *Vrĭhat-sanhitá,* c. 2.
[7] As. Res. 12.

sanhitá,[1] both relative to the ancient and actual position of the colures; and deemed parallel (though one be less precise than the other); since they are cited together as of the same author, and consequently as of like import, by the scholiast.[2] The text of the *Vrĭhat-sanhitá* is further authenticated by a quotation of it in the commentary of Prĭt'húdaca on Brahmegupta;[3] and the former position of the colures is precisely that which is described in the calendar appendant on the *Védas,*[4] and which is implied in a passage of Párásara, concerning the seasons, which is quoted by Bhattótpala.

The position of the colures, affirmed as actual in his time by Varáha-mihira, in the *Vrĭhat-sanhitá,* implies an antiquity of either 1216 or 1440 years before A. D. 1800, according to the origin of the ecliptic determined from the star *Chitrá* (Spica virginis) distant either 180° or 183° from it; or a still greater antiquity, if it be taken to have corresponded more nearly with the Grecian celestial sphere. The mean of the two numbers (disregarding the surmise of greater antiquity,) carries him to A. D. 472. If Varáha-mihira concurred with those Indian astronomers, who allow an oscillation of the equinox to 27° in 1800 years, or a complete oscillation of that extent both E. and W. in 7200 years, he must have lived soon after the year 3600 of the *Cali yuga,* or 421 *Sáca,* answering to A. D. 499; which is but six years from the date assigned to him by the astronomers of *Ujjayaní:* and twenty-seven from the mean before inferred.

It is probable, therefore, that he flourished about the close of the fifth century of the Christian era; and this inference is corroborated by the mention of an astrologer of this name in the *Panchatantra,* the *sanscrit* original of the fables of Pilpay translated in the reign of Nushirvan, King of Persia, in the latter part of the sixth century and beginning of the seventh.[5]

To that conclusion there is opposed an argument drawn from a passage of the *Bhásvatí-caraña;* in which the author of that treatise dated 1021 *Saca* (A. D. 1098) professes to have derived instruction from *Mihira,* meaning, as is supposed, oral instruction from Varáha-mihira; and the argument has been supported by computations which make the *Súrya-sidd'hánta* and *Játacárňava,* the latter ascribed to Varáha-mihira, to be both works of the same period, and as modern as the eleventh century.[6]

[1] C. 3. § 1 and 2.

[2] On *Vrĭ. Sanh.* c. 2.

[3] *Brahm. Sidd'h.* c. 11, § 54.

[4] As. Res. 8. 469.

[5] Pref. to the Sansc. *Hitópadésa.* Edit. Serampur.

[6] As. Res. 6. 572.

To this it has been replied, that the MIHIRA, from whom SATANANDA, author of the *Bháswatí*, derived instruction, is not the same person or personage with the author of the *Vrihat sanhitá;* if indeed SATÁNANDA's expression do intend the same name, VARÁHA.[1] That expression must be allowed to be a very imperfect designation, which omits half, and that the most distinctive half, of an appellation: and it is not such, as would be applied by a contemporary and auditor to an author and lecturer, whose celebrity could not yet be so generally diffused, as to render a part of his name a sufficient intimation of the remainder: without previous and well established association of the terms. But even conceding the interpretation, it would then be right to admit a third VARÁHA-MIHIRA, besides the two noticed by the chronologists of *Ujjayaní;* and the third will be an astronomer, contemporary with RÁJA BHÓJA-DÉVA; and the preceptor of SATÁNANDA; and author of the *Játacárnava*, supposing this treatise on nativities to be properly ascribed to an author bearing that name, and to be on sufficient grounds referred to the eleventh century.

There remains to be here noticed another treatise on casting of nativities, to which the same favourite name of a celebrated astrologer is affixed. It is a concise tract entitled *Laghu-játaca:* and its authenticity as a work of the astrologer of *Ujjayaní* is established by the verifying of a quotation of the scholiast BHATTÓTPALA; who cites a passage of his author's compendious treatise on the same subject (*swalpa játaca*) in course of expounding a rule of prognostication concerning the destination of a prince to the throne and his future character as a monarch (*Vrihat-játaca*, 11, 1.). That passage occurs in the *Laghu-játaca* (Misc. Chap.). It is hardly to be supposed, that the same writer can have given a third treatise on the same subject of nativities, entitled *Játacárnava*.

The question concerning the age of the *Súrya-sidd'hánta* remains for consideration. It is a very material one; as both VARÁHA-MIHIRA and BRAHMEGUPTA speak of a *Sáura* (or Solar) *sidd'hánta*, which is a title of the same import: and, unless a work bearing this title may have existed earlier than the age, which is assigned, for reasons to be at a future time examined, to the *Súrya-sidd'hánta*, the conclusions respecting the periods when they respectively wrote, are impeached in the degree in which those grounds of calculation may deserve confidence. Those grounds in detail will be discussed at a separate opportunity. But independently of this discussion of

[1] As. Res. 12. p. 224.

their merits, sufficient evidence does exist to establish, that more than one edition of a treatise of astronomy has borne the name of *Súrya* (with its synonyma) the sun. For LACSHMÍDÁSA cites one under the title of *Vrĭhat súrya-sidd'hánta*[1] (for a passage which the current solar *Sidd'hánta* does not exhibit;) in contradistinction to another more frequently cited by him without the distinctive epithet of *Vrĭhat:* and in these latter instances his quotations admit of verification. A reference of BHÁSCARA to a passage of the *Saura*, or, as explained by his own annotation, the *Súrya-sidd'hánta*, does not agree with the text of the received *Súrya-sidd'hánta.*[2] His commentators indeed do not unreservedly conclude from the discrepancy a difference of the work quoted, and that usually received under the same title. Yet the inference seems legitimate. At all events the quotation from the *Vrĭhat-súrya-sidd'hánta*, in the *Gańita-tatwa-chintámańi* of LACSHMÍDÁSA, proves beyond question, that in that commentator's opinion, and consistently with his knowledge, more than one treatise bearing the same name existed.

There is evidence besides of Arabian writers, that a system of astronomy bearing the equivalent title of *Árca* (Solar) was one of three, which were found by them current among the Hindus, when the Arabs obtained a knowledge of the Indian astronomy in the time of the Abbasside Khalifs, about the close of the eighth century or commencement of the ninth of the Christian era.[3] *Árcand,* the name by which the Arabs designate one of those three astronomical systems, assigning it as an Indian term, is the well known corruption of *Árca* in the common dialects, and is familiar in the application of the same word as a name of a plant (Asclepias Gigantea) which, bearing all the synonyma of the sun, is called vulgarly *Ácand*, or *Árcand.*

The *solar* doctrine of astronomy appears then to have been known by this name to the Arabians as one of three Indian astronomical systems a thousand years ago. The fact is that both the title and the system are considerably more ancient. Revisions of systems occasionally take place; like BRAHMEGUPTA's revisal of the *Brahma-sidd'hánta*, to adapt and modernise them; or, in other words, for the purpose, as BRAHMEGUPTA intimates, of reconciling computation and observation. The *Súrya* or *Árca-sidd'hánta,* no doubt, has undergone this process; and actually exhibits manifest indications of it.[4]

In every view, it is presumed, that any question concerning the age of the present text of the *Súrya-sidd'hánta,* or determination of that question,

[1] *Gán. tawat chint.* on Spherics of *Śirómańi*, ch. 4. Cons. of Sines.

[2] As. Res. 12.

[3] See Note N.

[4] As. Res. 2. 235.

will leave untouched the evidence for the age of the author of the *Vrihat-sanhitá*, VARÁHA-MIHIRA, son of ADITYADÁSA, an astrologer of *Ujjayaní*, who appears to have flourished at the close of the fifth, or beginning of the sixth century of the Christian era. He was preceded, as it seems, by another of the same name, who lived, according to the chronologists of *Ujjayaní*, at the close of the second century. He may have been followed by a third, who is said to have flourished at the Court of RÁJÁ BHÓJA-DÉVA of *Dhara*, and to have had SATÁNANDA, the author of the *Bhásvatí*, for his scholar.

L.

INTRODUCTION AND PROGRESS OF ALGEBRA AMONG THE ITALIANS.

LEONARDO of *Pisa* was unquestionably the first who made known the Arabian Algebra to Christian Europe. This fact was, indeed, for a time disputed, and the pretensions of the Italians to the credit of being the first European nation, which cultivated Algebra, were contested, upon vague surmises of a possible, and therefore presumed probable, communication of the science of Algebra, together with that of Arithmetic, by the Saracens of Spain to their Christian neighbours in the Peninsula, and to others alleged to have resorted thither for instruction. The conjecture, hazarded by WALLIS (Algebra historical and practical) on this point, was assisted by a strange blunder, in which BLANCANUS was followed by VOSSIUS and a herd of subsequent writers, concerning the age of LEONARDO, placed by them precisely two centuries too low. The claims of the Italians in his favour, and for themselves as his early disciples, were accordingly resisted with a degree of acrimony (GUA. Mem. de l'Acad. des Sc. 1741. p. 436.) which can only be accounted for by that disposition to detraction, which occasionally manifests itself in the literary, as in the idler, walks of society. The evidence of his right to acknowledgments for transplanting Arabian Algebra into Europe, was for a long period ill set forth: but, when diligently sought, and carefully adduced, doubt was removed and opposition silenced.[1]

[1] Montucla, 2d Ed. Addns.

The merit of vindicating his claim belongs chiefly to COSSALI.[1] A manuscript of LEONARDO's treatise on Arithmetic and Algebra, bearing the title of *Liber Abbaci compositus a Leonardo filio Bonacci Pisano in anno* 1202, was found towards the middle of the last century by TARGIONI TOZZETTI[2] in the Magliabecchian library at Florence, of which he had the care; and another work of that author, on square numbers, was afterwards found by the same person inserted in an anonymous compilation, treating of computation, (un trattato d'Abbaco), in the library of a royal hospital at the same place. A transcript of one more treatise of the same writer was noticed by TOZZETTI in the Magliabecchian collection, entitled *Leonardi Pisani de filiis Bonacci Practica Geometriæ composita anno* 1220. The subject of it is confined to mensuration of land; and, being mentioned by the author in his epistle prefixed to the revised Liber Abbaci, shows the revision to be of later date. It appears to be of 1228.[3] TOZZETTI subsequently met with a second copy of the Liber Abbaci in Magliabecchi's collection: but it is described by him as inaccurate and incomplete.[4] A third has been since discovered in the Riccardian collection, also at Florence: and a fourth, but imperfect one, was communicated by NELLI to COSSALI.[5] No diligence of research has, however, regained any trace of the volume which contained LEONARDO's treatise on square numbers: the library, in which it was seen, having been dispersed previously to COSSALI's inquiries.

It appears from a brief account of himself and his travels, and the motives of his undertaking, which LEONARDO has introduced into his preface to the Liber Abbaci, that he travelled into Egypt, Barbary, Syria, Greece, and Sicily; that being in his youth at *Bugia* in Barbary, where his father BONACCI held an employment of scribe at the Custom House by appointment from *Pisa*, for Pisan merchants resorting thither, he was there grounded in the Indian method of accounting by nine numerals; and that finding it more commodious, and far preferable to that which was used in other countries visited by him, he prosecuted the study,[6] and with some additions of his own, and taking some things from Euclid's geometry, he undertook the com-

[1] Origine, &c. dell'Algebra. Parma 1797.

[2] *Viaggi*, i and vi. Edit. 1751—1754.

[3] *Cossali*, Origine, &c. c. 1. § 5.

[4] *Viaggi*, ii. Edit. 1768.

[5] Origine, &c. dell' Algebra, c. 2. § 1.

[6] Quare amplectens strictius ipsum modum Yndorum, et actentius studens in eo, ex proprio sensu quædam addens, et quædam ex subtilitatibus Euclidis geometriæ artis apponens, &c.

position of the treatise in question, that "the Latin race might no longer be found deficient in the complete knowledge of that method of computation." In the epistle prefixed to the revision of his work he professes to have taught the complete doctrine of numbers according to the Indian method.[1]

His peregrinations then, and his study of the Indian computation through the medium of Arabic, in an African city, took place towards the close of the twelfth century; the earliest date of his work being A. C. 1202.

He had been preceeded by more than two centuries, in the study of arithmetic under Muhammedan instructors, by GERBERT (the Pope SILVESTER II.[2]), whose ardour for the acquisition of knowledge led him at the termination of a two years noviciate, as a Benedictine, to proceed by stealth into Spain, where he learnt astrology from the Saracens, and with it more valuable science, especially arithmetic. This, upon his return, he communicated to Christian Europe, teaching the method of numbers under the designation of *Abacus*, a name apparently first introduced by him, (rationes numerorum Abaci,[3]) by rules abstruse and difficult to be understood, as WILLIAM of *Malmesbury* affirms: *Abacum certe primus a Saracenis rapiens, regulas dedit, quæ a sudantibus Abacistis vix intelliguntur.*[4] It was probably owing to this obscurity of his rules and manner of treating the Arabian, or rather Indian arithmetic, that it made so little progress between his time and that of the Pisan.

LEONARDO's work is a treatise of Arithmetic, terminated, as Arabic treatises of computation are similarly,[5] by the solution of equations of the two first degrees. In the enumeration and exposition of the parts comprised in his fifteenth chapter, which is his last, he says, *Tertia erit super modum Algebræ et Almucabalæ;* and, beginning to treat of it, *Incipit pars tertia de solutione quarundam quæstionum secundum modum Algebræ et Almucabalæ, scilicet oppositionis et restaurationis.* The sense of the Arabic terms are here given in the inverse order, as has been remarked by COSSALI, and as clearly appears from LEONARDO's process of resolving an equation, which will be hereafter shown.

[1] Plenam numerorum doctrinam edidi Yndorum, quem modum in ipsa scientia præstantiorem elegi.

[2] Archbishop in 992; Pope in 999; died in 1003.

[3] Ep. prefixed to his Treatise De Numerorum Divisione. *Gerb. Ep.* 160. (Ed. 1611.)

[4] De Gestis Anglorum, c. 2.

[5] See Mr. *Strachey's* examination of the *Khulásatu'l hisáb*, As. Res. 12. Early History of Alg.

He premises the observation, that in number three considerations are distinguished: one simple and absolute, which is that of number in itself: the other two, relative; being those of root and of square. The latter, as he adds, is called *census*, which is the term he afterwards employs throughout.

It is the equivalent of the Arabic *Mál*, which properly signifies wealth, estate; and *census* seems therefore to be here employed by LEONARDO, on account of its correspondent acceptation; *(quicquid fortunarum quis habet.* Steph.) in like manner as he translates the Arabic *shai* by *res*, thing, as a designation of the root unknown.

He accordingly proceeds to observe, that the simple number, the root, and the square *(census)*, are equalled together in six ways: so that six forms of equality are distinguished: the three first of which are called simple; and the three others compound. The order, in which he arranges them, is precisely that which is copied by PACIOLO.[1] It differs by a slight transposition from the order in which they occur in the earliest Arabic treatises of Algebra;[2] and which, no doubt, was retained in the Italian version from the Arabic executed by GUGLIELMO DI LUNIS, and others who are noticed by COSSALI upon indications which are pointed out by him.[3] For PACIOLO cautions the reader not to regard the difference of arrangement, as this is a matter of arbitrary choice.[4] LEONARDO's six-fold distinction, reduced to the modern algebraic notation, is 1st, $x^2=p\,x$. 2d, $x^2=n$. 3d, $p\,x=n$. 4th, $x^2+p\,x=n$. 5th, $p\,x+n=x^2$. 6th, $x^2+n=p\,x$. In PACIOLO's abridged notation it is 1st, *c° e c*. 2d, *c° e n°*. 3d, *c e n°*, *&c.*[5] The Arabic arrangement, in the treatise of the *Khuwarezmite*, is, 1st, $x^2=p\,x$. 2d, $x^2=n$. 3d, $p\,x=n$. 4th, $x^2+p\,x=n$. 2d, $x^2+n=p\,x$. 3d, $p\,x+n=x^2$. Later compilations transfer the third of these to the first place.[6]

Like the Arabs, LEONARDO omits and passes unnoticed the fourth form of quadratic equations, $x^2+p\,x+n=o$. It could not, indeed, come within the Arabian division of equations into simple, between species and species, and compound, between one species and two:[7] quantity being either stated affirmatively, or *restored* in this Algebra to the positive form. PACIOLO expressly observes, that in no other but these six ways, is any equation be-

[1] Summa de Arithmetica, &c.

[2] See Note N.

[3] Origine, &c. dell'Alg.

[4] Summa, 8. 5. 5.

[5] Summa, 8. 5. 5.

[6] *Khulásatu'l hisáb.*

[7] *Khulásatu'l hisáb.*

tween those quantities possible: *Altramente che i questi 6 discorsi modi non e possible alcuna loro equatione.*

LEONARDO's resolution of the three simple cases of equation is not exhibited by COSSALI. It is, however, the same, no doubt, with that which is taught by PACIOLO; and which precisely agrees with the rules contained in the Arabic books.[1] To facilitate comparison, and obviate distant reference, PACIOLO's rules are here subjoined in fewer words than he employs.

1st, Divide the things by the squares [coefficient by coefficient], the quotient is the value of thing.

2d, Divide the number by the squares [by the coefficient of the square], the root of the quotient is the value of thing.

3d, Divide the number by the things [that is, by the coefficient], the quotient is the value of thing.[2]

The resolution of the three cases of compound equations is delivered by COSSALI from LEONARDO, contracting his rugged Latin into modern Algebraic form.

1st, Be $x^2+p\,x=n$. Then $x=-\frac{1}{2}p+\surd(\frac{1}{4}p^2+n)$.

2d, Be $x^2=p\,x+n$. Then $x=\frac{1}{2}p+\surd(\frac{1}{4}p^2+n)$.

3d, Be $x^2+n=p\,x$. Then, if $\frac{1}{4}p^2\angle n$, the equation is impossible. If $\frac{1}{4}p^2=n$, then $x=\frac{1}{2}p$. If $\frac{1}{4}p^2 \,7\, n$, then $x=\frac{1}{2}p-\surd(\frac{1}{4}p^2-n)$, or $=\frac{1}{2}p+\surd(\frac{1}{4}p^2-n)$.

He adds the remark: *Et sic, si non solvetur quæstio cum diminutione, solvetur cum additione.*

The rules are the same which are found in the Arabic treatises of algebra.[3] The same rules will be likewise found in the work of PACIOLO, expressed with his usual verboseness in his Italian text: to which, in this instance, he has added in the margin the same instructions delivered in a conciser form in Latin memorial verses. As they are given at length by MONTUCLA, it is unnecessary to cite them in this place. On the subject of the impossible case PACIOLO adds, as a *Notandum utilissimum*, '*Sel numero qual si trova in la detta equatione accompagnato con lo censo, sel non e minore o veramente equale al quadrato de la mita de le cose, el caso essere insolubile: e pur consequente detto agguagliamento non potere avvenire per alcun modo.*' Summa, 8. 4. 12.

Concerning the two roots of the quadratic equation in the other case,

[1] See Note N; and As. Res. 12. [2] Summa, 8. 5. 6. [3] See Note N.

under the same head, he thus expands the short concluding remark of LEONARDO: *Sia che l'uno e l'altro modo satisfa al tema: ma a le volte se havi la verita a l'uno modo, a le volte a l'altro;*[1] *el perche, se cavanda la radice del detto remanente de la mita de le cose non satisfacesse al tema, la detta radice aggiugni a la mita de le cose e averai el quesito: e mai fallera che a l'uno di lai modi non sia satisfatta al quesito, cioe giognendo le, ovvero cavando la del dimeciamento de le cose.* Summa, 8. 4. 12.

BOMBELLI remarks somewhat differently on the same point. *Nei quesiti alcuna volta, ben che di rado, il restante non servi, ma ben si la somma sempre.* Alg. 2. 262.

The rules for the resolution of compound equations are demonstrated by LEONARDO upon rectilinear figures; and in the last instance he has reference to Euclid.—Lib. 2. Th. 5. There is room then to surmise, that some of the demonstrations are among the additions which he professes to have made.

Among the many problems which he proceeds to resolve, two of which are selected by COSSALI for instances of his manner, it will be sufficient to cite one, in the resolution of which the whole thread of his operations is exhibited; substituting, however, the more compendious modern signs. His manner of conducting the algebraic process may be fully understood from this single instance.

Problem: To divide the number 10 into two parts, such that dividing one by the other, and adding 10 to the sum of the quotient, and multiplying the aggregate by the greater, the amount is finally 114.

Let the right line a be the greater of the parts sought; which I call *thing* (quam pono rem): and the right line $b\,g$ equal to 10: to which are joined in the same direction $g\,d$, $d\,e$, representing the quotients of division of the parts, one by the other. Since a multiplied by $b\ e$ is equal to 114, therefore $a\times b\ g+a\times g\ d+a+d\ e=114$; and taking from each side $a\times b\ g$, there will be $a\times g\ d+a\times d\ e=114-a\times b\ g$. Be $g\ d$ the quotient $\frac{10-a}{a}$, there will arise $10-a+a\times d\ e=114-a\times b\ g=114-10\,a$; since $b\,g$ is equal to 10. Whence $a\times d\ e=104-9\,a$. But $d\,e$ is the quotient $\frac{a}{10-a}$: wherefore $\frac{a^2}{10-a}=104-9\,a$. So that $a^2=1040-194\,a+9\,a^2$. Restore diminished things (restaura res di-

[1] Compare with Hindu Algebra. *Vij.-gan.* § 130 and 142.

minutas), and take one square from each side (et extrahe unum censum ab utraque parte), the remainder is $8a^2+1040=194a$; and, dividing by eight, $a^2+130=24\frac{1}{4}a$; and resolving this according to rule, $a=\frac{97}{8}-\sqrt{\left(\frac{97}{8}\right)^2-130}=\frac{97}{8}-\frac{33}{8}=8$: consequently $10-a=2$.

Besides his great work on arithmetic and algebra, LEONARDO was author of a separate treatise, as already intimated, on square numbers. Reference is formally made to it by PACIOLO, who drew largely from this source, and who mentions *Le quali domande* (Questions concerning square numbers) *sono difficillissime quanto ala demonstratione dela practica: comme sa chi ben la scrutinato. Maxime* Leonardo *Pisano in un particulare tractato che fa* de quadratis numeris *intitulato. Dove con grande sforzo se ingegna dare norma e regola a simili solutioni.* Summa 1. 4. 6.

The directions for the solution of such problems being professedly taken by PACIOLO chiefly from LEONARDO, and the problems themselves which are instanced by him being probably so, it can be no difficult task to restore the lost work of LEONARDO on this subject. The divination has accordingly been attempted by COSSALI, and with a considerable degree of success. (Origine, &c. dell' Algebra, c. 5.)

Among problems of this sort which are treated by PACIOLO after LEONARDO, several are found in the current Arabic treatises; others, which belong to the indeterminate analysis, occur in the algebraic treatises of the Hindus: some, which are more properly Diophantine, may have been taken from the Arabic translation, or commentary, of the work of Diophantus. LEONARDO's endeavour to reduce the solution of such problems to general rule and system, according to PACIOLO's intimation of his efforts towards that end, must have been purely his own: as nothing systematic to this effect is to be found in the Arabic treatises of Algebra; and as he clearly had no communication through his Arab instructors, nor any knowledge of the Hindu methods for the general resolution of indeterminate problems simple or quadratic.

MONTUCLA, who had originally underrated the performance of LEONARDO, seems to have finally conceded to it a merit rather beyond its desert, when he ascribes to that author the resolution of certain biquadratics as derivative equations of the second degree. The derivative rules were, according to CARDAN's affirmation, added to the original ones of LEONARDO by an uncertain author; and placed with the principal by PACIOLO. CARDAN's testimony in this respect is indeed not conclusive, as the passage, in which the subject is mentioned, is in other points replete with errors: attributing the invention of Algebra to MUHAMMED son of MUSA, and alleging the testimony of LEONARDO to that point; limiting LEONARDO's rules to four, and intimating that PACIOLO introduced the derivative rules in the same place with the principal: all which is unfounded and contrary to the fact. COSSALI, however, who seems to have diligently examined LEONARDO's remains, does not claim this honour for his author: but appears to admit CARDAN's position, that the derivative, or, as they are termed by PACIOLO, the proportional equations, and rules for the solution of them, were devised by an uncertain author; and introduced by PACIOLO into his compilation under a separate head: which actually is the case. (Summa 8, 6, 2, &c.)

In regard to the blunder, in which MONTUCLA copied earlier writers, respecting the time when LEONARDO of *Pisa* flourished, he has defended himself (2d edit. Additions) against the reprehension of COSSALI, upon the plea, that he was not bound to know of manuscripts existing in certain libraries of Italy, which served to show the age in which that author lived. The excuse is not altogether valid: for TARGIONI TOZZETTI had announced to the public the discovery of the manuscripts in question, with the date, and a sufficient intimation of the contents; several years before the first volumes of MONTUCLA's History of Mathematics appeared.[1]

I am withheld from further animadversion on the negligence of an author, who has in other respects deserved well of science, by the consideration, that equal want of research, and in the very same instance, has been manifested by more recent writers, and among our own countrymen. Even so lately as in the past year (1816) a distinguished mathematician, writing in the Encyclopædia which bears the national appellation,[2] has relied on obsolete authorities and antiquated disquisitions concerning the introduction of the denary numerals into Europe; and shown total unacquaintance with what was made public sixty years ago by TARGIONI TOZZETTI and amply discussed by COSSALI in a copious work on the progress of Algebra in Italy,

[1] TARGIONI TOZZETTI's first volume bears date 1751. His sixth, (the last of his first edition) 1754. MONTUCLA's first two volumes were published in 1758.

[2] Encycl. Brit. Supp. art. Arithmetic.

and in an earlier one on the origin of Arithmetic, published more than twenty years since: matter fully recognised by MONTUCLA in his second edition, and briefly noticed in common biographical dictionaries.[1]

In the article of the Encyclopædia, to which reference has been just made, the author is not less unfortunate in all that he says concerning the Hindus and their arithmetical knowledge. He describes the *Lílávatí* as "a short and meagre performance headed with a silly preamble and colloquy of the gods." (Where he got this colloquy is difficult to divine; the *Lílávatí* contains none). "The examples," he says, "are generally very easy, and only written on the margin with red ink." (Not so written in any one among the many copies collated or inspected.) "Of fractions," he adds, "whether decimal or vulgar, it treats not at all." (See Ch. 2. Sect. 3. and Ch. 4. Sect. 2. also § 138.)

He goes on to say, "the Hindus pretend, that this arithmetical treatise was composed about the year 1185 of the Christian era, &c." Every thing in that passage is erroneous. The date of the *Lílávatí* is 1150, at the latest. The uncertainty of the age of a manuscript does not, as suggested, affect the certainty of the date of the original composition. It is not true, as alleged, that the oriental transcriber is accustomed to incorporate without scruple such additions in the text as he thinks fit. Nor is it practicable for him to do so with a text arranged in metre, of which the lines are numbered: as is the case with *Sanscrit* text books in general. Collation demonstrates that no such liberty has been taken with the particular book in question.

The same writer affirms, that "the Persians, though no longer sovereigns of Hindustan, yet display their superiority over the feeble Gentoos, since they generally fill the offices of the revenue, and have the reputation of being the most expert calculators in the east." This is literally and precisely the reverse of the truth; as every one knows, who has read or heard any thing concerning India.

The author is not more correct when he asserts, that "it appears from a careful inspection of the manuscripts preserved in the different public libraries in Europe, that the Arabians were not acquainted with the denary numerals before the middle of the thirteenth century of the Christian era." LEONARDO of *Pisa* had learned the Indian numerals from Arabian instruc-

[1] Dict. Hist. par Chaudon and Delandine: art. Leonard de Pise. 7 Edit. (1789). Probably in earlier editions likewise.

tion in the twelfth century and taught the use of them in the second year of the thirteenth: and the Arabs were in possession of the Indian mode of computation by these numerals so far back as the eighth century of the Christian era.[1]

To return to the subject.

After LEONARDO of *Pisa*, and before the invention of the art of printing and publication of the first printed treatise on the science, by PACIOLO, Algebra was diligently cultivated by the Italian mathematicians; it was publicly taught by professors; treatises were written on it; and recurrence was again had to the Arabian source. A translation of "the Rule of Algebra" (La Regola dell' Argebra) from the Arabic into the language of Italy by GUGLIELMO DI LUNIS, is noticed at the beginning of the *Ragionamento di Algebra* by RAFFAELO CARACCI, the extant manuscript of which is considered by antiquarians to be of the fourteenth century.[2] A translation of the original treatise of MUHAMMED BEN MUSA the Khuwarezmite appears to have been current in Italy; and was seen at a later period by both CARDAN and BOMBELLI.[3] PAOLO DELLA PERGOLA, DEMETRIO BRAGADINI, and ANTONIO CORNARO, are named by PACIOLO as successively filling the professor's chair at Venice; the latter his own fellow-disciple. He himself taught Algebra publicly at *Peroscia* at two different periods. In the preceding age a number of treatises on Algorithm, some of them with that title; others like LEONARDO's, entitled De Abaco, and probably like his touching on Algebra as well as Arithmetic, were circulated. PAOLO DI DAGOMARI, in particular, a mathematician living in the middle of the fourteenth century, obtained the surname of *Dell' Abaco* for his skill in the science of numbers, and is besides said to have been conversant with equations (whether algebraic or astronomical may indeed be questioned;) as well as geometry.[4]

With the art of printing came the publication of PACIOLO; and the subsequent history of the inventions in Algebra by Italian masters is too well known to need to be repeated in this place.

[1] See Note N.

[2] COSSALI, Orig. &c. dell' Algebra, i. 7.

[3] Ibid. i. 9. CARDAN Ars Magn. 5.

[4] Ibid. i. 9.

M.

ARITHMETICS OF DIOPHANTUS.

Five copies of Diophantus, vizt. three in the Vatican *(Cossali, Orig. dell' Alg.* i. 4. § 2.); Xilander's, supposed (*Coss. ib.* § 5) to be the same with the Palatine inspected by Saumaise, though spoken of as distinct by Bachet, *(Epist. ad. lect.)*; and the Parisian used by Bachet himself *(ib.)*; all contain the same text. But one of the Vatican copies, believed to be that which Bombelli consulted, distributes a like portion of text into seven instead of six books. (*Coss. ib.* § 5.) In truth the division of manuscript books is very uncertain: and it is by no means improbable, that the remains of Diophantus, as we possess them, may be less incomplete and constitute a larger portion of the thirteen books announced by him (*Def.* 11.), than is commonly reckoned. His treatise on polygon numbers, which is surmised to be one, (and that the last of the thirteen,) follows, as it seems, the six (or seven) books in the exemplars of the work, as if the preceding portion were complete. It is itself imperfect: but the manner is essentially different from that of the foregoing books: and the solution of problems by equations is no longer the object, but rather the demonstration of propositions. There appears no ground, beyond bare surmise, to presume, that the author, in the rest of the tracts relative to numbers which fulfilled his promise of thirteen books resumed the Algebraic manner: or in short, that the Algebraic part of his performance is at all mutilated in the copies extant, which are considered to be all transcripts of a single imperfect exemplar. (Bachet *Ep. ad. lect.*)

It is indeed alleged, that the resolution of compound equations (two species left equal to one) which Diophantus promises (*Def.* 11.) to show subsequently, bears reference to a lost part of his work. But the author, after confining himself to cases of simple equations (one species equal to one species) in the first three books, passes occasionally to compound equations (two species equal to one: and even two equal to two species;) in the three following books. See iv. Q. 33; vi. Q. 6 and 19; and Bachet on Def. 11, and i. Q. 33. In various instances he pursues the solution of the problem, until he arrives at a final quadratic equation; and, as in the case of a simple equation, he then merely states the value inferrible, without specifying the steps by which he arrives at the inference. See iv. Q. 23; vi. Q. 7, 9 and 11. But, in other places, the steps are sufficiently indicated: particularly iv. Q. 33 and 45; v. Q. 13; vi. Q. 24: and his method of resolving the equation is the same with the second of BRAHMEGUPTA's rules for the resolution of quadratics (*Brahm.* 18. § 34). The first of the Hindu author's rules, the same with S'RID'HARA's quoted by BHÁSCARA (*Vij.-gań.* § 131. *Brahm.* 18. § 32.), differs from that of NUGNEZ (NONIUS) quoted by BACHET (on *Dioph.* i. 33), in dispensing with the preliminary step of reducing the square term to a single square: a preparation which the Arabs first introduced, as well as the distinction of three cases of quadratics: for it was practised neither by DIOPHANTUS, nor by the Hindu Algebraists.

DIOPHANTUS has not been more explicit, nor methodical, on simple, than on compound, equations. But there is no reason to conclude, that he returned to either subject in a latter part of his work, for the purpose of completing the instruction, or better explaining the method of conducting the resolution of those equations. Such does not seem to be the manner of his arithmetics, in which general methods and comprehensive rules are wanting. It is rather to be inferred, as COSSALI does, from the compendious way in which the principles of Algebra are delivered, or alluded to, by him, that the determinate analysis was previously not unknown to the Greeks; wheresoever they got it: and that DIOPHANTUS, treating of it cursorily as a matter already understood, gives all his attention to cases of indeterminate analysis, in which perhaps he had no Greek precursor. (*Coss. Orig. dell' Alg.* i. 4. § 10.) He certainly intimates, that some part of what he proposes to teach is new: ἴσως μὲν ἂν δοκεῖ το πρᾶγμα δυσχερέστερον ἐπειδὴ μήπω γνώριμόν ἐστι. While in other places (*Def.* 10) he expects the student to be previously exercised in the algorithm of Algebra. The seeming contradiction is reconciled by conceiving the principles to have been known; but the application of them to a certain class of problems concerning numbers to have been new.

Concerning the probable antiquity of the Diophantine Algebra; all that can be confidently affirmed is, that it is not of later date than the fourth century of Christ. Among the works of HYPATIA, who was murdered A. D. 415, as they are enumerated by SUIDAS, is a commentary on a work of a DIOPHANTUS, most likely this author. An epigram in the Greek anthologia (lib. 2. c. 22) is considered with probability to relate to him: but the age of its author Lucillius is uncertain. Bachet observes, that, so far as can be conjectured, Lucillius lived about the time of Nero. This, however, is mere conjecture.

Diophantus is posterior to Hypsicles, whom he cites in the treatise on polygon numbers. (Prop, 8.) This should furnish another fixt point. But the date of Hypsicles is not well determined. He is reckoned the author, or at least the reviser,[1] of two books subjoined to Euclid's elements, and numbered 14th and 15th. In the introduction, he makes mention of Apollonius, one of whose writings, which touched on the ratio of the dodecaedron and icosaedron inscribed in the same sphere, was considered by Basilides of Tyre, and by the father of him (Hypsicles) as incorrect, and was amended by them accordingly: but subsequently he (Hypsicles) met with another work of Apollonius, in which the investigation of the problem was satisfactory, and the demonstration of the proposition correct. Here again Bachet observes, that, so far as can be conjectured, from the manner in which he speaks of Apollonius, he must have lived not long after him. Cossali goes a little further; and concludes on the same grounds, that they were nearly contemporary. (*Orig. dell' Alg.* i. 4. § 4.) The grounds seem inadequate to support any such conclusion: and all that can be certainly inferred is, that Hypsicles of Alexandria was posterior to Apollonius, who flourished in the reign of Ptolomy Euergetis: two hundred years before Christ.

Several persons of the name of Diophantus are noticed by Greek authors; but none whose place of abode, profession, or avocations, seem to indicate any correspondence with those of the mathematician and Algebraist: one a prætor of Athens mentioned by Diodorus Siculus, Zenobius, and Suidas; another, secretary of king Herod, put to death for forgery, as noticed by Tzetzes; and a third, the instructor of Libanius in eloquence, named by Suidas in the article concerning that sophist and rhetorician.

The Armenian Abu'lfaraj places the Algebraist Diophantus under the Emperor Julian. But it may be questioned, whether he has any authority for that date, besides the mention by Greek authors of a learned person of the name, the instructor of Libanius, who was contemporary with that emperor.

[1] *Tárikhu'l hukmá* cited by Casiri, *Bibl. Arab. Hisp.* i. 346. The Arabian author uses the word *Ásleh* amended.

Upon the whole, however, it seems preferable to abide by the date furnished in a professed history, even an Arabic one, on a Grecian matter: and consider DIOPHANTUS as contemporary with the Emperor JULIAN, about A. D. 365. That date is consistent with the circumstance of HYPATIA writing a commentary on his works; and is not contradicted by any other fact; nor by the affirmation of any other writer besides BOMBELLI: on whose authority COSSALI neverthelefs relies.

BOMBELLI, when he announced to the public the existence of a manuscript of DIOPHANTUS in the Vatican, placed the author under the Emperor ANTONINUS PIUS without citing any grounds. His general accuracy is, however, impeached by his assertion, that the Indian authors are frequently cited by DIOPHANTUS. No such quotations are found in the very manuscript of that author's work, which he is known to have consulted: and which has been purposely reexamined. (*Coss.* i. 4. § 4.) BOMBELLI'S authority was, therefore, very properly rejected by BACHET; and should have been so by COSSALI.

N.

PROGRESS AND PROFICIENCY OF THE ARABIANS IN ALGEBRA.

IN the reign of the second Abbasside Khalif ALMANSÚR, and in the 156th year of the *Hejira* (A. D. 773), as is related in the preface to the Astronomical tables of BEN-AL-A'DAMÍ published by his continuator ALCA'SEM in 308 H. (A.D. 920), an Indian astronomer, well versed in the science which he professed, visited the court of the Khalif, bringing with him tables of the equations of planets according to the mean motions, with observations relative to both solar and lunar eclipses and the ascension of the signs; taken, as he affirmed, from tables computed by an Indian prince, whose name, as the Arabian author writes it, was PHÍGHAR. The Khalif, embracing the opportunity thus happily presented to him, commanded the book to be translated into Arabic, and to be published for a guide to the Arabians in matters pertaining to the stars. The task devolved on MUHAMMED *ben* IBRA'HÍM *Alfazárí;* whose version is known to astronomers by the name of the greater *Sind-hind* or *Hind-sind:* for the term occurs written both ways.[1] It signifies, according to the same author BEN-AL-ÁDAMÍ, the revolving ages, *Al dehr al dáher;* which CASIRI translates perpetuum æternumque.[2]

No *Sanscrit* term of similar sound occurs, bearing a signification reconcilable to the Arabic interpretation. If a conjecture is to be hazarded, the original word may have been *Sidd'hánta.* Other guesses might be proposed: partly combining sound with interpretation, and taking for a termination *sind'hu* ocean, which occurs in titles now familiar for works relative to the regulation of time, as *Cála-sind'hu, Samaya-sind'hu,* &c. or adhering exclusively to sound, as *Indu-sindhu,* or *Indu-sidd'hánta;* the last a title of the same import with *Sóma-siddhánta* still current. But whatever may have been the name, the system of astronomy, which was made known to the Arabs, and which is by them distinguished by the appellation in question, appears to have been that which is contained in the *Brahma-sidd'hánta,* and which is taught in BRAHMEGUPTA's revision of it. This fact is deducible from the number of elapsed days between the beginning of planetary motions and the commencement of the present age of the world, according to the Indian reckoning, as it is quoted by the astrologer of *Balkh* ABU-MÂSHAR, and which precisely agrees with BRAHMEGUPTA. The astrologer does not indeed specify which of the Indian systems he is citing. But it is distinctly affirmed by later Arabian authorities, that only one of the three Indian doctrines of astronomy was understood by the Arabs; and that they had no knowledge of the other two beyond their names.[3] Besides, ÁRYABHAṬṬA and the *Árca-sidd'hánta,* the two in question, would have furnished very different numbers.

The passage of ABU-MÂSHAR, to which reference has been now made, is remarkable, and even important; and, as it has been singularly misunderstood and grossly misquoted by BAILLY in his Astronomie Ancienne (p. 302), it may be necessary to cite it at full length in this place. It occurs at the end of the fourth tract (and not, as BAILLY quotes, the beginning of the fifth,) in ABU-MÂSHAR's work on the conjunctions of planets. The author there observes, that " the Indians reckoned the beginning [of the world] on

[1] *Bibl. Arab. Hisp.* citing *Bibl. Arab. Phil (Táríkhu'l hukmá)* i. 428. voce *Alphazárí.*

[2] Ibid. i. 426. voce Katka. *Sind* and *Hind* likewise signify, in the Arabian writers, the hither and remoter India. *D'Herbelot. Bibl. Orient.* 415.

[3] *Táríkhu'l hukmá,* cited by CASIRI, *Bibl. Arab. Hisp.* i. 426. voce Katka.

" Sunday at sunrise (or, to quote from the Latin version, Et estimaverunt " Indi quod principium fuit die dominica sole ascendente;) and between that " day and the day of the deluge (et est inter eos, s. inter illum diem et illum " diem diluvii) 720634442715 days equivalent to 1900340938* Persian years " and 344 days. The deluge happened on Friday (et fuit diluvium die " veneris) 27th day of *Rabe* 1st, which is 29 from *Cibat* and 14 from *Adris-* " *tinich*. Between the deluge and the first day of the year in which the " *Hejira* occurred (fuerunt ergo inter diluvium et primum diem anni in " quo fuit Alhegira) 3837 years and 268 days; which will be, according to " the years of the Persians, 3725 years and 348 days. And between the " deluge and the day of Jesdagir (YEZDAJERD) king of the Persians, from " the beginning of whose reign the Persians took their era, ... 3735 years, " 10 months, and 22 days." The author proceeds with the comparison of the eras of the Persians and Arabians, and those of ALEXANDER and PHILIP; and then concludes the treatise: completi sunt quatuor tractatus, deo adjuvante.

BAILLY's reference to this passage is in the following words. " ALBU-" MASAR[2] rapporte que selon les Indiens, il s'est écoulé 720634442715 jours " entre le déluge et l'époque de l'hégire. Il en conclud, on ne sait trop com-" ment, qu'il s'est écoulé 3725 ans dans cet intervalle: ce qui placeroit le " déluge 3103 ans avant J. C. précisément à l'époque chronologique et astro-" nomique des Indiens. Mais ALBUMASAR ne dit point comment il est " parvenu à égaler ces deux nombres de 3725 ans et de 720634442715 jours." *Ast. anc.* ecl. liv. i. § xvii.

Now on this it is to be observed, that BAILLY makes the antediluvian period between the Sunday on which the world began and the Friday on which the deluge took place, comprising 720634442715 days, to be the same with the postdiluvian period, from the deluge to the Hejira; and that he quotes the author, as unaccountably rendering that number equivalent to 3725 years, though the text expressly states more than 1900000000 years. The blunder is the more inexcusable, as BAILLY himself remarked the inconsistency, and should therefore have reexamined the text which he cited, to verify his quotation.

* There is something wanting in the number of years: which is deficient at the third place. Both editions of the translation (Augsburg 1489, Venice 1515) give the same words.

[2] De Magn. Conj. Traité v, au commencement.

Major WILFORD (As. Res. 10., 117.), relying on the correctness of BAILLY's quotation, concluded, that the error originated with either the transcriber or translator. But in fact the mistake rested solely with the citer: as he would have found if his attention had been drawn to the more correct quotation in ANQUETIL DU PERRON's letter prefixed to his *Rech. Hist. et Geog. sur l'Inde*, inserted in BERNOULLI's 2d vol. of *Desc. de l'Inde* (p. xx). But, though ANQUETIL is more accurate than BAILLY in quotation, he is not more successful in his inferences, guesses and surmises. For he strangely concludes from a passage, which distinctly proves the use of the great cycle of the *calpa* by the Indian astronomers to whom ABUMÂSHAR refers, that they were on the contrary unacquainted in those days with a less cycle, which is comprehended in it. So little did he understand the Indian periods, that he infers from a specified number of elapsed days and correspondent years, reckoned from the beginning of the great cycle which dates from the supposed moment of the commencement of the world, that they knew nothing of a subordinate period, which is one of the elements of that cycle. Nor is he nearer the truth, but errs as much the other way, in his conjecture, that the number of solar years stated by ABUMÂSHAR relates to the duration of a life of BRAHMÁ, comprising a hundred of that deity's years.

In short, ANQUETIL's conclusions are as erroneous as BAILLY's premises. The discernment of Mr. DAVIS, to whom the passage was indicated by Major WILFORD, anticipated the correction of this blunder of BAILLY, by restoring the text with a conjectural emendation worthy of his sagacity.[1]

The name of the Indian author, from whom ABUMÂSHAR derived the particulars which he has furnished, is written by BAILLY, *Kankaraf;* taken, as he says, from an ancient Arabic writer, whose work is subjoined to that of *Messala* published at Nuremberg by Joach. HELLER in 1648.[2] The Latin translation of *Messahala* (MÁ-SHÁÄ-ALLAH) was edited by Joachim HELLER at Nuremberg in 1549; but it is not followed, in the only copy accessible to me, by the work of any other Arabic author; and the quotation consequently has not been verified. D'HERBELOT writes the name variously; *Kankah* or *Cancah*, *Kenker* or *Kankar*, and *Kengheh* or *Kanghah;*[3]

[1] As. Res. 9. 242. Appendix to an Essay of Major WILFORD.

[2] Ast. Anc. 303.

[3] *Bibl. Or.* Art. Cancah al Hendi, and Kenker al Hendi. Also Ketab Menazel al Camar and Ketab al Keranat.

to which REISKE and SCHULTENS, from further research, add another variation, *Kengch;*[1] which is not of Arabic but Persian orthography. CASIRI, by a difference of the diacritical point, reads from the *Táríkhu'l hukmá*, and transcribes, *Katka.*[2] That the same individual is all along meant, clearly appears from the correspondence of the works ascribed to him; especially his treatise on the greater and less conjunctions of the planets, which was imitated by ABU-MÂSHAR.

Amidst so much diversity in the orthography of the word it is difficult to retrieve the original name, without too much indulgence in conjecture. *Canca*, which comes nearest to the Arabic corruption, is in Sanscrit a proper name among other significations: but it does not occur as the appellation of any noted astrologer among the Hindus. GARGA does; and, as the Arabs have not the soft guttural consonant, they must widely corrupt that sound: yet *Canghar* and *Cancah* seem too remote from it to allow it to be proposed as a conjectural restoration of the Indian name.

To return to the more immediate subject of this note. The work of *Alfazárí*, taken from the Hindu astronomy, continued to be in general use among the Muhammedans, until the time of ALMÁMÚN; for whom it was epitomized by MUHAMMED *ben* MUSA *Al Khuwárezmí;* and his abridgment was thenceforward known by the title of the less *Sind-hind.* It appears to have been executed for the satisfaction of ALMÁMÚN before this prince's accession to the Khelafet, which took place early in the third century of the Hejira and ninth of Christ. The same author compiled similar astronomical tables of his own; wherein he professed to amend the Indian tables which furnished the mean motions; and he is said to have taken, for that purpose, equations from the Persian astronomy; some other matters from PTOLOMY; and to have added something of his own on certain points. His work is reported to have been well received by both Hindus and Muhammedans: and the greater tables, of which the compilation was commenced in the following age by BEN AL ADAMÍ and completed by AL CASEM, were raised upon the like foundation of Indian astronomy: and were long in general use among the Arabs, and by them deemed excellent. Another and earlier set of astronomical tables, founded on the Indian system called *Sind-hind,* was compiled by HABASH an astronomer of *Baghdad;* who flourished in the

[1] *Bibl. Or.* (1777-79). iv. 725. Should be *Kengeh:* a like error occurs in p. 727, where *sharch* is put for *shareh.*

[2] *Bibl. Arab. Hisp.* i. 426.

time of the Khalif ALMÁMÚN.[1] Several others, similarly founded on the mean motions furnished by the same Indian system, were published in the third century of *Hejira* or earlier: particularly those of FAZL *ben* HÁTIM *Nárízí;* and AL HASAN *ben* MISBAH.[2]

It was no doubt at the same period, while the Arabs were gaining a knowledge of one of the Indian systems of Astronomy, that they became apprized of the existence of two others. No intimation at least occurs of any different specific time or more probable period, when the information was likely to be obtained by them; than that in which they were busy with the Indian astronomy according to one of the three systems that prevailed among the Hindus: as the author of the *Táríkhu'l hukmá* quoted by CASIRI affirms. This writer, whose compilation is of the twelfth century,[3] observes, that 'owing to the distance of countries and impediments to 'intercourse, scarcely any of the writings of the Hindus had reached the 'Arabians. There are reckoned, he adds, three celebrated systems *(Mazhab)* 'of astronomy among them; namely, *Sind* and *hind; Árjabahar,* and *Ár-'cand:* one only of which has been brought to us, namely, the *Sind-hind:* 'which most of the learned Muhammedans have followed.' After naming the authors of astronomical tables founded on that basis, and assigning the interpretation of the Indian title, and quoting the authority of BEN AL ADAMÍ, the compiler of the latest of those tables mentioned by him, he goes on to say, that 'of the Indian sciences no other communications have been re-'ceived by us (Arabs) but a treatise on music of which the title in *Hindí* is '*Biyáphar*, and the signification of that title "fruit of knowledge;"[4] the 'work entitled *Calílah* and *Damanah*, upon ethics: and a book of numerical 'computation, which ABU JÂFR MUHAMMED BEN MUSA *Al Khuwárezmí* 'amplified *(basat)* and which is a most expeditious and concise method, and 'testifies the ingenuity and acuteness of the *Hindus*.'[5]

The book, here noticed as a treatise on ethics, is the well-known collection of fables of *Pilpai* or *Bidpai* (Sans. *Vaidyapriya);* and was translated from

[1] *Tárikhu'l hukmá*, CASIRI, i. 426 and 428. ABULFARAJ; Pococke 161.

[2] *Ib.* i. 421 and 413.

[3] He flourished in 595 H. (A. D. 1198), as appears from passages of his work. M. S. MDCCLXXIII. Lib. Esc. p. 74 and 316. *Casiri*, ii. 332.

[4] *Sans. Vidyáphala*, fruit of science.

[5] CASIRI, i. 426 and 428. The *Cashfu'l zanún* specifies three astronomical systems of the Hindus under the same names.

the Pehleví version into Arabic, by command of the same Abbaside Khalif ALMANSÚR,[1] who caused an Indian Astronomical treatise to be translated into the Arabian tongue. The Arabs, however, had other communications of portions of Indian science, which the author of the *Táríkhu'l hukmá* has in this place overlooked: especially upon medicine, on which many treatises, general and particular, were translated from the Indian tongue. For instance, a tract upon poisons by SHANAC, (Sansc. *Characa?*) of which an Arabic version was made for the Khalif ALMÁMÚN, by his preceptor ABBAS *ben* SÂÏD *Jóharí*. Also a treatise on medicine and on materia medica in particular, which bears the name of SHASHURD (Sansc. ŚUŚRUTA); and numerous others.[2]

The Khuwarezmite MUHAMMED BEN MUSA, who is named as having made known to the Arabians the Indian method of computation, is the same who is recognized by Arabian authors with almost a common consent (ZACARIA of *Casbín*, &c.) as the first who wrote upon Algebra. His competitor for the honour of priority is ABU KÁMIL SHUJAÂ *ben* ASLAM, surnamed the Egyptian arithmetician, *(Hásib al Miśrí,)*; whose treatise on Algebra was commented by ÂLI *ben* AHMED *Al Âmrání* of *Muśella;*[3] and who is said by D'HERBELOT to have been the first among learned Muslemans, that wrote upon this branch of mathematics.[4] The commentator is a writer of the tenth century; the date of his decease being recorded as of 344 H.[5] (A. D. 955.) The age, in which his author flourished, or the date of his text, is not furnished by any authority which has been consulted: and unless some evidence be found, showing that he was anterior to the *Khuwárezmí*, we may abide by the historical authority of ZACARIA of *Casbín;* and consider the *Khuwárezmí* as the earliest writer on Algebra in Arabic. Next was the celebrated Alchindus (ABU YUSEF ALKENDÍ) contemporary with the astrologer ABU-MÂSHER in the third century of the Hejira and ninth of the Christian era,[6] an illustrious philosopher versed in the sciences of Greece, of

[1] Introd. Rem. *Hitópadéśa.* Sansc. ed. 1804.

[2] D'HERBELOT, Bibl. Orient. Ketab al samoun, Ketab Sendhaschat, Ketab al sokkar, Ketab Schaschourd al Hendi, Ketab Rai al Hendi, Ketab Noufschal al Hendi, Ketab al akakir, &c.

[3] *Táríkhu'l hukmá*, CASIRI, i. 410.

[4] Bibl. Orient. 482. Also 226 and 494. No grounds are specified. EBN KHALCÁN and HÁJÍ KHALFAH, whom he very commonly follows, have been searched in vain for authority on this point.

[5] *Tár.* CASIRI, i. 410.

[6] *Abulfaraj;* Pococke, 179.

India, and of Persia, and author of several treatises upon numbers. In the prodigious multitude of his writings upon every branch of science, one is specified as a tract on Indian computation *(Hisábu'l hindí)*: others occur with titles which are understood by CASIRI to relate to Algebra, and to the 'finding of hidden numbers:' but which seem rather to appertain to other topics.[1] It is, however, presumable, that one of the works composed by him did treat of Algebra as a branch of the science of computation. His pupil AHMED *ben* MUHAMMED of *Sarkhasi* in Persia, (who flourished in the middle of the third century of the Hejira, for he died in 286 H.) was author of a complete treatise of computation embracing Algebra with Arithmetic. About the same time a treatise of Algebra was composed by ABU HANIFAH *Dainawarí*, who lived till 290 H. (A. D. 903.)

At a later period ABU'LWAFÁ *Buzjání*, a distinguished mathematician, who flourished in the fourth century of the *Hejira*, between the years 348 when he commenced his studies, and 388 the date of his demise, composed numerous tracts on computation, among which are specified several commentaries on Algebra: One of them on the treatise of the Khuwarezmite upon that subject: another on a less noted treatise by ABU YAHYA, whose lectures he had attended: an interpretation (whether commentary or paraphrase may perhaps be doubted) of the work of DIOPHANTUS: demonstrations of the propositions contained in that work: a treatise on numerical computation in general: and several tracts on particular branches of this subject.[2]

A question has been raised, as just now hinted, whether this writer's interpretation of DIOPHANTUS is to be deemed a translation or a commentary. The term, which is here employed in the *Táríkhu'l hukmá*, (*tafsir*, paraphrase,) and that which ABULFARAJ uses upon the same occasion (*fasr*, interpreted,) are ambiguous. Applied to the relation between works in the same language, the term, no doubt, implies a gloss or comment; and is so understood in the very same passage where an interpretation of the Khuwarezmite's treatise, and another of ABU YAHYA's, were spoken of. But, where a difference of language subsists, it seems rather to intend a version, or at least a paraphrase, than mere scholia; and is employed by the same author in a passage before cited,[3] where he gives the Arabic signification of a *Hindí* term. That *Buzjání*'s performance is to be deemed a

[1] *Táríkhu'l hukmá*; CASIRI, i. 353—360. [2] Ib. i. 433. [3] Ib. i. 426. Art. Katka.

translation, appears to be fairly inferrible from the separate mention of the demonstration of the propositions in DIOPHANTUS, as a distinct work: for the latter seems to be of the nature of a commentary; and the other consequently is the more likely to have been a version, whether literal or partaking of paraphrase. Besides, there is no mention, by any Arabian writer, of an earlier Arabic translation of DIOPHANTUS; and the *Buzjání* was not likely to be the commentator in Arabic of an untranslated Greek book. D'HERBELOT then may be deemed correct in naming him as the translator of the Arithmetics of DIOPHANTUS; and COSSALI, examining a like question, arrives at nearly the same conclusion; namely, that the *Buzjání* was the translator, and the earliest, as well as the expositor, of DIOPHANTUS.—(*Orig. dell' Alg.* i. 175.) The version was probably made soon after the date, which ABULFARAJ assigns to it, 348 H. (A. D. 969), which more properly is the date of the commencement of the translator's mathematical studies.

From all these facts, joined with other circumstances to be noticed in progress of this note, it is inferred, 1st, that the acquaintance of the Arabs with the Hindu astronomy is traced to the middle of the second century of the Hejira, in the reign of ALMANSÚR; upon authority of Arabian historians citing that of the preface of ancient astronomical tables: while their knowledge of the Greek astronomy does not appear to have commenced until the subsequent reign of HÁRÚN ALRASHÍD, when a translation of the Almagest is said to have been executed under the auspices of the Barmacide YAHYA *ben* KHÁLED, by ABA HIÁN and SALAMÁ employed for the purpose.[1] 2dly, That they were become conversant, in the Indian method of numerical computation, within the second century; that is, before the beginning of the reign of ALMÁMÚN, whose accession to the Khelafet took place in 205 H. 3dly, That the first treatise on Algebra in Arabic was published in his reign; but their acquaintance with the work of DIOPHANTUS is not traced by any historical facts collected from their writings to a period anterior to the middle of the fourth century of the *Hejira*, when ABU'LWAFÁ *Buzjání* flourished. 4thly, That MUHAMMED *ben* MUSA *Khuwárezmí*, the same Arabic author, who, in the time of ALMÁMÚN, and before his accession, abridged an earlier astronomical work taken from the Hindus, and who published a treatise on the Indian method of numerical computation, is the first also who furnished

[1] CASIRI, i. 349.

the Arabs with a knowledge of Algebra, upon which he expressly wrote, and in that Khalif's reign: as will be more particularly shown, as we proceed.

A treatise of Algebra bearing his name, it may be here remarked, was in the hands of the Italian Algebraists, translated into the Italian language, not very long after the introduction of the science into that country by LEONARDO of *Pisa*. It appears to have been seen at a later period both by CARDAN and by BOMBELLI. No manuscript of that version is, however, now extant; or at least known to be so.

Fortunately a copy of the Arabic original is preserved in the Bodleian collection. It is the manuscript marked CMXVIII Hunt. 214. fo. and bearing the date of the transcription 743 H. (A. D. 1342.) The rules of the library, though access be readily allowed, preclude the study of any book which it contains, by a person not enured to the temperature of apartments unvisited by artificial warmth. This impediment to the examination of the manuscript in question has been remedied by the assistance of the under librarian Mr. Alexander NICOLL; who has furnished ample extracts purposely transcribed by him from the manuscript. This has made it practicable to ascertain the contents of the book, and to identify the work as that in which the *Khuwárezmí* taught the principles of Algebra; and consequently to compare the state of the science, as it was by him taught, with its utmost progress in the hands of the Muhammedans, as exhibited in an elementary work of not very ancient date, which is to this time studied among Asiatic Muslemans.

I allude to the *Khulásetu'l hisáb* of BEHAU'LDIN; an author, who lived between the years 953 and 1031 H. The Arabic text, with a Persian commentary, has been printed in Calcutta; and a summary of its contents had been previously given by Mr. STRACHEY in his "Early History of Algebra," in which, as in his other exertions for the investigation of Hindu and Arabian Algebra, his zeal surmounted great difficulties; while his labours have thrown much light upon the subject.[1]

The title page of the manuscript above described, as well as a marginal note on it, and the author's preface, all concur in declaring it the work of MUHAMMED *ben* MUSA *Khuwárezmí*: and the mention of the Khalif ALMA-

[1] See *Bija Ganita*, or Algebra of the Hindus; London, 1813. HUTTON's Math. Dict. Ed. 1815. Art. Algebra: and As. Res. 12. 159.

MÚN in that preface, establishes the identity of the author, whose various works, as is learned from Arabian historians, were composed by command, or with encouragement, of that Khalif, partly before his accession, and partly during his reign.

The preface, a transcript of which was supplied by the care of Mr. NICOLL, has been examined at my request, by Colonel John BAILLIE. After perusing it with him, I am enabled to affirm, that it intimates " encouragement from the *Imám* ALMÁMÚN Commander of the Faithful, to compile a compendious treatise of calculation by Algebra;" terms, which amount not only to a disclaimer of any pretensions to the invention of the Algebraic art; but which would to my apprehension, as to that of the distinguished Arabic scholar consulted, strongly convey the idea of the preexistence of ampler treatises upon Algebra in the same language (Arabic), did not the marginal note above cited distinctly assert this to be " the first treatise composed upon Algebra among the faithful;" an assertion corroborated by the similar affirmation of ZACARIA of *Casbín*, and other writers of Arabian history. Adverting, however, to that express affirmation, the author must be here understood as declaring that he compiled (*alaf* is the verb used by him) the treatise upon Algebra from books in some other language: doubtless then in the Indian tongue; as it has been already shown, that he was conversant with Hindu astronomy, and Hindu computation and account.

It may be right to notice, that the title of the manuscript denominates the author " ABU ABDULLAH MUHAMMED *ben* MUSA *al Khuwárezmí*, differing in the first part of the name from the designation, which occurs in one passage of the *Tárikhu'l hukmá*, quoted by CASIRI, where the *Khuwárezmí* MUHAMMED *ben* MUSA is called ABU-JÂFR.[1] But that is not a sufficient ground for questioning the sameness of persons and genuineness of the work, as the *Khuwárezmí* is not usually designated by either of those additions, or by any other of that nature taken from the name of offspring: and error may be presumed; most probably on the part of the Egyptian author of the *Tárikhu'l hukmá*, since the addition, which he introduces, that of ABU-JÂFR, belongs to MUHAMMED *ben* MUSA *ben* SHAKER, a very different person; as appears from another passage of the same Egyptian's compilation.[2]

[1] CASIRI, i. 428. [2] CASIRI, i. 418.

The following is a translation of the *Khuwárezmí's* directions for the solution of equations: simple and compound: a topic, which he enters upon at no great distance from the commencement of the volume: having first treated of unity and number in general.

'I found, that the numbers, of which there is need in computation by restoration and comparison,[1] are of three kinds; namely, roots and squares, and simple number relative to neither root nor square. A root is the whole of thing multiplied by [root] itself, consisting of unity, or numbers ascending, or fractions descending. A square is the whole amount of root multiplied into itself. And simple number is the whole that is denominated by the number without reference to root or square.

'Of these three kinds, which are equal, some to some, the cases are these: for instance, you say "squares are equal to roots;" and "squares are equal to numbers;" and "roots are equal to numbers."

'As to the case in which squares are equal to roots; for example, "a square is equal to five roots of the same:" the root of the square is five; and the square is twenty-five: and that is equivalent to five times its root.

'So you say "a third of the square is equal to four roots:" the whole square then is equal to twelve roots; and that is a hundred and forty-four; its root is twelve.

'Another example: you say "five squares are equal to ten roots." Then one square is equal to two roots: and the root of the square is two; and the square is four.

'In like manner, whether the squares be many or few, they are reduced to a single square: and as much is done to the equivalent in roots; reducing it to the like of that to which the square has been brought.

'Case in which squares are equal to numbers: for instance, you say, "the square is equal to nine." Then that is the square, and the root is three. And you say "five squares are equal to eighty:" then one square is a fifth of eighty; and that is sixteen. And, if you say, "the half of the square is equal to eighteen:" then the square is equal to thirty-six; and its root is six.

'In like manner, with all squares affirmative and negative, you reduce them to a single square. If there be less than a square, you add thereto, until the square be quite complete. Do as much with the equivalent in numbers.

[1] *Hisábu'l jebr wa al mukábalah.*

'Case in which roots are equal to number: for instance, you say "the root equals three in number." Then the root is three; and the square, which is raised therefrom, is nine. And, if you say "four roots are equal to twenty;" then a single root is equal to five; and the square, that is raised therefrom, is twenty-five. And, if you say "the half of the root is equal to ten:" then the [whole] root is equal to twenty; and the square, which is raised therefrom, is four hundred.

'I found, that, with these three kinds, namely, roots, squares, and number compound, there will be three compound sorts [of equation]; that is, squares and roots equal to number; squares and number equal to roots; and roots and number equal to squares.

'As for squares and roots, which are equal to number: for example, you say "square, and ten roots of the same, amount to the sum of thirty-nine." Then the solution of it is: you halve the roots; and that in the present instance yields five. Then you multiply this by its like, and the product is twenty-five. Add this to thirty-nine: the sum is sixty-four. Then take the root of this, which is eight, and subtract from it half the roots, namely, five; the remainder is three. It is the root of the square which you required; and the square is nine.

'In like manner, if two squares be specified, or three, or less, or more, reduce them to a single square; and reduce the roots and number therewith to the like of that to which you reduced the square.

'For example, you say "two squares and ten roots are equal to forty-eight *dirhems:*" and the meaning is, any two [such] squares, when they are summed and unto them is added the equivalent of ten times the root of one of them, amount to the total of forty-eight *dirhems*. Then you must reduce the two squares to a single square: and assuredly you know, that one of two squares is a moiety of both. Then reduce the whole thing in the instance to its half: and it is as much as to say, a square and five roots are equal to twenty-four *dirhems;* and the meaning is, any [such] square, when five of its roots are added to it, amounts to twenty-four. Then halve the roots, and the moiety is two and a half. Multiply that by its like, and the product is six and a quarter. Add this to twenty-four, the sum is thirty *dirhems* and a quarter. Extract the root, it is five and a half. Subtract from this the moiety of the roots; that is, two and a half: the remainder is three. It is the root of the square: and the square is nine.

'In like manner, if it be said "half of the square and five roots are equal to twenty-eight *dirhems.*" It signifies, that, when you add to the moiety of any [such] square the equivalent of five of its roots, the amount is twenty-eight *dirhems.* Then you desire to complete your square so as it shall amount to one whole square; that is, to double it. Therefore double it, and double what you have with it; as well as what is equal thereunto. Then a square and ten roots are equal to fifty-six *dirhems.* Add half the roots multiplied by itself, twenty-five, to fifty-six; and the sum is eighty-one. Extract the root of this, it is nine. Subtract from this the moiety of the roots; that is, five: the remainder is four. It is the root of the square which you required: and the square is sixteen; and its moiety is eight.

'Proceed in like manner with all that comes of squares and roots; and what number equals them.

'As for squares and number, which are equal to roots; for example, you say, "a square and twenty-one are equal to ten of its roots:" the meaning of which is, any [such] square, when twenty-one *dirhems* are added to it, amounts to what is the equivalent of ten roots of that square: then the solution is, halve the roots; and the moiety is five. Multiply this by itself, the product is twenty-five. Then subtract from it twenty-one, the number specified with the square: the remainder is four. Extract its root; which is two. Subtract this from the moiety of the roots; that is, from five; the remainder is three. It is the root of the square which you required: and the square is nine. Or, if you please, you may add the root to the moiety of the roots: the sum is seven. It is the root of the square which you required; and the square is forty-nine.

'When a case occurs to you, which you bring under this head, try its answer by the sum: and, if that do not serve, it certainly will by the difference. This head is wrought both by the sum and by the difference. Not so either of the others of three cases requiring for their solution that the root be halved. And know, that, under this head, when the roots have been halved, and the moiety has been multiplied by its like, if the amount of the product be less than the *dirhems* which are with the square, then the instance is impossible: and, if it be equal to the *dirhems* between them, the root of the square is like the moiety of the roots, without either addition or subtraction.

'In every instance where you have two squares, or more or less, reduce to a single square, as I explained under the first head.

'As for roots and number, which are equal to squares: for example, you say, "three roots and four in number are equal to a square:" the solution of it is, halve the roots; and the moiety will be one and a half. Multiply this by its like, [the product is two and a quarter. Add it to four, the sum is six and a quarter. Extract the root, which is two and a half. To this add the moiety of the roots. The sum is four. It is the root of the square which you required: and the square is sixteen.]'

The author returns to the subject in a distinct chapter, which is entitled "On the six cases of Algebra." A short extract from it may suffice.

'The first of the six cases. For example, you say, "you divide ten into two parts, and multiply one of the two parts by the other: then you multiply one of them by itself, and the product of this multiplication into itself is equal to four times that of one of the parts by the other."

'Solution. Make one of the two parts *thing*, and the other ten less *thing:* then multiply *thing* by ten less *thing*, and the product will be ten *things* less a square. Multiply by four: for you said "four times." It will be four times the product of one part by the other; that is, forty *things* less four squares. Now multiply *thing* by *thing*, which is one of the parts by itself: the result is, square equal to forty *things* less four squares. Then *restore* it in the four squares, and add it to the one square. There will be forty *things* equal to five squares; and a single square is equal to eight roots. It is sixty-four; and its root is eight: and that is one of the two parts, which was multipled into itself: and the remainder of ten is two; and that is the other part. Thus has this instance been solved under one of the six heads: and that is the case of squares equal to roots.

'The second case. "You divide ten into two parts, and multiply the amount of a part into itself. Then multiply ten into itself; and the product of this multiplication of ten into itself, is equivalent to twice the product of the part taken into itself, and seven ninths: or it is equivalent to six times and a quarter the product of the other part taken into itself.

'Solution. Make one of the parts *thing*, and the other ten less *thing*. Then you multiply *thing* into itself: it is a square. Next by two and seven ninths: the product will be two squares, and seven ninths of a square. Then multiply ten into itself, and the product is a hundred. Reduce it to a single square, the result is nine twenty-fiths; that is, a fifth and four fifths of a fifth. Take a fifth of a hundred and four fifths of a fifth, the quotient is thirty-six, which is equal to one square. Then extract the root, which is six. It is one of the two parts; and the other is undoubtedly four. Thus you solve this instance under one of the six heads: and that is "squares equal to number."

These extracts may serve to convey an adequate notion of the manner, in which the *Khuwárezmí* conducts the resolution of equations simple and compound, and the investigation of problems by their means. If a comparison be made with the *Khulásetu'l hisáb*, of which a summary by Mr. STRACHEY will be found in the researches of the Asiatic society,[1] it may be seen, that the Algebraic art has been nearly stationary in the hands of the Muhammedans, from the days of MUHAMMED of *Khuwárezm*[2] to those of BEHÁU'LDÍN of *Aamul*,[3] notwithstanding the intermediate study of the arithmetics of DIOPHANTUS, translated and expounded by MUHAMMED of *Buzján*. Neither that comparison, nor the exclusive consideration of the *Khuwárezmí's* performance, leads to any other conclusion, than, as before intimated, that, being conversant with the sciences of the Hindus, especially with their astronomy and their method of numerical calculation, and being the author of the earliest Arabic treatise on Algebra, he must be deemed to have learnt from the Hindus the resolution of simple and quadratic equations, or, in short, Algebra, a branch of their art of computation.

The conclusion, at which we have arrived, may be strengthened by the coincident opinion of COSSALI, who, after diligent research and ample disquisition, comes to the following result.[4]

'Concerning the origin of Algebra among the Arabs, what is certain is, that MUHAMMED *ben* MUSA the *Khuwárezmite* first taught it to them. The *Casbinian*, a writer of authority affirms it; no historical fact, no opinion, no reasoning, opposes it.

'There is nothing in history respecting MUHAMMED *ben* MUSA individually, which favours the opinion, that he took from the Greeks, the Algebra, which he taught to the Muhammedans.

'History presents in him no other than a mathematician of a country most distant from Greece and contiguous to India; skilled in the Indian tongue; fond of Indian matters: which he translated, amended, epitomised, adorned: and he it was, who was the first instructor of the Muhammedans in the Algebraic art.[5]

[1] Vol. 12. [2] On the Oxus. [3] A district of Syria; not *Ámal* a town in *Khurásán*. COM.
[4] *Orig. dell' Alg.* i. 216. [5] *Orig. dell' Alg.* i. 219.

'Not having taken Algebra from the Greeks, he must have either invented it himself, or taken it from the Indians. Of the two, the second appears to me the most probable.'[1]

O.

COMMUNICATION OF THE HINDUS WITH WESTERN NATIONS ON ASTROLOGY AND ASTRONOMY.

THE position, that Astrology is partly of foreign growth in India; that is, that the Hindus have borrowed, and largely too, from the astrology of a more western region, is grounded, as the similar inference concerning a different branch of divination,[2] on the resemblance of certain terms employed in both. The mode of divination, called *Tájaca,* implies by its very name its Arabian origin. Astrological prediction by configuration of planets, in like manner, indicates even by its Indian name a Grecian source. It is denominated *Hórá,* the second of three branches which compose a complete course of astronomy and astrology:[3] and the word occurs in this sense in the writings of early Hindu astrologers. VARÁHA-MIHIRA, whose name stands high in this class of writers, has attempted to supply a Sanscrit etymology; and in his treatise on casting nativities derives the word from *Ahórátra,* day and night, a nycthemeron. This formation of a word by dropping *both* the first and last syllables, is not conformable to the analogies of Sanscrit etymology. It is more natural then to look for the origin of the term in a foreign tongue: and that is presented by the Greek ὥρα and its derivative ὡροσκόπος, an astrologer, and especially one who considers the natal *hour,* and hence predicts events.[4] The same term *hórá* occurs again in the writings of the Hindu astrologers, with an acceptation (that of hour[5]) which more exactly conforms to the Grecian etymon.

The resemblance of a single term would not suffice to ground an inference of common origin, since it might be purely accidental. But other words are also remarked in Hindu astrology, which are evidently not Indian. An in-

[1] See his reasons at large.
[2] As. Res. 9. 376.
[3] See Note K.
[4] *Hesych.* and *Suid.*
[5] As. Res. 5. 107.

stance of it is *dréshcáńa,*[1] used in the same astrological sense with the Greek δεκανος and Latin *decanus:* words, which, notwithstanding their classic sound, are to be considered as of foreign origin (Chaldean or Egyptian) in the classic languages, at least with this acceptation.[2] The term is assuredly not genuine Sanscrĭt; and hence it was before[2] inferred, that the particular astrological doctrine, to which it belongs, is exotic in India. It appears, however, that this division of the twelve zodiacal signs into three portions each, with planets governing them, and pourtrayed figures representing them, is not implicitly the same among the Hindu astrologers, which it was among the Chaldeans, with whom the Egyptians and Persians coincided. Variations have been noticed.[3] Other points of difference are specified by the astrologer of *Balkh;*[4] and they concern the allotment of planets to govern the *decani* and *dréshcánas,* and the figures by which they are represented. Abu-Mâshar is a writer of the ninth century;[5] and his notice of this astrological division of the zodiac as received by Hindus, Chaldeans, and Egyptians, confirms the fact of an earlier communication between the Indians and the Chaldeans, perhaps the Egyptians, on the subject of it.

With the sexagesimal fractions, the introduction of which is by Wallis ascribed to Ptolomy among the the Greeks,[6] the Hindus have adopted for the minute of a degree, besides a term of their own language, *calá,* one taken from the Greek λεπτα scarcely altered in the Sanscrĭt *liptá.* The term must be deemed originally Greek, rather than Indian, in that acceptation, as it there corresponds to an adjective λεπτος, slender, minute: an import which precisely agrees wth the Sanscrĭt *calá* and Arabic *dakík,* fine, minute; whence, in these languages respectively, *calá* and *dakík* for a minute of a degree. But the meanings of *liptá* in Sanscrĭt[7] are, 1st, smeared; 2d, infected with poison; 3d, eaten: and its derivative *liptaca* signifies a poisoned arrow, being derived from *lip,* to smear: and the dictionaries give no interpretation of the word that has any affinity with its special acceptation as a technical term in astronomy and mathematics. Yet it occurs so employed in the work of Brahmegupta.[8]

By a different analogy of the sense and not the sound, the Greek μοῖρα, a

[1] As. Res. 9. 367. [2] Ibid. Vide *Salm.* Exerc. Plin. [3] Ibid. 9. 374.
[4] Lib. intr. in Ast. Albumasis Abalachi, 5. 12 and 13.
[5] Died in 272 H. (885 C.) aged a hundred.
[6] Wallis. Alg. c. 7. [7] *Am. Cósh.* [8] C. 1. § 6, et passim.

part, and specially a degree of a circle, is in Sanscrĭt *anśa, bhága,* and other synonyma of part, applied emphatically in technical language to the 360th part of the periphery of a circle. The resemblance of the radical sense, in the one instance, tends to corroborate the inference from the similarity of sound in the other.

Céndra is used by BRAHMEGUPTA and the *Súrya-sidd'hánta,* as well as other astronomical writers (BHÁSCARA, &c.), and by the astrologers VARÁHA-MIHIRA and the rest, to signify the equation of the centre.[1] The same term is employed in the Indian mensuration for the centre of a circle;[2] also denoted by *med'hya,* middle. It comes so near in sound, as in signification, to the Greek κέντρον, that the inference of a common origin for these words is not to be avoided. But in Sanscrĭt it is exclusively technical; it is unnoticed by the vocabularies of the language; and it is not easily traced to a Sanscrĭt root. In Greek, on the contrary, the correspondent term was borrowed in mathematics from a familiar word signifying a goad, spur, thorn, or point; and derived from a Greek theme κέντεω.

The other term, which has been mentioned as commonly used for the centre of a circle, namely *med'hya,* middle, is one of the numerous instances of radical and primary analogy between the Sanscrĭt and the Latin and Greek languages. It is a common word of the ancient Indian tongue; and is clearly the same with the Latin *medius;* and serves to show that the Latin is nearer to the ancient pronunciation of Greek, than μέσος; from which SIPONTINUS derives it; but which must be deemed a corrupted or softened utterance of an ancient term coming nearer to the Sanscrĭt *med'hyas* and Latin *medius.*

On a hasty glance over the *Játacas* or Indian treatises upon horoscopes, several other terms of the art have been noticed, which are not Sanscrĭt, but apparently barbarian. For instance *anapha, sunapha, durud'hara,* and *cémadruma,* designating certain configurations of the planets. They occur in both the treatises of VARÁHA-MIHIRA; and a passage, relative to this subject, is among those quoted from the abridgment by the scholiast of the greater treatise, and verified in the text of the less.[3] The affinity of those

[1] *Brahm. sidd'h.* c. 2. *Súr. Sidd'h.* c. 2. *Vrĭhat* and *Laghu Játacas.* [2] *Súr.* on *Lίl.* § 207.

[3] See p. xlix. Another passage so quoted and verified uses the term *céndra* in the sense above-mentioned.

terms to words of other languages used in a similar astrological sense, has not been traced: for want, perhaps, of competent acquaintance with the terminology of that silly art. But it must not be passed unremarked, that VARÁHA-MIHIRA, who has in another place praised the *Yavanas* for their proficiency in astrology (or astronomy; for the term is ambiguous;) frequently quotes them in his great treatise on horoscopes and his scholiast marks a distinction between the ancient *Yavanas*, whom he characterises as "a race of barbarians conversant with *(hórá)* horoscopes," and a known Sanscrit author bearing the title of YAVANÉSWARA, whose work he had seen and repeatedly cites; but the writings and doctrine of the ancient *Yavanas*, he acknowledges, had not been seen by him, and were known to him only by this writer's and his own author's references.

No argument, bearing upon the point under consderation, is built on BHÁSCARA's use of the word *dramma* for the value of 64 cowry-shells (*Lil.* § 2.) in place of the proper Sanscrit term *pramáńa*, which ŚRÍD'HARA and other Hindu authors employ; nor on the use of *dínára*, for a denomination of money, by the scholiast of BRAHMEGUPTA (12 § 12.) who also, like BHÁSCARA, employs the first mentioned word (12. § 14.): though the one is clearly analogous to the Greek *drachma*, a word of undoubted Grecian etymology, being derived from δρασσομαι; and the other apparently is so, to the Roman *denarius* which has a Latin derivation. The first has not even the Sanscrit air; and is evidently an exotic, or, in short, a barbarous term. It was probably received mediately through the Muhammedans, who have their *dirhem* in the like sense. The other is a genuine Sanscrit word, of which the etymology, presenting the sense of 'splendid,' is consistent with the several acceptations of a specific weight of gold; a golden ornament or breast-piece; and gold money: all which senses it bears, according to the ancient vocabularies of the language.[1]

The similarity seems then to be accidental in this instance; and the Muhammedans, who have also a like term, may have borrowed it on either hand: not improbably from the Hindus, as the *dínár* of the Arabs and Persians is a gold coin like the Indian; while the Roman *denarius* is properly a silver one. D'HERBELOT assigns as a reason for deriving the Arabic *dínár* from the Roman *denarius*, that this was of gold. The nummus aureus some-

[1] *Amera-cósha*, &c.

times had that designation; and we read in Roman authors of golden as well as silver *denarii*.[1] But it is needless to multiply references and quotations to prove, that the Roman coin of that name was primarily silver, and so denominated because it was equal in value to ten copper *as*;[2] that it was all along the name of a silver coin;[3] and was still so under the Greek empire, when the δηνάριον was the hundredth part of a large silver coin termed ἀργυροῦς.[4]

[1] *Plin.* 33. § 13, and 37 § 3. Petron. Satyr. 106. 160.

[2] *Plin.* 33. 13 *Vitr.* 3. 1. *Volus. Mæcianus.* Didymus.

[3] Vitr. and Vol. Mæc.

[4] Epiphanius, cum multis aliis.

INDIAN
Arithmetic and Algebra.

CHAPTER I.

INTRODUCTION.

1. HAVING bowed to the deity, whose head is like an elephant's;[1] whose feet are adored by gods; who, when called to mind, relieves his votaries from embarrassment; and bestows happiness on his worshippers; I propound this easy process of computation,[2] delightful by its elegance,[3] perspicuous with words concise, soft and correct, and pleasing to the learned.

AXIOMS.

[CONSISTING IN DEFINITIONS OF TECHNICAL TERMS.]

[*Money by Tale.*]

2. Twice ten cowry shells[4] are a *cácińí*; four of these are a *pańa;* sixteen of which must be here considered as a *dramma*; and in like manner, a *nishca,* as consisting of sixteen of these.

[1] GAŃÉSA, represented with an elephant's head and human body.

[2] *Pátí-gańita; pátí, paripátí,* or *vyacta-gańita,* arithmetic.

[3] *Lílávatí* delightful: an allusion to the title of the book. See notes on § 13 and 277.

[4] Cypræa moneta. *Sans.* Varáṭaca, capardí; *Hind.* Caurí.

[*Weights.*]

3. A *gunja*[1] (or seed of Abrus) is reckoned equal to two barley-corns; a *valla*, to two *gunjas;* and eight of those are a *d'haraṅa;* two of which make a *gadyáṅaca*. In like manner one *d'hátaca* is composed of fourteen *vallas*.

4. Half ten *gunjas* are called a *másha*,[2] by such as are conversant with the use of the balance: a *carsha* contains sixteen of what are termed *máshas;* a *pala*, four *carshas*. A *carsha* of gold is named *suverṅa*.

[*Measures.*]

5—6. Eight breadths of a barley-corn[3] are here a finger; four times six fingers, a cubit;[4] four cubits, a staff;[5] and a *crósa* contains two thousand of these; and a *yójana*, four *crósas*.

So a bambu pole consists of ten cubits; and a field (or plane figure) bounded by four sides, measuring twenty bambu poles, is a *nivartana*.[6]

7. A cube,[7] which in length, breadth and thickness measures a cubit, is termed a solid cubit: and, in the meting of corn and the like, a measure,

[1] A seed of Abrus precatorius: black or red; the one called *crĭshṅala;* the other *racti*, *racticá* or *ratticá;* whence *Hind.* ratti.

[2] Physicians reckon seven *gunjas* to the *másha;* lawyers, seven and a half. The same weight is intended; and the difference of description arises only from counting by heavier or lighter seeds of Abrus: in like manner as the earth is the same, whether rated at 3300 *yójanas;* or, with the *Śirómaṅi*, 4967; or, according to others, 6522. GAṄ.

[3] Eight barley-corns (*yava*) by breadth, or three grains of rice by length, are equal to one finger (*angula*). GAṄ.

[4] *Hasta*, *cara* and synonyma of hand or fore arm. According to the commentator GAṄÉŚA, this intends the practical cubit as received by artisans, and vulgarly called *gaj* [or *gaz*]. It is nearer to the yard than to the true cubit: but the commentator seems to have no sufficient ground for so enlarging the cubit.

[5] *Daṅda*, a staff: directed to be cut nearly of man's height. MENU, 2. 46.

[6] A superficial measure or area containing 400 square poles. SUR.

[7] *Dwádaśásri*, lit. dodecagon, but meaning a parallelopipedon; the term *asra*, corner or angle, being here applied to the edge or line of incidence of two planes. See CHATURVÉDA on BRAHMEGUPTA, § 6.

which contains a solid cubit, is a *c'hári* of *Magad'ha*[1] as it is denominated in science.

8. A *dróńa* is the sixteenth part of a *c'hári;* an *ád'haca* is a quarter of a *dróńa;* a *prast'ha* is a fourth part of an *ád'haca;* and a *cuḋaba*[2] is by the ancients[3] termed a quarter of a *prast'ha.*[4]

The rest of the axioms, relative to time[5] and so forth, are familiarly known.[6]

[1] The country or province situated on the *Sónebhadrá* river.—GAŃ. It is South Bihar. See, concerning other *c'hári* measures, a note on § 236.

[2] 'In the *Cuťapa,* the depth is a finger and a half; the length and breadth, each, three.' SRÍD'HARA ÁCHÁRYA cited by GANGÁD'HARA and SÚRYADÁSA. 'The *cuťapa* or *cuḋaba* is a wooden measure containing 13½ cubic *angulas;* the *prast'ha,* (four times as many) 54; the *ád'haca,* 216; the *dróńa,* 864; the *c'hári,* 13824.'—GANG. and SÚR. See As. Res. vol. 5, p. 102.

[3] By SRÍD'HARA and the rest. SUR.

[4] Another stanza, (an eighth, on the subject of weights and measures,) occurs in one copy of the text; and that number is indicated in the *Manóranjana.* But the commentaries of GAŃÉSA and SÚRYADÁSA specify seven, and GANGÁD'HARA alone expounds the additional stanza. It is therefore to be rejected as spurious, and interpolated: not being found in other copies of the text. The subject of it is the *mańa (mań)* of forty *sétas (sér);* which, as a measure of corn by weight, is ascribed to the *Turushcas* or Muhammedans of India; the people of *Yávana-désa,* as the commentator terms them.

"The *séta** is here reckoned at twice seven *tancas,* each equal to three-fourths of a *gadyánaca:* and a *mańa,* at forty *sétas.* The name is in use among the *Turushcas,* for a weight of corn and like articles." See notes on § 97 and 236.

[5] The author has himself explained the measures of time in the astronomical part of his treatise. *(Sidd'hánta-śirómańi,* § 16-18.*)* GANG. and SÚR.

[6] Concerning weights and measures, see *Gańita-sára* of SRÍD'HARA, § 4—8; and PRÍT'HÚDACA SWÁMÍ CHATURVÉDA on BRAHMEGUPTA's arithmetic, § 10-11.

* The copy of GANGÁD'HARA's commentary writes *śaura.* But the exemplar of the text, containing the passage, has *séta.*

CHAPTER II.

SECTION I.

Invocation.[1]

9. SALUTATION to GAŃÉŚA, resplendent as a blue and spotless lotus; and delighting in the tremulous motion of the dark serpent, which is perpetually twining within his throat.

Numeration.

10—11. Names of the places of figures have been assigned for practical use by ancient writers,[2] increasing regularly[3] in decuple proportion: namely, unit, ten, hundred, thousand, myriad, hundred thousands, million, ten millions, hundred millions, thousand millions, ten thousand millions, hundred thousand millions, billion, ten billions, hundred billions, thousand billions, ten thousand billions, hundred thousand billions.[4]

[1] A reason of this second introductory stanza is, that the foregoing definitions of terms are not properly a part of the treatise itself; none such having been premised by A'RYA-BHAṬṬA and other ancient authors to their treatises of arithmetic. GAŃ. and *Manó*.

[2] According to the *Hindus*, numeration is of divine origin; 'the invention of nine figures *(anca)*, with the device of places to make them suffice for all numbers, being ascribed to the beneficent Creator of the universe,' in BHÁSCARA's *Vásaná* and its gloss; and in CRĪSHṆA's commentary on the *Vīja-gańita*. Here nine figures are specified; the place, when none belongs to it, being shown by a blank *(śúnya)*; which, to obviate mistake, is denoted by a dot or small circle.

[3] From the right, where the first and lowest number is placed, towards the left hand. GAŃ. &c.

[4] Sans. *éca*, *daśa*, *śata*, *sahasra*, *ayuta*, *lacsha*, *prayuta*, *cóṭi*, *arbuda*, *abja* or *padma*, *c'harva*, *nic'harva*, *mahápadma*, *śancu*, *jalad'hi* or *samudra*, *antya*, *mad'hya*, *parárd'ha*.

A passage of the *Véda*, which is cited by SU'RYA-DÁSA, contains the places of figures. 'Be these the milch kine before me, one, ten, a hundred, a thousand, ten thousand, a hundred thousand, a million, ... Be these milch kine my guides in this world.'

GAŃÉŚA observes, that numeration has been carried to a greater number of places by ŚRÍD'HARA and others; but adds, that the names are omitted on account of the numerous contradictions and the little utility of those designations. The text of the *Gańita-sára* or abridgment of ŚRÍD'HARA does not correspond with this reference: for it exhibits the same eighteen places, and no more. *Gań-sár*. § 2—3.)

SECTION II.

Eight Operations[1] *of Arithmetic.*

12. Rule of addition and subtraction:[2] half a stanza.

The sum of the figures according to their places is to be taken in the direct or inverse order:[3] or [in the case of subtraction] their difference.

13. Example. Dear intelligent LÍLÁVATÍ,[4] if thou be skilled in addition and subtraction, tell me the sum of two, five, thirty-two, a hundred and ninety-three, eighteen, ten, and a hundred, added together; and the remainder, when their sum is subtracted from ten thousand.

Statement, 2, 5, 32, 193, 18, 10, 100.
[Answer.] Result of the addition, 360.
Statement for subtraction, 10000, 360.
[Answer.] Result of the subtraction, 9640.[5]

14—15. Rule of multiplication:[6] two and a half stanzas.

Multiply the last[7] figure of the multiplicand by the multiplicator, and

[1] *Paricarmáshťaca*, eight operations, or modes of process: logistics or algorism.

[2] *Sancalana, sancalita, miśrana, yuti, yóga*, &c. summation, addition. *Vyavacalana, vyavacalita, sód'hana, patana*, &c. subtraction. *Antara*, difference, remainder.

[3] From the first on the right, towards the left; or from the last on the left, towards the right. GANG.

[4] Seemingly the name of a female to whom instruction is addressed. But the term is interpreted in some of the commentaries, consistently with its etymology, "Charming."—See § 1. and 277.

[5] Mode of working addition as shown in the *Manóranjana*:

Sum of the units,	2, 5, 2, 3, 8, 0, 0, ...	20
Sum of the tens,	3, 9, 1, 1, 0, ...	14
Sum of the hundreds,	1, 0, 0, 1, ...	2
Sum of the sums	. . . .	360

[6] *Guńana, abhyása;* also *hańana* and any term implying a tendency to destroy. It is denominated *pratyutpanna* by BRAHMEGUPTA, § 3; and by ŚRÍD'HARA, § 15—17.

Guńya multiplicand. *Guńaca* multiplicator. *Gháta* product.

[7] The digit standing last towards the left. The work may begin either from the first or the last digit, according to ŚRÍD'HARA. *Gańita-sára*, § 15.

next the penult, and then the rest, by the same repeated. Or let the multiplicand be repeated under the several parts of the multiplicator, and be multiplied by those parts: and the products be added together. Or the multiplier being divided by any number which is an aliquot part of it, let the multiplicand be multiplied by that number and then by the quotient, the result is the product. These are two methods of subdivision by form. Or multiply separately by the places of figures, and add the products together. Or multiply by the multiplicator diminished or increased by a quantity arbitrarily assumed; adding, or subtracting, the product of the multiplicand taken into the assumed quantity.[1]

16. Example. Beautiful and dear LÍLÁVATÍ, whose eyes are like a fawn's! tell me what are the numbers resulting from one hundred and thirty-five, taken into twelve? if thou be skilled in multiplication by whole or by parts, whether by subdivision of form or separation of

[1] The author teaches six methods, according to the exposition of SU'RYADA'SA, &c.; but seven, as interpreted by GANGA'D'HARA: and those, combined with the four of SCANDASE'NA and S'RÍD'HARA, (one of which at the least is unnoticed by BHA'SCARA,) make eight distinct ways. The mode of multiplication by parts (*c'haṅda-pracára*) is distinguished into *rúpa-vibhága* and *st'hána-vibhága*, or subdivision of the form and severance of the digits: the first is again divided into multiplication by integrant or by aliquot parts: the second in like manner furnishes two ways, according as the digits of the multiplier or of the multiplicand are severed. These then are four methods, deduced from two of SCANDASE'NA and S'RÍD'HARA; to which two others are added by BHA'SCARA, consisting in the increase or decrease of the multiplier by an arbitrary quantity, and taking the sum or difference of the products. To those six must be joined the *Tatst'ha* of the older authors, and their *Capáṭasand hi*; if indeed this be not (conformably with GANGA'D'HARA's opinion,) intended by BHA'SCARA's first method. It is wrought by repeating or moving the multiplier over (according to GANGA'D'HARA, or under, as directed by the *Manóranjana*,) every digit of the multiplicand; and, according to the explanation of GAN'E'SA, it proceeds obliquely, joining products along compartments. The *tatst'ha*, so named because the multiplier is stationary, appears from GAN'E'SA's gloss to be cross multiplication. 'After setting the multiplier under the multiplicand,' he directs to 'multiply unit by unit, and note the result underneath. Then, as in cross multiplication,* multiply unit by ten, and ten by unit, add together, and set down the sum in a line with the foregoing result. Next multiply unit by hundred, and hundred by unit, and ten by ten; add together, and set down the result as before: and so on, with the rest of the digits. This being done, the line of results is the product of the multiplication.' The commentator considers this method as 'difficult, and not to be learnt by dull scholars without oral instruction.' He adds, that 'other modes may be devised by the intelligent.' See Arithm. of BRAHM. § 55, *Gaṅ.-sár.* § 15—17.

* *Vajrábhyása.* See *Víja-gańita*, § 77.

digits.[1] Tell me, auspicious woman, what is the quotient of the product divided by the same multiplier?

Statement, Multiplicand 135. Multiplicator 12.

Product (multiplying the digits of the multiplicand successively by the multiplicator) 1620.

Or, subdividing the multiplicator into parts, as 8 and 4; and severally multiplying the multiplicand by them; adding the products together: the result is the same, 1620.

Or, the multiplicator 12 being divided by three, the quotient is 4; by which, and by 3, successively multiplying the multiplicand, the last product is the same, 1620.

Or, taking the digits as parts, viz. 1 and 2; the multiplicand being multiplied by them severally, and the products added together, according to the places of figures, the result is the same, 1620.

Or, the multiplicand being multiplied by the multiplicator less two, viz. 10, and added to twice the multiplicand, the result is the same, 1620.

Or, the multiplicand being multiplied by the multiplicator increased by eight, viz. 20, and eight times the multiplier being subtracted, the result is the same, 1620.

[1] The following scheme of the process of multiplication is exhibited in GAŃÉSA's commentary.

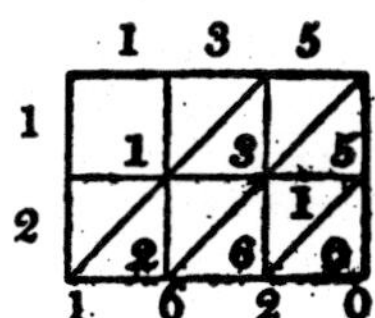

Or the process may be thus ordered, according to GANGÁD'HARA,

12	12	12
1	3	5
12		60
	36	
	1620	

Or, in this manner,

135	135
1	2
	270
	135
	1620

Or in the subjoined modes taken from CHATURVÉDA, &c.

135	1	135
135	2	270
		1620

135	8	1080
135	4	540
		1620

135	20	2700
135	8	1080
		1620

17. Rule of division.[1] One stanza.

That number, by which the divisor being multiplied balances the last digit of the dividend [and so on[2]], is the quotient in division: or, if practicable, first abridge[3] both the divisor and dividend by an equal number, and proceed to division.

[Example.] Statement of the number produced by multiplication in the foregoing example, and of its multiplicator for a divisor: Dividend 1620. [Divisor 12.]

Quotient 135; the same with the original multiplicand.[4]

Or both the dividend and the divisor, being reduced to least terms by the common measure three, are 540 and 4; or by the common measure four, they become 405 and 3. Dividing by the respective reduced divisors, the quotient is the same, 135.

18—19. Rule for the square[5] of a quantity: two stanzas.

The multiplication of two like numbers together is the square. The square of the last[6] digit is to be placed over it; and the rest of the digits, doubled and multiplied by that last, to be placed above them respectively; then repeating the number, except the last digit, again [perform the like operation]. Or twice the product of two parts, added to the sum of the squares of the parts, is the square [of the whole number].[7] Or the product of the sum and

[1] *Bhága-hára, bhájana, haraṅa, ch'hédana:* division. *Bhájya,* dividend. *Bhájaca, hara,* divisor. *Labd'hi,* quotient.

[2] Repeating the divisor for every digit, like the multiplier in multiplication. GANG.

[3] *Apavartya,* abridging. See note on § 249.

[4] The process of long division is exhibited in the *Manóranjana* thus: The highest places of the proposed dividend, 16, being divided by 12, the quotient is 1; and 4 over. Then 42 becomes the highest remaining number, which divided by 12 gives the quotient 3, to be placed in a line with the preceding quotient (1): thus 13. Remains 60, which, divided by 12, gives 5: and this being carried to the same line as before, the entire quotient is exhibited: viz. 135. *Manór.*

[5] *Varga, crĭti,* a square number.

[6] The process may begin with the first digit: as intimated by the author, § 24.

[7] Let the portions, or quantities comprising the first and last figures, be represented by the first letters of the alphabet, says the commentator on the *Vásaná*: Then, proceeding by the rule of multiplication, there results *a v* 1, *a. á g* 1, *a. á g* 1, *á v* 1; and, adding together like terms, *a v* 1, *a. á g* 2, *á v* 1. RANG.

difference of the number and an assumed quantity, added to the square of the assumed quantity, is the square.[1]

20. Example. Tell me, dear woman, the squares of nine, of fourteen, of three hundred less three, and of ten thousand and five, if thou know the method of computing the square.

Statement, 9, 14, 297, 10005.

[Answer.] Proceeding as directed, the squares are found: 81, 196, 88209, 100,100,025.

Or, put 4 and 5, parts of nine. Their product doubled 40, added to the sum of their squares 41, makes 81.

So, taking 10 and 4, parts of fourteen. Their product 40, being doubled, is 80; which, added to 116, the sum of the squares 100 and 16, makes the entire square, 196.

Or, putting 6 and 8. Their product 48, doubled, is 96; which, added to the sum of the squares 36 and 64, viz. 100, makes the same, 196.

Again, 297, diminished by three, is 294; and, in another place, increased by the same, is 300. The product of these is 88200; to which adding the square of three 9, the sum is as before the square, 88209.

21. Rule for the square-root:[2] one stanza.

Having deducted from the last of the odd digits[3] the square number,

The proposed quantity may be divided into three parts instead of two; and the products of the first and second, first and third, and second and third, being added together and doubled, and added to the sum of the squares of the parts, the total is the square sought. GAṄ.

[1] Another method is hinted in the author's note on this passage; consisting in adding together the product of the proposed quantity by any assumed one, and its product by the proposed less the assumed one. RANG.

[2] *Varga-múla* root of the square: *Múla*, *pada*, and other synonyma of root.

[3] Every uneven place is to be marked by a vertical line, and the intermediate even digits by a horizontal one. But, if the last place be even, it is joined with the contiguous odd digit. Ex. $\overset{|}{8}\,\bar{8}\,\overset{|}{2}\,\bar{0}\,\overset{|}{9}$.

From the last uneven place 8, deduct the square 4, remains $4\,\bar{8}\,\overset{|}{2}\,\bar{0}\,\overset{|}{9}$. Double the root 2, and divide by that (4) the subsequent even digit $4\bar{8}$; quotient nine [a higher one cannot be taken for the root of the foregoing digit would become greater than 2:] the remainder is $1\,2\,\overset{|}{2}\,\bar{0}\,\overset{|}{9}$. From the uneven place [with the residue] $1\,2\,\overset{|}{2}$, subtract the square of the quotient 9, viz. 81, the remainder is $4\,1\,\bar{0}\,\overset{|}{9}$. The double of the quotient 18 is to be placed in a line with the former double number double its root; and by that dividing the subsequent even digit, and subtracting the square of the quotient from the next uneven place, note in a line [with the preceding double number] the double of the quotient. Divide by the [number as noted in a] line the next even place, and deduct the square of the quotient from the following uneven one, and note the double of the quotient in the line. Repeat the process [until the digits be exhausted.] Half the [number noted in the] line is the root.

22. Example. Tell me, dear woman, the root of four, and of nine, and those of the squares before found, if thy knowledge extend to this calculation.

Statement, 4, 9, 81, 196, 88209, 100100025.
Answer. The roots are 2, 3, 9, 14, 297, 10005.

23—25. Rule for the cube[1]: three stanzas.

The continued multiplication of three like quantities is a cube. The cube of the last [digit] is to be set down; and next the square of the last multiplied by three times the first; and then the square of the first taken into the last and tripled; and lastly the cube of the first: all these, added together according to their places, make the cube. The proposed quantity [consisting of more than two digits] is distributed into two portions, one of which is then taken for the last [and the other for the first]; and in like manner repeatedly [if there be occasion.[2]] Or the same process may be begun from the first place of figures, either for finding the cube, or the square. Or three times the proposed number, multiplied by its two parts, added to the sum of the cubes of those parts, give the cube. Or the square root of the proposed number being cubed, that, multiplied by itself, is the cube of the proposed square.[3]

4; thus, 58. By this divide the even place 41$\bar{0}$; the quotient is 7, and remainder 4$\bar{9}$; to which uneven digit the square of the quotient 49 answers without residue. The double of the quotient 14 is put in a line with the preceding double number 58, making 594. The half of which is the root sought, 297. *Manó.* and GANG.

[1] *C'hana*, a cube. (*Lit.* solid.)

[2] The subdivision is continued until it comes to single digits.

GANÉSA confines it to the places of figures *(st'hána-vibhága,)* not allowing the portioning of the number *(rúpa-vibhága;)* because the addition is to be made according to the places.

[3] This carries an allusion to the raising of quantities to higher powers than the cube. GANÉSA

26. Example. Tell me, dear woman, the cube of nine, and the cube of the cube of three, and the cube of the cube of five, and the cube-roots of these cubes, if thy knowledge be great in computation of cubes.

Statement, 9, 27, 125.
Answer: The cubes in the same order, are 729, 19683, 1953125.[1]

The proposed number being nine and its parts 4 and 5. Then 9 multiplied by them and by three is 540; which, added to the sum of the cubes 64 and 125, viz. 189, makes the cube of nine 729.

The entire number being 27, its parts are 20 and 7: by which, the number being successively multiplied, and then tripled, is 11340; and this, added to the sum of the cubes of the parts 8343, makes the cube 19683.

The proposed number being a square, as 4. Its root 2 cubed is 8. This, taken into itself, gives 64 the cube of four. So nine being proposed, its square root 3, cubed, is 27; the square of which 729 is the cube of 9. In short the square of the cube is the same with the cube of the square.

specifies some of them. The product of four like numbers multiplied together is the square of a square, *varga-varga.* Continued multiplication up to six is the cube of a square, or square of a cube, *varga-g'hana* or *g'hana-varga.* Continued to eight, it is the square of a square's square, *varga-varga-varga.* Continued to nine, it is the cube of a cube, *g'hana-g'hana.* Intermediately are, the fifth power, *varga-g'hana-ghâta;* and the seventh, *varga-varga-g'hana-ghâta.*

[1] The number proposed being 125, distributed into two parts 12 and 5; and the first of these again into two portions, 1 and 2:

Then 1, cubed, is	1
1, square of 1, tripled and multiplied by 2,	6
4, square of 2, tripled and multiplied by 1,	12
2, cubed,	8
	1728

Now 12, cubed as above, is	1728
144, square of 12, tripled and multiplied by 5,	2160
25, square of 5, tripled and multiplied by 12,	900
5, cubed,	125
	1953125

Man6.

27—28. Rule for the cube-root[1]: two stanzas.

The first [digit] is a cube's place; and the two next uncubic; and again, the rest in like manner. From the last cubic place take the [nearest] cube, and set down its root apart. By thrice the square of that root divide the next [or uncubic] place of figures, and note the quotient in a line [with the quantity before found.] Deduct its square taken into thrice the last [term,] from the next [digit;] and its cube from the succeeding one. Thus the line [in which the result is reserved] is the root of the cube. The operation is repeated [as necessary.][2]

Example. Statement of the foregoing cubes for extraction of the root: 729, 19683, 1953125.[3]

Answer. The cube-roots respectively are 9, 27, 125.

[1] *G'hana-múla;* root of the cube.

[2] The same rule is taught by BRAHMEGUPTA, § 7, and ŚRÍD'HARA, § 29—31.

[3] The mode of conducting the work is shown in the *Manóranjana*, viz. 1953125. Here the last cubic digit is 1. Subtracting 1 the cube of the number 1, the remainder is 953125; and the root obtained is 1, which is to be set down in two places. Dividing the next digit by three times the square of that, the quotient taken is 2 [for 3 would soon appear to be too great;] and the residue is 353125; and the quotient 2, put in a line with 1, makes 12. Subtract the square of this 2, tripled and multiplied by the last term, viz. 12, from the next digit, the remainder is 233125; and the cube of the quotient 2, viz. 8, being taken from the succeeding digit, the residue is 225125. Again, the reserved root 12, being squared and tripled, gives 432. The next place of figures, divided by this, yields the quotient 5 and remainder 9125; and the quotient is set down in the line, which becomes 125. The square of that 5, viz. 25, multiplied by the last term 12, is 300, and tripled 900; which subtract from the next place, and the residue is 125. Take the cube of the quotient 5, viz. 125, from the succeeding digit, and the remainder is 0. Thus the root is found 125.

SECTION III.

FRACTIONS.[1]

FOUR RULES FOR THE ASSIMILATION OR REDUCTION OF FRACTIONS TO A COMMON DENOMINATOR.[2]

Simple Reduction of Fractions.[3]

29. Rule. The numerator and denominator[4] being multiplied reciprocally by the denominators of the two quantities,[5] they are thus reduced to the same denomination. Or both numerator and denominator may be multiplied by the intelligent calculator into the reciprocal denominators abridged by a common measure.

30. Example. Tell me the fractions reduced to a common denominator

[1] *Bhinna* a fraction; *lit.* a divided quantity, or one obtained by division.—GAŃ. An incomplete quantity or non-integer *(apurńa)*.—GANG. A proper or improper fraction, including a quantity, to which a part, as a moiety, a quarter, &c. is added; or from which such a part is deducted.—GAŃ.

[2] *Bhága-játi-chatushťaya, Játi-chatushťaya,* or four modes of assimilation or process for reducing, to a common denomination, fractions having dissimilar denominators: preliminary to addition and subtraction, and other arithmetical operations upon fractions. BRAHMEGUPTA's commentator CHATURVÉDA carries to six the number of rules termed *játi,* assimilation, or reduction to uniformity; and ŚRÍD'HARA has no less than eight; including rules answering to BHÁSCARA's for the arithmetic of fractions *(Lil.* § 36—43*)*, and for the solution of certain problems *(Lil.* § 52—54, and § 94—95.*)* See BRAHMEGUPTA's Arithmetic, § 1, note, (and § 2—5. § 8—9.) and *Gań. sár.* § 32—57.

[3] *Bhága-játi* or *Anśa-savarńa,* assimilation of fractions; or rendering fractions homogeneous: reduction of them to uniformity.

[4] *Bhága, anśa, vibhága, lava,* &c. a part or fraction: the numerator of a fraction.

Hara, hára, ch'héda, &c. the divisor; the denominator of a fraction. That, which is to be divided, is the part *(anśa)*; and that, by which it is to be divided, is *(hara)* the divisor. GAŃ. and SúR.

[5] *Rási* a quantity, § 36. It here intends one consisting of two terms; a part and a divisor, or numerator and denominator. GANG.

which answer to three and a fifth, and one-third, proposed for addition; and those which correspond to a sixty-third and a fourteenth offered for subtraction.

Statement: $\frac{3}{1}\ \frac{1}{5}\ \frac{1}{3}$.*

Answer. Reduced to a common denominator $\frac{45}{15}\ \frac{3}{15}\ \frac{5}{15}$. Sum $\frac{53}{15}$.

Statement of the 2d example: $\frac{1}{63}\ \frac{1}{14}$.

Answer. The denominators being abridged, or reduced to least terms, by the common measure seven, the fractions become $\frac{1}{9}\ \frac{1}{2}$.

Numerator and denominator, multiplied by the abridged denominators, give respectively $\frac{2}{126}$ and $\frac{9}{126}$.

Subtraction being made, the difference is $\frac{7}{126}$.

Reduction of subdivided Fractions.

31. Rule: half a stanza.

The numerators being multiplied by the numerators, and the denominators by the denominators, the result is a reduction to homogeneous form in subdivision of fractions.

32. Example. The quarter of a sixteenth of the fifth of three-quarters of two-thirds of a moiety of a *dramma* was given to a beggar by a person, from whom he asked alms: tell me how many cowry shells the miser gave,

* Among astronomers and other arithmeticians, oral instruction has taught to place the numerator above and denominator beneath. GAN'.

No line is interposed in the original: but in the version it is introduced to conform to the practice of European arithmetic. BHA'SCARA subsequently directs (§ 36) an integer to be written as a fraction by placing under it unity for its denominator. The same is done by him in this place in the text. It corresponds with the directions of S'RI'D'HARA and of BRAHMEGUPTA's commentator. *Gan'. sár.* § 32; *Brahm.* § 5.

[2] *Prabhága-játi* assimilation of sub-fractions, or making uniform the fraction of a fraction: it is a sort of division of fractions. GAN'.

Prabhága a divided fraction, or fraction of a fraction: as a part of a moiety, and so forth. GANG.

CHATURVÉDA terms this operation *Pratyutpanna-játi*; assimilation consisting in multiplication, or reduction to homogeneousness by multiplication. *Brahm.* § 8.

if thou be conversant, in arithmetic, with the reduction termed subdivision of fractions.

Statement: $\frac{1}{1}$ $\frac{1}{2}$ $\frac{2}{3}$ $\frac{3}{4}$ $\frac{1}{5}$ $\frac{1}{16}$ $\frac{1}{4}$.
Reduced to homogeneousness $\frac{6}{7680}$, or in least terms $\frac{1}{1280}$.
Answer. A single cowry shell was given.[1]

Reduction of Quantities increased or decreased by a Fraction.[2]

33. Rule: A stanza and a half.

The integer being multiplied by the denominator, the numerator is made positive or negative,[3] provided parts of an unit be added or be subtractive. But, if indeed the quantity be increased or diminished by a part of itself, then, in the addition and subtraction of fractions, multiply the denominator by the denominator standing underneath,[4] and the numerator by the same augmented or lessened by its own numerator.

[1] For a cowry shell is in the tale of money the 1280th part of a *dramma*, § 2.

[2] *Bhágánuband'ha-játi*; assimilation of fractional increase; reduction to uniformity of an increase by a fraction, or the addition of a part: from *anuband'ha* junction—GAŃ., union—SÚR., addition—GANG.

Bhágápaváha-játi; assimilation of fractional decrease, reduction to uniformity of a decrease by a fraction, or the subtraction of a part: derived from *apaváha* deduction, lessening, or subtraction.

These, as remarked by GAŃÉŚA, are sorts of addition and subtraction.

The fractions may be parts of an integer, or proportionate parts of the proposed quantity itself. Hence two sorts of each, named by the commentators (GANG. and SÚR.) *Rúpa-bhágánuband'ha*, addition of the fraction of an unit; *Rúpa-bhágápaváha*, subtraction of the fraction of an unit; *Ráśi-bhágánuband'ha*, addition of a fraction of the quantity; *Ráśi-bhágápaváha*, subtraction of a fraction of the quantity.

[3] And added or subtracted accordingly. See explanation of positive and negative quantity (*d'hana* and *rína*) in *Víja-gańita*, § 3.

[4] Indian arithmeticians write fractions under the quantities to which they are additive, or from which they are subtractive. Accordingly, 'the numerators and denominators are put in their order, one under the other. Then multiply the denominator which stands above, by that which stands below; and the upper numerator, by a multiplier consisting of the same denominator with its own numerator added or deducted. Repeat the operation till the up and down line contain but two quantities.'—SÚR.

It must have originally contained three terms or numbers, at the least, in examples of the first rule; and four, in those of the last.—GANG.

34. Example. Say, how much two and a quarter, and three less a quarter, are, when reduced to uniformity, if thou be acquainted with fractional increase and decrease.

Statement:

2	3
$\frac{1}{4}$	$\cdot\frac{1}{4}$

Answer: Reduced to homogeneousness, they become $\frac{9}{4}$ and $\frac{11}{4}$.

35. Example. How much is a quarter added to its third part, with a quarter of the sum? and how much are two-thirds, lessened by one-eighth of them, and then diminished by three-sevenths of the residue? Tell me, likewise, how much is half less its eighth part, added to nine-sevenths of the residue, if thou be skilled, dear woman, in fractional increase and decrease?

Statement:

$\frac{1}{4}$*	$\frac{2}{3}$	$\frac{1}{2}$
$\frac{1}{3}$	$\cdot\frac{1}{8}$	$\cdot\frac{1}{8}$
$\frac{1}{2}$	$\cdot\frac{3}{7}$	$\frac{9}{7}$

Answer: Reduced to uniformity, the results are $\frac{1}{2}$ $\frac{1}{3}$ $\frac{1}{1}$.

The Eight Rules of Arithmetic applied to Fractions.[1]

36. Rule for addition and subtraction of fractions:[2] half a stanza. The sum or [in the case of subtraction] the difference of fractions having

* Multiply the upper denominator 4, by the one beneath, 3; the product is 12. Then, by the same denominator 3 added to its numerator 1, making 4, multiply the upper numerator; the product is 4. Again multiply the denominator as above found by the lower denominator 2, the product is 24; and by the same added to its numerator, making 3, multiply the numerator before found, viz. 4, the product is 12. The result, therefore, is $\frac{12}{24}$; which, abridged by the common divisor six, gives $\frac{1}{2}$ or a moiety.—*Manó.*

[1] *Bhinna-paricarmáshtaca*; the eight modes of process, as applicable to fractions; the preceding section being relative to those arithmetical processes as applicable to whole terms (*abhinna-paricarmáshtaca*).

[2] *Bhinna-sancalita*, addition of fractions. *Bhinna-vyavacalita*, subtraction of fractions.

a common denominator, is [taken]. Unity[1] is put denominator of a quantity[2] which has no divisor.[3]

37. Example. Tell me, dear woman, quickly, how much a fifth, a quarter, a third, a half, and a sixth, make, when added together. Say instantly what is the residue of three, subtracting those fractions?

Statement: $\frac{1}{5}\ \frac{1}{4}\ \frac{1}{3}\ \frac{1}{2}\ \frac{1}{6}$.
Answer: Added together the sum is $\frac{29}{20}$.
[Statement $\frac{3}{1}\ \frac{1}{5}\ \frac{1}{4}\ \frac{1}{3}\ \frac{1}{2}\ \frac{1}{6}$.]
Subtracting those fractions from three, the remainder is $\frac{31}{20}$.

38. Rule for multiplication of fractions:[4] half a stanza.
The product of the numerators, divided by the product of denominators, [gives a quotient, which] is the result of multiplication of fractions.

39. Example. What is the product of two and a seventh, multiplied by two and a third? and of a moiety multiplied by a third? tell, if thou be skilled in the method of multiplication of fractions.

Statement: $\begin{matrix}2\\ \frac{1}{3}\end{matrix}\ .\ \begin{matrix}2\\ \frac{1}{7}\end{matrix}$ (or reduced $\frac{7}{3}\ .\ \frac{15}{7}$) $\frac{1}{2}\ .\ \frac{1}{3}$.
Answer: the products are $\frac{5}{1}$ and $\frac{1}{6}$.

40. Rule for division of fractions:[5] half a stanza.
After reversing the numerator and denominator of the divisor, the remaining process for division of fractions is that of multiplication.

41. Example. Tell me the result of dividing five by two and a third;

[1] *Rúpa,* the species or form; any thing having bounds.—GANG. Discrete quantity. In the singular, the arithmetical unit; in the plural, integer number. See *Víja-gańita,* § 4.

[2] *Ráśi,* a congeries; a heap of things, of which one is the scale of numeration; a quantity or number. See *Víja-gań.* ib.

[3] That is, it is put denominator of an integer.

[4] *Bhinna-guńana,* multiplication of fractions.

[5] *Bhinna-bhága-hara;* division of fractions.

and a sixth by a third; if thy understanding, sharpened into confidence, be competent to the division of fractions.[1]

Statement: $\begin{matrix} 2 \\ \frac{1}{3} \end{matrix}$ $(\frac{7}{3})$ $\frac{6}{1}$ $\frac{1}{3}$ $\frac{1}{6}$.

Answer: Proceeding as directed, the quotients are $\frac{15}{7}$ and $\frac{1}{2}$.

42. Rule for involution and evolution of fractions:[2] half a stanza.

If the square be sought, find both squares; if the cube be required, both cubes: or, to discover the root [of cube or square,] extract the roots of both [numerator and denominator].

43. Example. Tell me quickly what is the square of three and a half; and the square root of the square; and the cube of the same; and the cube root of that cube: if thou be conversant with fractional squares and roots?

Statement: $\begin{matrix} 3 \\ \frac{1}{2} \end{matrix}$ or reduced $\frac{7}{2}$.

Answer. Its square is $\frac{49}{4}$; of which the square root is $\frac{7}{2}$. The cube of it is $\frac{343}{8}$; of which again the cube root is $\frac{7}{2}$.

[1] GAŃÉSA omits the latter half of the stanza. GANGÁD'HARA gives it entire.

[2] *Bhinna-varga*, square of a fraction; *Bhinna-g'hana*, cube of a fraction, &c.

SECTION IV.

CIPHER.[1]

44—45. Rule for arithmetical process relative to cipher: two couplets.

In addition, cipher makes the sum equal to the additive.[2] In involution and [evolution][3] the result is cipher. A definite quantity,[4] divided by cipher, is the submultiple of nought.[5] The product of cipher is nought: but it must be retained as a multiple of cipher,[6] if any further operation impend. Cipher having become a multiplier, should nought afterwards become a divisor, the definite quantity must be understood to be unchanged. So likewise any quantity, to which cipher is added, or from which it is subtracted, [is unaltered.]

46. Example. Tell me how much is cipher added to five? and the square of cipher? and its square root? its cube? and cube-root? and five multiplied by cipher? and how much is ten, subtracting cipher? and what

[1] *Śúnya, c'ha,* and other synonyma of vacuum or etherial space: nought or cipher; a blank or the privation of specific quantity.—CRĬSHṆ. on *Vija-gańita.*

The arithmetic of cipher is briefly treated by BRAHMEGUPTA in his chapter on Algebra, § 19—24. See CH. on Arithm. of BRAHM. § 13, note.

[2] *Cshépa*; that which is cast or thrown in *(cshipyaté)*: additive. GANG.

[3] Involution, &c. That is, square and square-root; cube and cube-root. GANG.

[4] *Rási.* See § 36.

[5] *C'ha-hara,* a fraction with cipher for its denominator. According to the remark of GAŃÉSA, an indefinite, unlimited, or infinite quantity: since it cannot be determined how great it is. Unaltered by addition or subtraction of finite quantities: since, in the preliminary operation of reducing both fractional expressions to a common denominator, preparatory to taking their sum or difference, both numerator and denominator of the finite quantity vanish. RANGANÁT'HA affirms, that it is infinite, because the smaller the divisor is, the greater is the quotient: now cipher, being in the utmost degree small, gives a quotient infinitely great. See *Vija-gańita,* § 14.

[6] *C'hagúńa,* a quantity which has cipher for its multiplier. Cipher is set down by the side of the multiplicand, to denote it. GAŃ.

number is it, which multiplied by cipher, and added to half itself, and multiplied by three, and divided by cipher, amounts to the given number sixty-three?

Statement: 0. Cipher added to five makes 5. Square of cipher, 0. Square-root, 0. Cube of cipher, 0. Cube-root, 0.

Statement: 5. This, multiplied by cipher makes 0.

Statement: 10. This, divided by cipher, gives $\frac{10}{0}$.

Statement: An unknown quantity; its multiplier, 0; additive, $\frac{1}{2}$; multiplicator, 3; divisor, 0; given number, 63; assumption, 1. Then, either by inversion or position, as subsequently explained (§ 47 and 50), the number is found, 14. This mode of computation is of frequent use in astronomical calculation.

CHAPTER III.

MISCELLANEOUS RULES.[1]

SECTION I.

INVERSION.

47—48. Rule of inversion:[2] two stanzas.

To investigate a quantity, one being given,[3] make the divisor a multiplicator: and the multiplier, a divisor; the square, a root; and the root, a square;[4] turn the negative into positive; and the positive into negative. If a quantity was to be increased or diminished by its own proportionate part, let the [lower[5]] denominator, being increased or diminished by its numerator, become the [corrected[5]] denominator, and the numerator remain unchanged; and then proceed with the other operations of inversion, as before directed.

49. Example. Pretty girl with tremulous eyes, if thou know the correct method of inversion, tell me, what is the number, which multiplied by

[1] *Pracírńa* miscellaneous. The rules, contained in the five first sections of this chapter, have none answering to them in the Arithmetic of BRAHMEGUPTA and ŚRÍD'HARA. Some of the examples, however, serving to illustrate the reduction of fractions (as § 51—54.) do correspond. Compare § 54 with *Gań. sár.* § 52.

[2] *Vilóma-vid'hi, Vilóma-criyá, Vyasta-vid'hi,* inversion.

[3] *Drĭśya;* the quantity or number, which is visible; the one known by the enunciation of the problem: the given quantity.

[4] And the cube, a cube-root; and the cube-root, a cube. GAŃ.

[5] GANGÁD'HARA.

three, and added to three quarters of the quotient, and divided by seven, and reduced by subtraction of a third part of the quotient, and then multiplied into itself, and having fifty-two subtracted from the product, and the square root of the remainder extracted, and eight added, and the sum divided by ten, yields two?[1]

Statement: Multiplier 3. Additive $\frac{3}{4}$. Divisor 7. Decrease $\frac{1}{3}$. Square —. Subtractive 52. Square-root —. Additive 8. Divisor 10. Given number 2.

Answer. Proceeding as directed, the result is 28; the number sought.

[1] All the operations are inverted. The known number 2, multiplied by the divisor 10 converted into a multiplicator, makes 20; from which the additive 8, being subtracted, leaves 12; the square whereof (extraction of the root being directed) is 144; and adding the subtractive 52, becomes 196: the root of this (squaring was directed) is 14; added to its half, 7, it amounts to 21; and multiplied by 7, is 147. This again divided by 7 and multiplied by 3, makes 63; which, subtracted from 147, leaves 84: and this, divided by 3, gives 28. *Manó.*

SECTION II.

SUPPOSITION.

50. Rule of supposition:[1] one stanza.

Or any number, assumed at pleasure, is treated as specified in the particular question; being multiplied and divided, raised or diminished by fractions: then the given quantity, being multiplied by the assumed number and divided by that [which has been found,] yields the number sought. This is called the process of supposition.[2]

51. Example.[3] What is that number, which multiplied by five, and having the third part of the product subtracted, and the remainder divided by ten, and one-third, a half and a quarter of the original quantity added, gives two less than seventy?

Statement: Mult. 5. Subtractive $\frac{1}{3}$ of itself. Div. 10. Additive $\frac{1}{3}\,\frac{1}{2}\,\frac{1}{4}$ of the quantity. Given 68.

Putting 3; this, multiplied by 5, is 15; less its third part, is 10; divided by ten, yields 1. Added to the third, half and quarter of the assumed number three, viz. $\frac{3}{3}\,\frac{3}{2}\,\frac{3}{4}$, the sum is $\frac{17}{4}$. By this divide the given number 68 taken into the assumed one 3; the quotient is 48.

The answer is the same with any other assumed number, as one, &c.

Thus, by whatever number the quantity is multiplied or divided in any example, or by whatever fraction of the quantity, it is increased or diminished; by the same should the like operations be performed on a number

[1] *Ishṫa-carman:* operation with an assumed number. It is the rule of false position, supposition, and trial and error.

[2] In this method, multiplication, division, and fractions only are employed. GAŃ.

[3] Reduction of a given number with affirmative fractions is the subject of this example; as reduction of a number given, with negative fractions, is that of the next. SŪR.

In the rule of position or reduction appertaining to it, are comprehended reduction of given quantity (with fractions affirmative or negative), reduction of fractions of residues, and reduction of differences of fractions. GANG.

arbitrarily assumed: and by that, which results, divide the given number taken into the assumed one; the quotient is the quantity sought.

52. Example of reduction of a given quantity.[1] Out of a heap of pure lotus flowers, a third part, a fifth and a sixth, were offered respectively to the gods ŚIVA, VISHŃU and the Sun; and a quarter was presented to BHAVÁNÍ. The remaining six lotuses were given to the venerable preceptor. Tell quickly the whole number of flowers.

Statement: $\frac{1}{3}\ \frac{1}{5}\ \frac{1}{6}\ \frac{1}{4}$; known 6.

Putting one for the assumed number, and proceeding as above, the quantity is found 120.

53. Example of reduction of residues:[2] A traveller, engaged in a pilgrimage, gave half his money at *Prayága;* two-ninths of the remainder at *Cáśi;* a quarter of the residue in payment of taxes on the road; six-tenths of what was left at *Gáyá;* there remained sixty-three *nishcas;* with which he returned home. Tell me the amount of his original stock of money, if you have learned the method of reduction of fractions of residues.

Statement: $\frac{1}{2}\ \frac{2}{9}\ \frac{1}{4}\ \frac{6}{10}$; known 63.

Putting one for the assumed number; subtracting the numerator from its denominator, multiplying denominators together, and in other respects proceeding as directed, the remainder is found $\frac{7}{60}$. By this dividing the given number 63 taken into the assumed quantity, the original sum comes out 540.

Or it may be found by the method of reduction of fractional decrease [§ 33]. Statement: $\frac{1}{1}\ \frac{1}{2}\cdot\ \frac{2}{9}\cdot\ \frac{1}{4}\cdot\ \frac{6}{10}\cdot$. Being reduced to homogeneous form, the result is $\frac{7}{60}$: whence the sum is deduced 540.

Or this may also be found by the rule of inversion [§ 47.]

54. Example of reduction of differences.[3] Out of a swarm of bees, one-

[1] *Drísya-játi;* assimilation of the visible; reduction of the given quantity with fractions affirmative or negative: here, with negative; in the preceding example, with affirmative.

[2] *Śésha-játi,* assimilation of residue; reduction of fractions of residues or successive fractional remainders.

[3] *Viślésha-játi,* assimilation of difference; reduction of fractional differences.

fifth part settled on a blossom of *Cadamba*;[1] and one-third on a flower of *Silínd'hrí*:[2] three times the difference of those numbers flew to the bloom of a *Cutaja*.[3] One bee, which remained, hovered and flew about in the air, allured at the same moment by the pleasing fragrance of a jasmin and pandanus. Tell me, charming woman, the number of bees.[4]

Statement: $\frac{1}{5}\ \frac{1}{3}\ \frac{2}{15}$; known quantity, 1; assumed, 30.

A fifth of the assumed number is 6; a third is 10; difference 4; multiplied by three gives 12; and the remainder is 2. Then the product of the known quantity by the assumed one, being divided by this remainder, shows the number of bees 15.

Here also putting unit for the assumed quantity, the number of the swarm is found 15.

So in other instances likewise.[5]

[1] *Cadamba*, Nauclea orientalis or N. Cadamba.

[2] *Śilínd'hrí*, a plant resembling the *Cachóra*. CRĬSHN. on *Víja-gań.*

[3] Echites antidysenterica.

[4] See the same example in *Víja-gańita*, § 108.

[5] The *Manóranjana* introduces one more example, which is there placed after the second. It is similar to one which occurs in ŚRÍD'HARA's *Gańita-sára*, § 50; and is here subjoined:—"The third part of a necklace of pearls, broken in an amorous struggle, fell to the ground: its fifth part rested on the couch; the sixth part was saved by the wench; and the tenth part was taken up by her lover: six pearls remained strung. Say, of how many pearls the necklace was composed."

Statement: $\frac{1}{3}\ \frac{1}{5}\ \frac{1}{6}\ \frac{1}{10}$; Rem. 6. Answer 30.

SECTION III.[1]

55. Rule of concurrence: half a stanza.

The sum with the difference added and subtracted, being halved, gives the two quantities. This is termed concurrence.[2]

56. Example. Tell me the numbers, the sum of which is a hundred and one; and the difference, twenty-five; if thou know the rule of concurrence, dear child.

Statement: Sum 101; diff. 25.—Answer: the two numbers are 38 and 63.

57. Rule of dissimilar operation:[3] half a stanza.

The difference of the squares, divided by the difference of the radical quantities,[4] gives their sum: whence the quantities are found in the mode before directed.

58. Example. Tell me quickly, skilful calculator, what numbers are they, of which the difference is eight, and the difference of squares four hundred?

Statement: Diff. of the quantities 8. Diff. of the squares 400.

Answer. The numbers are 21 and 29.

[1] The rules comprised in this section are treated under the same titles (*Sancramańa* and *Vishama-carman)* by BRAHMEGUPTA, in his chapter on Algebra, or, as by him termed, lecture on the pulverizer, § 25. See CHATURVÉDA on Arithm. of *Brahm.* § 66.

[2] *Sancramańa,* concurrence or mutual penetration in the shape of sum and difference.—GANG. Investigation of two quantities concurrent or grown together in form of sum and difference.—GAN. Calculation of quantities latent within those exhibited.—SÚR. The same term signifies transition (or transposition). See BRAHMEGUPTA, Arithm. § 12.

[3] *Vishama-carman:* the finding of the quantities, when the difference of their squares is given, and either the sum or the difference of the quantities.—GAŃ. A species of concurrence.—GANG. See below *Lílávatí,* § 135. *Víja-gań.* § 148.

[4] Or divided by their sum, gives their difference.—GAŃ.

SECTION IV.

Problem concerning Squares.[1]

A certain problem relative to squares is propounded in the next instance.

59—60 Rule: The square of an arbitrary number, multiplied by eight and lessened by one, then halved and divided by the assumed number, is one quantity: its square, halved and added to one, is the other. Or unity, divided by double an assumed number and added to that number, is a first quantity; and unity is the other. These give pairs of quantities, the sum and difference of whose squares, lessened by one, are squares.

Tell me, my friend, numbers, the sum and difference of whose squares, less one, afford square roots: which dull smatterers in algebra labor to excruciate, puzzling for it in the six-fold method of discovery there taught.[2]

To bring out an answer by the first rule, let the number put be $\frac{1}{2}$. Its square, $\frac{1}{4}$, multiplied by eight, is 2; which, lessened by one, is 1. This halved is $\frac{1}{2}$, and divided by the assumed number ($\frac{1}{2}$) gives $\frac{1}{1}$ for the first quantity. Its square halved is $\frac{1}{2}$; which, added to one, makes $\frac{3}{2}$. Thus the two quantities are $\frac{1}{1}$ and $\frac{3}{2}$.

So, putting one for the assumed number, the numbers obtained are $\frac{7}{2}$, and $\frac{57}{8}$. With the supposition of two, they are $\frac{31}{4}$ and $\frac{993}{32}$.

By the second method, let the assumed number be 1. Unity divided by the double of it is $\frac{1}{2}$, which, added to the assumed number, makes $\frac{3}{2}$. The first quantity is thus found. The second is unity, 1. With the supposition of two, the quantities are $\frac{9}{4}$ and $\frac{1}{1}$. Putting three, they are $\frac{19}{6}$ and $\frac{1}{1}$.

61. Another Rule:[3] The square of the square of an arbitrary number, and

[1] *Varga-carman.* Operation relative to squares. An indeterminate problem; admitting innumerable solutions.

[2] This question, found in some copies of the text, and interpreted by GANGÁD'HARA and the *Manóranjana*, is unnoticed by other commentators.

[3] To bring out answers in whole numbers: the two preceding solutions giving fractions.

E 2 GAN'. and SUR.

the cube of that number, respectively multiplied by eight, adding one to the first product, are such quantities;[1] equally in arithmetic and in algebra.

Put $\frac{1}{2}$. The square of the square of the assumed number is $\frac{1}{16}$; which, multiplied by eight, makes $\frac{1}{2}$. This, added to one, is $\frac{3}{2}$; and is the first quantity. Again put $\frac{1}{2}$. Its cube is $\frac{1}{8}$; which, multiplied by eight, gives the second quantity $\frac{1}{1}$. Next, supposing one, the two quantities are 9 and 8. Assuming two, they are 129 and 64. Putting three, they are 649 and 216. And so on, without end, by means of various suppositions, in the several foregoing methods.

"Algebraic solution, similar to arithmetical rules, appears obscure; but is not so, to the intelligent: nor is it sixfold, but manifold."

[1] The greater quantity is to be taken such, that the square of it may consist of three portions, whereof one shall be unity; and the remaining two be squares; and twice the product of the roots of those squares constitute a square, the root of which will be the second quantity.

RANG.

SECTION V.

62—63. Rule for assimilation of the root's coefficient:[1] two stanzas.

The sum or difference of a quantity and of a multiple of its square-root being given, the square of half the coefficient[2] is added to the given number; and the square root of their sum [is extracted: that root,] with half the coefficient added or subtracted, being squared, is the quantity sought by the interrogator.[3] If the quantity have a fraction added, or subtracted, divide the number given and the multiplicator of the root, by unity increased or lessened by the numerator, and the required quantity may be then discovered, proceeding with those quotients as above directed.

A quantity, increased or diminished by its square-root multiplied by some number, is given. Then add the square of half the multiplier of the root to the given number: and extract the square-root of the sum. Add half the multiplier, if the difference were given; or subtract it, if the sum were so. The square of the result will be the quantity sought.

64. Example (the root subtracted, and the difference given). One pair out of a flock of geese remained sporting in the water, and saw seven times the half of the square-root of the flock proceeding to the shore tired of the diversion. Tell me, dear girl, the number of the flock.

[1] *Múla-játi*, *mula-guńaca-játi* or *Ishta-mulánśa-játi*, assimilation and reduction of the root's coefficient with a fraction.

[2] *Guńa*, multiplicator; *mula-guńa*, root's multiplier; the coefficient of the root.

[3] The quantity sought consists of two portions; one the square-root taken into its multiplicator; the other the given number. The number given too is the quantity required less the root taken into its multiplicator: and the quantity sought is the square of that root. Therefore the number given is one that consists of two portions; viz. the square of the root less the root taken into its multiplier. Now the root taken into its multiplier is equivalent to twice the product of the root by half the multiplicator. By adding then the square of half the multiplicator to the given number, a quantity results of which the root may be taken; and this root is the root (of the quantity sought) less half the multiplier. Therefore that added to half the multiplier is the root (of the quantity required); and its square, of course, is the number sought. RANG.

Statement: Coeff. $\frac{7}{2}$. Given 2. Half the coefficient is $\frac{7}{4}$. Its square $\frac{49}{16}$; added to the given number, makes $\frac{81}{16}$; the square root of which is $\frac{9}{4}$. Half the coefficient being added, the sum is $\frac{16}{4}$; or, reduced to least terms, 4. This squared is 16; the number of the flock, as required.

65. Example (the root added and the sum given). Tell me what is the number, which, added to nine times its square-root, amounts to twelve hundred and forty?

Statement: Coeff. 9. Given 1240. Answer 961.

66. Example (the root and a fraction both subtracted). Of a flock of geese, ten times the square-root of the number departed for the *Mánasa* lake,[1] on the approach of a cloud: an eighth part went to a forest of *St'halapadminís*:[2] three couples were seen engaged in sport, on the water abounding with delicate fibres of the lotus. Tell, dear girl, the whole number of the flock.

Statement: Coeff. 10. Fraction $\frac{1}{8}$. Given 6.

By the [second] rule (§ 63); unity, less the numerator of the fraction, is $\frac{7}{8}$; and the coefficient and given number, being both divided by that, become $\frac{80}{7}$ and $\frac{48}{7}$; and the half coefficient is $\frac{40}{7}$. With these, proceeding by the [first] rule (§ 62), the number of the flock is found 144.

67. Example.[3] The son of PRIT'HA,[4] irritated in fight, shot a quiver of arrows to slay CARŃA. With half his arrows, he parried those of his antagonist; with four times the square-root of the quiver-full, he killed his

[1] Wild geese are observed to quit the plains of India, at the approach of the rainy season; and the lake called *Mánasaróvar* (situated in the *Ún-* or *Hún-dés* is covered with water-fowl, geese especially, during that season. The *Hindus* suppose the whole tribe of geese to retire to the holy lake at the approach of rain. The bird is sacred to BRAHMÁ.

[2] The plant intended is not ascertained. The context would seem to imply that it is arboreous: as the term signifies forest.

[3] This example is likewise inserted in the *Víja-gańita*, § 133.

[4] ARJUNA, surnamed PÁRT'HA: his matronymic from PRÍT'HÁ or KUNT'HÍ.

horses; with six arrows, he slew ŚALYA;[1] with three he demolished the umbrella, standard and bow; and with one, he cut off the head of the foe. How many were the arrows, which ARJUNA let fly?

Statement: Fraction $\frac{1}{2}$. Coeff. 4. Given 10.

The given number and coefficient being divided (by unity less the fraction) become 20 and 8; and proceeding by the rule (§ 63), the number of arrows comes out 100.

68. Example.[2] The square-root of half the number of a swarm of bees is gone to a shrub of jasmin;[3] and so are eight-ninths of the whole swarm: a female is buzzing to one remaining male that is humming within a lotus, in which he is confined, having been allured to it by its fragrance at night.[4] Say, lovely woman, the number of bees.

Here eight-ninths of the quantity, and the root of its half, are negative [and consequently subtractive] from the quantity: and the given number is two of the specific things. The negative quantity and the number given, being halved, bring out half the quantity sought.[5] Thus,

Statement: Fraction $\frac{8}{9}$. Coeff. $\frac{1}{2}$. Given $\frac{1}{1}$.

A fraction of half the quantity is the same with half the fraction of the quantity: the fraction is therefore set down [unaltered].

Here, proceeding as above directed, there comes out half the quantity, 36; which, being doubled, is the number of bees in the swarm, 72.

69. Example. Find quickly, if thou have skill in arithmetic, the quantity,

[1] One of the *Cauravas*, and charioteer of CARŃA.

[2] Inserted also in the *Víja-gańita*. § 132.

[3] *Málatí*, Jasminum grandiflorum.

[4] The lotus being open at night and closed in the day, the bee might be caught in it. GAŃ.

[5] In such questions, it is necessary to observe whether the coefficient of the root be so of the root of the whole number, or of that of its part. For that quantity is found, of whose root the coefficient is used. But, in the present case, the root of half the quantity is proposed; and accordingly, the half of the quantity will be found by the rule. The number given, however, belongs to the entire quantity. Therefore, taking half the given number, half the required number is to be brought out by the process before directed.

Manó. and SÚR.

which added to its third part and eighteen times its square root, amounts to twelve hundred.

Statement: Fraction $\frac{1}{3}$. Coeff. 18. Given 1200.

Here, dividing the coefficient and given number by unity added to the fraction [§ 63] and proceeding as before directed, the number is brought out, *576*.

SECTION VI.

RULE of PROPORTION.[1]

70. Rule of three terms.[2]

The first and last terms, which are the argument and requisition, must be of like denomination; the fruit, which is of a different species, stands between them: and that, being multiplied by the demand and divided by the first term, gives the fruit of the demand.[3] In the inverse method, the operation is reversed.[4]

71. Example. If two and a half *palas* of saffron be obtained for three-sevenths of a *nishca*; say instantly, best of merchants, how much is got for nine *nishcas*?

Statement: $\frac{3}{7}$ $\frac{5}{2}$ $\frac{9}{1}$. Answer: 52 *palas* and 2 *carshas.*

72. Example. If one hundred and four *nishcas* are got for sixty-three *palas* of best camphor, consider and tell me, friend, what may be obtained for twelve and a quarter *palas*?

Statement: 63 104 $\frac{49}{4}$. Answer: 20 *nishcas,* 3 *drammas,* 8 *pańas,* 3 *cácinís,* 11 cowryshells and $\frac{1}{9}$th part.

73. Example. If a *c'hárí* and one eighth of rice, may be procured for two *drammas,* say quickly what may be had for seventy *pańas*?

[1] The rule of proportion, direct and inverse, simple and compound, including barter, has been similarly treated by BRAHMEGUPTA, Arithm. § 10—13; and by ŚRÍD'HARA (adding, however, as a distinct article, the sale of live animals and slaves, which BHÁSCARA places under the rule of three inverse). *Gań. sár.* § 58—90.

[2] *Trairáśica,* calculation belonging to a set of three terms.—GANG. Rule of three.

The first term is *pramáńa,* the measure or argument; the second is its fruit, *phala,* or produce of the argument; the third is *ich'há,* the demand, requisition, desire or question. GAŃ.

[3] *Ich'há-phala,* produce of the requisition, or fruit of the question: it is of the same denomination or species with the second term.

[4] See § 74.

Statement (reducing *drammas* to *pańas*): 32 $\frac{9}{8}$ 70.
Answer: 2 *c'háris*, 7 *drónas*, 1 *ád'haca*, 2 *prast'has.*

74. Rule of three inverse.[1]

If the fruit diminish as the requisition increases, or augment as that decreases, they, who are skilled in accounts, consider the rule of three terms to be inverted.[2]

When there is diminution of fruit, if there be increase of requisition, and increase of fruit if there be diminution of requisition, then the inverse rule of three is [employed]. For instance,

75. When the value of living beings[3] is regulated by their age; and in the case of gold, where the weight and touch[4] are compared; or when heaps[5] are subdivided; let the inverted rule of three terms be [used].

76. Example. If a female slave sixteen years of age, bring thirty-two [*nishcas*], what will one aged twenty cost? If an ox, which has been worked a second year, sell for four *nishcas*, what will one, which has been worked six years, cost?

1st Qu. Statement: 16 32 20. Answer: $25\frac{3}{5}$ *nishcas.*
2d Qu. Statement: 2 4 6. Answer: $1\frac{1}{3}$ *nishca.*

77. Example. If a *gadyánaca* of gold of the touch of ten may be had

[1] *Vyasta-trairásica* or *Vilóma-trairásica*, rule of three terms inverse.

[2] The method of performing the inverse rule has been already taught (§ 70). "In the inverse method, the operation is reversed." That is, the fruit is to be multiplied by the argument and divided by the demand. SU'R.

When fruit increases or decreases, as the demand is augmented or diminished, the direct rule (*crama-trairásica*) is used. Else the inverse. GAŃ.

[3] Slaves and cattle. The price of the older is less; of the younger, greater. GANG. and SU'R.

[4] Colour on the touchstone. See Alligation, § 101.

[5] See Chap. 10. When heaps of grain, which had been meted with a small measure, are again meted with a larger one, the number decreases; and when those, which had been meted with a large measure, are again meted with a smaller one, there is increase of number. GANG and SUR.

for one *nishca* [of silver], what weight of gold of fifteen touch may be bought for the same price?

Statement: 10 1 15. Answer $\frac{2}{3}$.

78. Example. A heap of grain having been meted with a measure containing seven *áḋhacas*, if a hundred such measures were found, what would be the result with one containing five *áḋhacas*?

Statement: 7 100 5. Answer 140.

79. Rule of compound proportion.[1]

In the method of five, seven, nine or more[2] terms, transpose the fruit and divisors;[3] and the product of multiplication of the larger set of terms, being divided by the product of the less set of terms,[4] the quotient is the produce [sought].

[1] This, which is the compound rule of three, comprises, according to the remark of GAŃÉSA, two or more sets of three terms *(trairásica)*; or two or more proportions *(anupáta)*, as SÚRYADÁSA observes. "Thus the rule of five *(pancha-rásica)* comprises two proportions; that of seven *(sapta-rásica)*, three; that of nine *(nava-rásica)*, four; and that of eleven *(écádaśa-rásica)*, five."

[2] Meaning eleven.—*Manó.* Eleven or more.—SÚR. It is a rule for finding a sixth term, five being given; (or, from seven known terms, an eighth; from nine, a tenth; from eleven, a twelfth).

[3] GAŃÉSA and the commentator of the *Vásaná* understand this last word (*ch'hid* divisor) as relating to denominators of fractions; and the transposing of them (if any there be) is indeed right: accordingly the author gives, under this rule, an example of working with fractions (§ 81). But the *Manóranjana* and SÚRYADÁSA explain it otherwise; and the latter cites an ancient commentary entitled *Gańita-caumudí* (also quoted by RANGANÁT'HA) in support of his exposition. 'There are two sets of terms; those which belong to the argument; and those which appertain to the requisition. The fruit, in the first set, is called produce of the argument; that, in the second, is named divisor of the set. They are to be transposed, or reciprocally brought from one set to the other. That is, put the fruit in the second set; and place the divisor in the first. Would it not be enough to say transpose the fruits of both sets? The author of the *Caumudí* replies "the designation of divisor serves to indicate, that, after transposition, the fruit of the second set, being included in the product of the multiplication of the less set of terms, the product of the greater set is to be divided by it." 'Some, however, interpret it as relative to fractions ["transpose denominators, if any there be."—GANG.] But that is wrong: for the word would be superfluous.'

[4] *Bahu-rási (pacsha)*, set of many terms: the one which is most numerous. (That, to which the fruit is brought, is the larger set.—GANG. Or, if there be fruit on both sides, that, in which the fruit of the requisition is.—GAŃ.) *Laghu-rási*, set of fewer terms; that, which is less numerous.

80. Example. If the interest of a hundred for a month be five, say what is the interest of sixteen for a year? Find likewise the time from the principal and interest; and knowing the time and produce, tell the principal sum.

Statement: $\begin{array}{cc} 1 & 12 \\ 100 & 16 \\ 5 & \end{array}$ [1]Answer: the interest is $9\frac{3}{5}$.

To find the time; Statement: $\begin{array}{cc} 1 & \\ 100 & 16 \\ 5 & \frac{48}{5} \end{array}$ [2]Answer: months 12.

To find the principal; Statement: $\begin{array}{cc} 1 & 12 \\ 100 & \\ 5 & \frac{48}{5} \end{array}$ [3]Answer: principal 16.

81. Example. If the interest of a hundred for a month and one-third, be five and one fifth, say what is the interest of sixty-two and a half for three months and one fifth?

Statement: $\begin{array}{cc} \frac{4}{3} & \frac{16}{5} \\ \frac{100}{1} & \frac{125}{2} \\ \frac{26}{5} & \end{array}$ [4]Answer: interest $7\frac{4}{5}$.

[1] Transposing the fruit, $\begin{array}{cc} 1 & 12 \\ 100 & 16 \\ & 5 \end{array}$ Product of the larger set, 960 of the less set, 100. Quotient, $\frac{960}{100}$ or $\frac{48}{5}$.

[2] Transposing both fruits, $\begin{array}{cc} 1 & \\ 100 & 16 \\ \frac{48}{5} & 5 \end{array}$ and the denominator, $\begin{array}{cc} 1 & \\ 100 & 16 \\ 48 & 5 \\ & 5 \end{array}$

Product of the larger set, 4800 of the less set, 400. Quotient, 12.

[3] Transposing both fruits, $\begin{array}{cc} 1 & 12 \\ 100 & \\ \frac{48}{5} & 5 \end{array}$ and the denominator, $\begin{array}{cc} 1 & 12 \\ 100 & \\ 48 & 5 \\ & 5 \end{array}$

Product of the larger set 4800 of the less set 300. Quotient, 16.

[4] Transposing the fruit, $\begin{array}{cc} \frac{4}{3} & \frac{16}{5} \\ \frac{100}{1} & \frac{125}{2} \\ \frac{26}{5} & \end{array}$ and the denominators, $\begin{array}{cc} 4 & 16 \\ 5 & 3 \\ 100 & 125 \\ 2 & 1 \\ 5 & 26 \end{array}$ Abridging by correspondent reduction on both sides* $\begin{array}{cc} 1 & 4 \\ 5 & 3 \\ 4 & 5 \\ 2 & 1 \\ 5 & 26 \end{array}$, and by further reduction, $\begin{array}{cc} 1 & 1 \\ 1 & 3 \\ 1 & 1 \\ 1 & 1 \\ 5 & 13 \end{array}$ Answer $\frac{39}{5}$ or $7\frac{4}{5}$

* The *Manóranjana* teaches to abridge the work by reduction of terms on both sides by their common divisors.

82. Example of the rule of seven: If eight, best, variegated, silk scarfs, measuring three cubits in breadth and eight in length, cost a hundred [*nishcas*]; say quickly, merchant, if thou understand trade, what a like scarf, three and a half cubits long and half a cubit wide, will cost.

Statement:
3 ½
8 ½
8 1
100

[1]Answer: *Nishca* 0, *drammas* 14, *paṅas* 9, *cácini* 1, cowryshells 6⅔.

83. Example of the rule of nine: If thirty benches, twelve fingers thick, square of four wide, and fourteen cubits long, cost a hundred [*nishcas*]; tell me, my friend, what price will fourteen benches fetch, which are four less in every dimension?

Statement:
12 8
16 12
14 10
30 14
100

[2]Answer: *Nishcas* 16⅔.

84. Example of the rule of eleven: If the hire of carts to convey the benches of the dimensions first specified, a distance of one league (*gavyúti*,)[3] be eight *drammas*; say what should be the cart-hire for bringing the benches last mentioned, four less in every dimension, a distance of six leagues?

[1] Transposing fruit and denominators,
3 1
2
8 7
2
8 1
100
Product of larger set, 700; of less set, 768. Quotient, 0 14 9 1 6⅔.

[2] Transposing fruit,
12 8
16 12
14 10
30 14
100
abridging by correspondent reduction on both sides;
1 1
2 1
1 1
3 1
100
Product of larger set, 100; of less set, 6. Quotient, 16⅔.

[3] *Gavyúti*; two *crósas* or half a *yójana*: it contains 4000 *daṅdas* or fathoms; about 8000 yards; and is about 3, 8 B. miles: the *crósa* being 1, 9 B. m. See As. Res. 5. 105.

Statement:

12	8
16	12
14	10
30	14
1	6
8	

[1]Answer: *Drammas* 8.

85. Rule of barter;[2] half a stanza.

So in barter likewise, the same process is [followed;] transposing both prices, as well as the divisors.[3]

86. Example. If three hundred mangoes be had in this market for one *dramma*, and thirty ripe pomegranates for a *paṅa*, say quickly, friend, how many should be had in exchange for ten mangoes?

Statement:

16	1
300	30
10	

[4]Answer: 16 pomegranates.

[1] Transposing the fruit, abridging by correspondent reduction on both sides, and by further reduction,

Transposing the fruit		abridging by correspondent reduction on both sides		and by further reduction	
12	8	1	1	1	1
16	12	2	1	1	1
14	10	1	1	1	1
30	14	3	1	1	1
1	6	1	6	1	2
	8		8		4

Product of larger set, 8; of less set, 1. Quotient, 8.

[2] *Bhánda-prati-bhándaca* commodity for commodity; computation of the exchange of goods (*vastu-vinimaya-gańita*,—GANG.): barter.

[3] GANGÁD'HARA, SU'RYADÁSA and the *Manóranjana* so read this passage: *haráni-cha maulyé.* But GAṄÉSA and RANGANÁT'HA have the affirmative adverb *sadá-hi*, in place of the words "and the divisors;" *haráni-cha.* At all events, the transposition of denominators takes place, as usual; and so does that of the lower term, as in the rule of five; to which, as SU'RYADÁSA remarks, this is analogous. It comprises two proportions, thus stated by him from the example in the text. "If for one *paṅa*, thirty pomegranates may be had, how many for sixteen? Answer, 480. Again, if for three hundred mangoes, four hundred and eighty pomegranates may be had, how many for ten? Answer, 16. Here thirty is first multiplied by sixteen and then divided by one; and then multiplied by ten and divided by three hundred. For brevity, the prices are transposed, and the result is the same" SU'R.

[4] Transposing the prices, and, transferring the fruit,

Transposing the prices		transferring the fruit	
1	16	1	16
300	30	300	30
10			10

Then product of the larger set, 4800; of the less set, 300. Quotient, 16. Or, by correspondent reduction, and further

by correspondent reduction		and further	
1	16	1	16
10	1	1	1
	10		1

Whence products 16 and 1 and quotient 16.

CHAPTER IV.

INVESTIGATION OF MIXTURE.[1]

SECTION I.

INTEREST.

87—88. RULE: a stanza and a half.[2]

The argument[3] multiplied by its time, and the fruit multiplied by the mixt quantity's time, being severally set down, and divided by their sum and multiplied by the mixt quantity, are the principal and interest [composing the quantity]. Or the principal being found by the rule of supposition, that, taken from the mixt quantity, leaves the amount of interest.

89. Example. If the principal sum, with interest at the rate of five on the hundred by the month, amount in a year to one thousand, tell the principal and interest respectively.

Statement: $\begin{matrix} 1 & 12 \\ 100 & 1000. \\ 5 & \end{matrix}$ [4]Answer: Principal, 625; Interest, 375.

Or, by the rule of position, put one; and proceeding according to that rule (§ 50), the interest of unity is $\frac{3}{5}$; which, added to one, makes $\frac{8}{5}$. The

[1] *Miśra-vyavahára*, investigation of mixture, ascertainment of composition, as principal and interest joined, and so forth.—GAŃ. It is chiefly grounded on the rule of proportion.—*Ibid.* The rules in this chapter bear reference to the examples which follow them. Generally they are quæstiones otiosæ; problems for exercise.

[2] To investigate a mixt amount of principal and interest.—GAŃ. The first rule agrees with SRÍD'HARA'S (*Gań. sár.* § 91). The second answers to one deduced from BRAHMEGUPTA by his Commentator. Arithm. of BRAHM. § 14.

[3] *Pramáńa* argument; and *phala* fruit (§ 70): principal and interest.

[4] 100 multiplied by 1 is 100; 5 by 12 is 60. Their sum 160 is the divisor. The first number, 100, multiplied by 1000, and divided by 160, is 625. The second 60, multiplied by 1000 and divided by 160, gives 375. GANG.

given quantity 1000, multiplied by unity, and divided by that, shows the principal 625. This, taken from the mixt amount, leaves the interest 375.[1]

90. Rule:[2] The arguments taken into their respective times are divided by the fruit taken into the elapsed times; the several quotients, divided by their sum and multiplied by the mixt quantity, are the parts as severally lent.

91. Example: The sum of six less than a hundred *nishcas* being lent in three portions at interest of five, three and four per cent. an equal interest was obtained on all three portions, in seven, ten, and five months respectively. Tell, mathematician, the amount of each portion.[3]

Statement:	1	7	1	10	1	5	Mixt amount 94.[4]
	100		100		100		
	5		3		4		

Answer: the portions are 24, 28 and 42. The equal amount of interest $8\frac{2}{5}$.

92. Rule: half a stanza.[5]

1 Or the principal being known, the interest may be found by the rule of five. SU'R.

2 For determining parts of a compound sum. SU'R.

3 Since the amount of interest on all the portions is the same, put unity for its arbitrarily assumed amount: whence corresponding principal sums are found by the rule of five. For instance, if a hundred be the capital, of which five is the interest for a month, what is the capital of which unity is the interest for seven months? and, in like manner, the other principal sums are to be found. Thus, a compound proportion being wrought, the time is multiplied by the argument to which it appertains, and divided by the fruit taken into the elapsed time. Then, as the total of those principal sums is to them severally, so is the given total to the respective portions lent. They are thus severally found by the rule of three. GAN'.

4 Multiplying the argument and fruit by the times, and dividing one product by the other, there result the fractions $\frac{100}{35}$ $\frac{100}{30}$ $\frac{100}{20}$ or $\frac{20}{7}$ $\frac{20}{6}$ $\frac{20}{4}$; which reduced to a common denominator and summed, make $\frac{1880}{168}$ or $\frac{235}{21}$; multiplied by the mixt amount 94, they are $\frac{1880}{7}$, $\frac{1880}{6}$, $\frac{1880}{4}$; and then divided by the sum $\frac{1880}{168}$, they give $\frac{168}{7}$, $\frac{168}{6}$, $\frac{168}{4}$, or 24, 28, 42. *Manó.*

To find the interest, employ the rule of five;
1	7
100	24.
5	

Answer, $8\frac{2}{5}$. By the same method, with all three portions, the interest comes out the same. SU'R.

5 The capital sums, their aggregate amount, and the sum of the gains being given; to apportion the gains.—GAN'. The rule is taken from BRAHMEGUPTA, Arithm. § 16. It answers to SRÍD'HARA'S, *Gan. sár.* § 109.

The contributions,[1] being multiplied by the mixt amount and divided by the sum of the contributions, are the respective fruits.[2]

93. Example. Say, mathematician, what are the apportioned shares of three traders, whose original capitals were respectively fifty-one, sixty-eight, and eighty-five; which have been raised by commerce conducted by them on joint stock, to the aggregate amount of three hundred?

Statement: 51, 68, 85. Sum: 204. Mixt amount: 300.

Answer: 75, 100, 125. These, less the capital sums, are the gains: viz. 24, 32, 40.

Or the mixt amount, less the sum of aggregate capital, is the profit on the whole: viz. 96. This being multiplied by the contributions, and divided by the sum of the contributions, gives the respective gains; viz. 24, 32, and 40.

[1] *Pracshépaca*, that which is thrown in or mixt.—GAN'. Joined together.—SU'R.

[2] The principle of the rule is obvious, being simply the rule of three.—GAN'. 'If by this sum of contributions, this contribution be had, then by the compound sum what will be? The numbers thus found, less the contributions, are the gains.' *Vásaná* by RANG.

SECTION II.

FRACTIONS.

94. Rule :[1] half a stanza.

Divide denominators by numerators; and then divide unity by those quotients added together. The result will be the time of filling [a cistern by several fountains.][2]

95. Example. Say quickly, friend, in what portion of a day will [four] fountains, being let loose together, fill a cistern, which, if severally opened, they would fill in one day, half a day, the third, and the sixth part, respectively?

Statement: $\frac{1}{1}\,\frac{1}{2}\,\frac{1}{3}\,\frac{1}{6}$.
Answer: $\frac{1}{12}$th part of a day.

[1] To apportion the time for a mixture of springs to fill a well or cistern.—GAŃ. To solve an instance relative to fractions.—SÚR. A similar problem occurs in BRAHMEGUPTA's Arithmetic, § 8.

[2] The rule is grounded on a double proportion, according to GAŃÉŚA and RANGANÁT'HA; but on the rule of three inverse, according to SÚRYADÁSA and the *Manóranjana:* "if by one fountain's time one day be had; then, by all the fountains' times in portions of days, summed together, what is had?" Or, "If, by this portion of a day one cistern be filled, how many by a whole day?" Then, after adding together the number of full cisterns, "if, by so many, one day be had, then by one cistern what will be?"

SECTION III.

PURCHASE and SALE.

96. Rule.[1] By the [measure of the] commodities,[2] divide their prices taken into their respective portions [of the purchase]; and by the sum of the quotients divide both them and those portions severally multiplied by the mixt sum: the prices and quantities are found in their order.[3]

97. Example:[4] If three and a half *mánas*[5] of rice may be had for one *dramma*, and eight of kidney-beans[6] for the like price, take these thirteen *cácinis*, merchant, and give me quickly two parts of rice with one of kidney-beans; for we must make a hasty meal and depart, since my companion will proceed onwards.

Statement: $\begin{matrix} \frac{2}{1} & \frac{1}{1} \\ \frac{1}{1} & \frac{1}{1} \\ \frac{7}{2} & \frac{8}{1} \end{matrix}$ Mixt sum $\frac{13}{64}$.

[1] For a case where a mixture of portions, and composition of things, are given.—GAŃ. Concerning measure of grain, &c.—SU'R. See SRÍD'HARA, § 116.

[2] *Pańya:* the measure of the grain or other commodity procurable for the current price in the market. SU'R. and *Manó.*

[3] Founded on the rule of proportion: 'if by this measure of goods this price be obtained, then by this portion of goods what will be?' So for the second commodity. Then, summing the prices so found, 'if by this sum, these several prices, then by this mixt amount what prices?' and, 'if by this sum, these portions, then by this mixt amount what quantities?' RANG.

[4] See *Víja-gańita*, § 115; which is word for word the same.

[5] *Mána* or *Mánaca* a measure; seemingly intending a particular one; the same with the *máńicá*, according to the *Manóranjana*, if a passage in the margin of that commentary be genuine. The *Máńicá* is the quarter of the *c'hárí*. See CHATURVÉDA on BRAHMEGUPTA, § 11. But, according to GAŃÉSA, the *mańa* (apparently the same with the *máńicá*) is at most an eighth of a *c'hárí;* being a cubic span. See note to § 236. A spurious couplet (see note on § 2.) makes it the modern measure of weight containing forty *sérs.*

[6] *Mudga:* Phaseolus mungo; sort of kidney-bean.

The prices, $\frac{1}{1}$ $\frac{1}{1}$, multiplied by the portions $\frac{2}{1}$ $\frac{1}{1}$, and divided by the goods $\frac{7}{2}$ $\frac{8}{1}$, make $\frac{4}{7}$ $\frac{1}{8}$; the sum of which is $\frac{39}{56}$. By this divide the same fractions ($\frac{4}{7}$ $\frac{8}{1}$) taken into the mixt sum ($\frac{13}{64}$); and the portions ($\frac{2}{1}$ $\frac{1}{1}$) taken into that mixt sum ($\frac{13}{64}$). There result the prices of the rice and kidney-beans, $\frac{1}{5}$ and $\frac{7}{192}$ of a *dramma;* or 10 *cácinís* and 13$\frac{1}{3}$ shells for the rice, and 2 *cácinís* and 6$\frac{2}{3}$ shells for the kidney-beans; and the quantities are $\frac{7}{12}$ and $\frac{7}{24}$ of a *mána* of rice and kidney-beans respectively.

98. Example. If a *pala* of best camphor may be had for two *nishcas,* and a *pala* of sandal-wood[1] for the eighth part of a *dramma,* and half a *pala* of aloe-wood[2] also for the eighth of a *dramma,* good merchant, give me the value of one *nishca* in the proportions of one, sixteen and eight: for I wish to prepare a perfume.

Statement: 32 $\frac{1}{8}$ $\frac{1}{8}$ Mixt sum 16.
1 1 $\frac{1}{2}$
1 16 8

Answer: Prices: *drammas* 14$\frac{2}{9}$ $\frac{8}{9}$ $\frac{8}{9}$.
Quantities: *palas* $\frac{4}{9}$ $\frac{64}{9}$ $\frac{32}{9}$.

1 *Chandana:* Santalum album.

2 *Aguru:* Aquillaria Agallochum.

SECTION IV.

99. Rule. Problem concerning a present of gems.[1]

From the gems subtract the gift multiplied by the persons; and any arbitrary number being divided by the remainders, the quotients are numbers expressive of the prices. Or the remainders being multiplied together, the product, divided by the several reserved remainders, gives the values in whole numbers.[2]

100. Example. Four jewellers, possessing respectively eight rubies, ten sapphires, a hundred pearls, and five diamonds, presented, each from his own stock, one apiece to the rest in token of regard and gratification at meeting: and they thus become owners of stock of precisely equal value. Tell me, severally, friend, what were the prices of their gems respectively?

Statement: Rub. 8; sapph. 10; pearls 100; diam. 5. Gift 1. Persons 4.

Here, the product of the gift 1 by the persons 4, viz. 4, being severally subtracted, there remain rubies 4; sapphires 6; pearls 96; diamond 1. Any number arbitrarily assumed being divided by these remainders, the quotients are the relative values. Taking it at random, they may be fractional values; or by judicious selection, whole numbers: thus, put 96; and the prices thence deduced are 24, 16, 1, 96; and the equal stock 233.

Or the remainders being multiplied together, and the product severally divided by those remainders, the prices are 576, 384, 24, 2304: and the equal amount of stock (after interchange of presents) is 5592.

[1] The problem is an indeterminate one. The solution gives relative values only.

[2] SÚRYADÁSA cites the *Víja-gańita* for the solution of the problem. (See *Víja-gań.* § 11. where the same example occurs.) The principle is explained by RANGANÁT'HA without reference to algebra. It is founded on the axiom, that "equality continues, if addition or subtraction of equal things be made to or from equal things." After interchange of presents, each person has one of every sort of gem, and a certain further number of one sort. Deducting then one of each sort from the equalized stock of every person, remains a number of a single sort equal in value one to the other. Put an arbitrary number for that value; and make the proportion; 'as this number of gems is to this equal value, so is one gem to its price.' RANG.

SECTION V.

ALLIGATION.[1]

101. Rule.[2] The sum of the products of the touch[3] and [weight of several parcels][4] of gold being divided by the aggregate of the gold, the touch of the mass is found. Or [after refining] being divided by the fine gold, the touch is ascertained; or divided by the touch, the quantity of purified gold is determined.[5]

102—103. Example. Parcels of gold weighing severally ten, four, two and four *máshas*, and of the fineness of thirteen, twelve, eleven and ten respectively, being melted together, tell me quickly, merchant, who art conversant with the computation of gold, what is the fineness of the mass? If the twenty *máshas* above described be reduced to sixteen by refining, tell me instantly the touch of the purified mass. Or, if its purity when refined be sixteen, prithee what is the number to which the twenty *máshas* are reduced?

Statement:	Touch	13	12	11	10.
	Weight	10	4	2	4.

Answer:[6] After melting, fineness 12.
Weight 20.

[1] *Suverńa-gańita;* computation of gold; that is, of its weight and fineness. Alligation medial. Śríd'hara has similar rules, § 99—108. The topic is unnoticed by Brahmegupta; but the omission is supplied by his commentator. See Chaturvéda on Brahmegupta's Arithm. note to Sect. 2.

[2] To find the fineness produced by mixture of parcels of gold; and, after refining, the weight, if the fineness be known; and the fineness, if the weight of refined gold be given. Gań.

[3] *Varńa,* colour of gold on the touchstone. Fineness of gold determined by that toucn. See § 77. " The degrees of fineness increase as the weight is reduced by refining."—Gań.

[4] Gang.

[5] The solution of the problem is grounded on the rule of supposition, together with the rule of three inverse: as shown at large by Rangana'at'ha and Gańésa under this and § 77.

[6] Products 130, 48, 22, 40. Their sum 240; divided by 20, gives 12: divided by 16, gives 15.

After refining, the weight being sixteen *máshas;* touch 15. The touch being sixteen; weight 15.

104. Rule.[1] From the acquired fineness of the mixture, taken into the aggregate quantity of gold, subtract the sum of the products of the weight and fineness [of the parcels, the touch of which is known,] and divide the remainder by the quantity of gold of unknown fineness; the quotient is the degree of its touch.[2]

105. Example. Eight *máshas* of ten, and two of eleven by the touch, and six of unknown fineness, being mixed together, the mass of gold, my friend, became of the fineness of twelve; tell the degree of unknown fineness.

Statement:	10	11		Fineness of the
	8	2	6	mixture 12.

Answer: Degree of the unknown fineness 15.

106. Rule.[3] The acquired fineness of the mixture being multiplied by the sum of the gold [in the known parcels], subtract therefrom the aggregate products of the weight and fineness [of the parcels]: divide the remainder by the difference between the fineness of the gold of unknown weight and that of the mass, the quotient is the weight of gold that was unknown.

107. Example. Three *máshas* of gold of the touch of ten, and one of the fineness of fourteen, being mixt with some gold of the fineness of sixteen, the degree of purity of the mixture, my friend, is twelve. How many *máshas* were there of the fineness of sixteen?

Statement:	10	14	16	Fineness of the
	3	1		mixture 12.

Answer: *Másha* 1.

[1] To discover the fineness of a parcel of unknown degree of purity mixed with others of which the touch was known.
GAŃ.

[2] The rule being the converse of the preceding, the principle of it is obvious. RANG.

[3] To find the weight of a parcel of known fineness, but unknown weights, mixt with other parcels of known weight and fineness.
GAŃ.

108. Rule.[1] Subtract the effected fineness from that of the gold of a higher degree of touch, and that of the one of lower touch from the effected fineness; the differences, multiplied by an arbitrarily assumed number, will be the weight of gold of the lower and higher degrees of purity respectively.[2]

109. Example. Two ingots of gold, of the touch of sixteen and ten respectively, being mixt together, the gold became of the fineness of twelve: tell me, friend, the weight of gold in both lumps.

Statement: 16, 10. Fineness resulting 12.

Putting one, and multiplying by that; and proceeding as directed; the weights of gold are found, *máshas* 2 and 4. Assuming two, they are 4 and 8. Taking half, they come out 1 and 2. Thus, manifold answers are obtained by varying the assumption.

1 To find the weight of two parcels of given fineness and unknown weight.—GAŃ. and SU'R. A rule of alligation alternate in the simplest case. The problem is an indeterminate one: as is intimated by the author.

2 By as much as the higher degree of fineness exceeds the fineness effected, so much is the measure of the weight of less pure gold; and by as much as the lower degree of purity is under the standard of the mixture, so much is the weight of the purer gold. SU'R.

SECTION VI.

PERMUTATION and COMBINATION.

110—112. Rule:[1] three stanzas.

Let the figures from one upwards, differing by one, put in the inverse order, be divided by the same [arithmeticals] in the direct order; and let the subsequent be multiplied by the preceding, and the next following by the foregoing [result]. The several results are the changes, ones, twos, threes, &c.[2] This is termed a general rule.[3] It serves in prosody, for those versed therein, to find the variations of metre; in the arts [as in architecture] to compute the changes upon apertures [of a building]; and [in music] the scheme of musical permutations;[4] in medicine, the combinations of different savours. For fear of prolixity, this is not [fully] set forth.

113. A single example in prosody: In the permutations of the *gáyatrí* metre,[5] say quickly, friend, how many are the possible changes of the verse? and tell severally, how many are the combinations with one, [two, three,] &c. long syllables?

Here the verse of the *gáyatrí* stanza comprises six syllables. Wherefore,

[1] To find the possible permutations of long and short syllables in prosody; combinations of ingredients in pharmacy; variations of notes, &c. in music; as well as changes in other instances. GAŃ.

[2] According to GAŃÉSA, there is no demonstration of the rule, besides acceptation and experience. RANGANÁT'HA delivers an explanation of the principle of it grounded on the summing of progressions.

[3] Commentators appear to interpret this as a name of the rule here taught; *sád'háraṅa*, or *sád'háraṅa-ch'handó-gaṅita*, general rule of prosodian permutation: subject to modification in particular instances; as in music, where a special method *(asád'háraṅa)* must be applied. GANG. SÚR.

[4] *C'haṅda-méru:* a certain scheme.—GAŃ. It is more fully explained by other commentators: but the translator is not sufficiently conversant with the theory of music to understand the term distinctly.

[5] The *Gáyatrí* metre in sacred prosody is a triplet comprising twenty-four syllables: as in the famous prayer containing the Brahmenical creed, called *gáyatrí*, (See As. Res. vol. 10, p. 463). But, in the prosody of profane poetry, the same number of syllables is distributed in a tetrastic: and the verse consequently contains six syllables. (As. Res. vol. 10, p. 469.)

the figures from one to six are set down, and the statement of them, in direct and inverse order is $\begin{smallmatrix}6&5&4&3&2&1\\1&2&3&4&5&6\end{smallmatrix}$. Proceeding as directed, the results are, changes with one long syllable, 6; with two, 15; with three, 20; with four, 15; with five, 6; with six, 1; with all short, 1. The sum of these is the whole number of permutations of the verse, 64.

In like manner, setting down the numbers of the whole tetrastic, in the mode directed, and finding the changes with one, two, &c. and summing them, the permutations of the entire stanza are found: 16777216.

In the same way may be found the permutations of all varieties of metre, from *Uctá* [which consists of monosyllabic verses] to *Utcrĭti* [the verses of which contain twenty-six syllables.][1]

114. Example: In a pleasant, spacious and elegant edifice, with eight doors,[2] constructed by a skilful architect, as a palace for the lord of the land, tell me the permutations of apertures taken one, two, three, &c.[3] Say, mathematician, how many are the combinations in one composition, with ingredients of six different tastes, sweet, pungent, astringent, sour, salt and bitter,[4] taking them by ones, twos, or threes, &c.

Statement [1st Example]: $\begin{smallmatrix}8&7&6&5&4&3&2&1\\1&2&3&4&5&6&7&8\end{smallmatrix}$.

Answer: the number of ways in which the doors may be opened by ones, twos, or threes, &c. is $\underset{1}{8}, \underset{2}{28}, \underset{3}{56}, \underset{4}{70}, \underset{5}{56}, \underset{6}{28}, \underset{7}{8}, \underset{8}{1}$. And the changes on the apertures of the octagon* palace amount to 255.

Statement 2d example: $\begin{smallmatrix}6&5&4&3&2&1\\1&2&3&4&5&6\end{smallmatrix}$.

Answer: the number of various preparations with ingredients of divers tastes is $\underset{1}{6}, \underset{2}{15}, \underset{3}{20}, \underset{4}{15}, \underset{5}{6}, \underset{6}{1}$.†

[1] Asiat. Res. vol. 10, p. 468—473.

[2] *Múc'há*, aperture for the admission of air: a door or window; (same with *gavácsha*;—GAŃ.) a portico or terrace, (*bhúmi-viśésha*;—GANG. and SU'R.)

[3] The variations of one window or portico open (or terrace unroofed) and the rest closed; two open, and the rest shut; and so forth.

[4] *Amera-cósha* 1. 3. 18.

* An octagon building, with eight doors (or windows; porticos or terraces;) facing the eight cardinal points of the horizon, is meant. See GAŃ.

† Total number of possible combinations, 63. GANG.

CHAPTER V.

PROGRESSIONS.[1]

SECTION I.

ARITHMETICAL PROGRESSION.

115. RULE:[2] Half the period,[3] multiplied by the period added to unity, is the sum of the arithmeticals one, &c. and is named their addition.[4] This, being multiplied by the period added to two, and being divided by three, is the aggregate of the additions.[5]

[1] *Śrēd'hí*, a term employed by the older authors for any set of distinct substances or other things put together.—GAŃ. It signifies sequence or progression. *Śrēd'hí-vyavahára*, ascertainment or determination of progressions.

[2] To find the sums of the arithmeticals.—GAŃ.

[3] *Pada* the place.—GAŃ. Any one of the figures, or digits; being that of which the sum is required.—SU'R. The last of the numbers to be summed.—*Manó*. See below: note to § 119.

[4] *Sancalitá*, the first sum, or addition of arithmeticals. *Sancalitaicya*, aggregate of additions, summed sums, or second sum.

[5] The first figure is unity. The sum of that and the period being halved, is the middle figure. As the figures decrease behind it, so they increase before it: wherefore the middle figure, multiplied by the period, is the sum of the figures one, &c. continued to the period. The only proof of the rule for the aggregate of sums, is acceptation.—GAŃ. It is a maxim, that 'a number multiplied by the next following arithmetical, and halved, gives the sum of the preceding:' wherefore, &c.—SU'R. CAMALÁCARA is quoted by RANGANÁT'HA for a demonstration grounded on placing the numbers of the progression in the reversed order under the direct one: where it becomes obvious, that each pair of terms gives the like sum: wherefore this sum, multiplied by the number of terms, is twice the sum of the progression.

116. Example: Tell me, quickly, mathematician, the sums of the several [progressions of] numbers one, &c. continued to nine; and the summed sums of those numbers.

Statement:	Arithmeticals:	1	2	3	4	5	6	7	8	9.
Answer:	Sums:	1	3	6	10	15	21	28	36	45.
	Summed sums:	1	4	10	20	35	56	84	120	165.

117. Rule:[1] Twice the period added to one and divided by three, being multiplied by the sum [of the arithmeticals], is the sum of the squares. The sum of the cubes of the numbers one, &c. is pronounced by the ancients equal to the square of the addition.

118. Example: Tell promptly the sum of the squares, and the sum of the cubes, of those numbers, if thy mind be conversant with the way of summation.

Statement: 1 2 3 4 5 6 7 8 9.
Answer: Sum of squares 285. Sum of the cubes 2025.[2]

119. Rule:[3] The increase multiplied by the period less one, and added to the first quantity, is the amount of the last.[4] That, added to the first,

[1] To find the sums of squares and of cubes. SU'R. and GAN'.

[2] Sums of the squares, 1 5 14 30 55 91 140 204 285.
Sums of the cubes, 1 9 36 100 225 441 784 1296 2025.

[3] Where the increase is arbitrary.—GANG. In such cases, to find the last term, mean amount, and sum of the progression.—SU'R. From first term, common difference and period, to find the whole amount, &c.—GAN'.

[4] *Ádi*, and *múc'ha*, *vadana*, *vactra*, and other synonyma of face; the initial quantity of the progression; (that, from which as an origin the sequence commences.—SU'R.) the first term.

Chaya, *prachaya* or *uttara*; the more (*ád'hica.*—SU'R.) or augment (*vrīdd'hi.*—GANG.) by which each term increases: the common increase or difference of the terms.

Antya; the last term.

Mad'hya; the middle term, or the mean of the progression.

Pada or *gach'ha*; the period (so many days as the sequence reaches.—SU'R.) the number of terms.

Sarva-d'hana, *Śrĕd'hi-phala* or *Gańita*; the amount of the whole; the sum of the progression. 'It is called *gańita*, because it is found by computation (*gańaná*).' GAN'.

and halved, is the amount of the mean: which, multiplied by the period, is the amount of the whole, and is denominated *(gańita)* the computed sum.[1]

120. Example: A person, having given four *drammas* to priests on the first day, proceeded, my friend, to distribute daily alms at a rate increasing by five a day. Say quickly how many were given by him in half a month?

Statement: Initial quant. 4; Com. diff. 5; Period 15.
Here, First term 4. Middle term 39. Last term 74. Sum 585.

121. Example:[2] The initial term being seven, the increase five, and the period eight, tell me, what are the nnmbers of the middle and last amounts? And what is the total sum?

Statement: First term 7; Com. diff. 5; Period 8.
Answer: Mean amount $\frac{49}{2}$. Last term 42. Sum 196.
Here, the period consisting of an even number of days, there is no middle day: wherefore the half of the sum of the days preceding and following the mean place, must be taken for the mean amount: and the rule is thus proved.

122. Rule:[3] half a stanza. The sum of the progression being divided by the period, and half the common difference multiplied by one less than the number of terms, being subtracted, the remainder is the initial quantity.[4]

123. We know the sum of the progression, one hundred and five; the number of terms, seven; the increase, three; tell us, dear boy, the initial quantity.

[1] The rule is founded on the proportion; as one day is to the increase of one day, or common difference, so is the number of increasing terms to the total increase: which, added to the initial quantity, gives the last term. SU'R. &c.

[2] To exhibit an instance of an even number of terms; where there can consequently be no middle term [but a mean amount]. GAŃ.

[3] The difference, period and sum being given, to find the first term. GAŃ. SUR.

[4] The rule is converse of the preceding. GAŃ. and SU'R.

Statement: First term? Com. diff. 3; Period 7; Sum 105.
Answer: First term, 6.

Rule:[1] half a stanza.[2] The sum being divided by the period, and the first term subtracted from the quotient, the remainder, divided by half of one less than the number of terms, will be the common difference.[3]

124. Example: On an expedition to seize his enemy's elephants, a king marched two *yójanas* the first day. Say, intelligent calculator, with what increasing rate of daily march did he proceed, since he reached his foe's city, a distance of eighty *yójanas*, in a week?

Statement: First term 2; Com. diff.? Period 7; Sum 80.
Answer: Com. diff. $\frac{22}{7}$.

125. Rule:[4] From the sum of the progression multiplied by twice the common increase, and added to the square of the difference between the first term and half that increase, the square root being extracted, this root less the first term and added to the [above-mentioned] portion of the increase, being divided by the increase, is pronounced[5] to be the period.

126. Example: A person gave three *drammas* on the first day, and continued to distribute alms increasing by two [a day]; and he thus bestowed on the priests three hundred and sixty *drammas*: say quickly in how many days?

Statement: First term 3; Com. diff. 2; Period? Sum 360.
Answer: Period 18.

[1] The first term, period and sum being known, to find the common difference which is unknown. GAŃ.

[2] Second half of one, the first half of which contained the preceding rule. § 22.

[3] This rule also is converse of the foregoing. GAŃ.

[4] The first term, common difference and sum being known, to find the period which is unknown. GAŃ.

[5] By BRAHMEGUPTA and the rest.—GAŃ. See BRAHM. c. 12, § 18. and *Gań. sár.* of ŚRÍD'H. § 123. The rules are substantially the same; the square being completed for the solution of the quadratic equation in the manner taught by ŚRÍD'HARA (cited in *Víja-gańita* § 131) and by BRAHMEGUPTA c. 8. § 32—33.

SECTION II.

GEOMETRICAL PROGRESSION.

127. Rule:[1] a couplet and a half. The period being an uneven number, subtract one, and note "multiplicator;" being an even one, halve it, and note "square:" until the period be exhausted. Then the produce arising from multiplication and squaring [of the common multiplier] in the inverse order from the last,[2] being lessened by one, the remainder divided by the common multiplier less one, and multiplied by the initial quantity, will be the sum of a progression increasing by a common multiplier.[3]

128. Example: A person gave a mendicant a couple of cowry shells first; and promised a two-fold increase of the alms daily. How many *nishcas* does he give in a month?

Statement: First term, 2; Two-fold increase, 2; Period, 30.

Answer, 2147483646 cowries; or 104857 *nishcas*, 9 *drammas*, 9 *paṅas*, 2 *cácinís*, and 6 shells.

[1] To find the sum of a progression, the increase being a multiplier.—GAṄ. That is, the sum of an increasing geometrical progression. The rule agrees with PRÍT'HÚDACA'S. (See Com. on BRAHMEGUPTA, c. 12, § 17.) It is borrowed from prosody (ibid).

[2] The last note is of course "multiplicator:" for in exhausting the number of the period, you arrive at last, at unity an uneven number. The proposed multiplier [the common multiplicator of the progression] is therefore put in the last place; and the operations of squaring and multiplying by it, are continued in the inverse order of the line of the notes. GAṄ.

[3] The effect of squaring and multiplying, as directed, is the same with the continued multiplication of the multiplier for as many times as the number of the period. For dividing by the multiplier the product of the multiplication, continued to the uneven number, equals the product of multiplication continued to one less than the number; and the extraction of the square root of a product of multiplication, continued to the even number, equals continued multiplication to half that number. Conversely, squaring and multiplying equals multiplication for double and for one more time. GAṄ.

129. Example: The initial quantity being two, my friend; the daily augmentation, a three-fold increase; and the period, seven; say what is in this case the sum?

Statement: First term, 2; three-fold increase, 3; Period, 7.
Answer: 2186.

130--131. Rule:[1] a couplet and an half.
The number of syllables in a verse being taken for the period, and the increase two-fold, the produce of multiplication and squaring [as above directed § 127] will be the number [of variations] of like verses. [2] Its square, and square's square, less their respective roots, will be [the variations] of alternately similar, and of dissimilar verses [in tetrastics].[3]

132. Example: Tell me directly the number [of varieties] of like, alternately like, and dissimilar verses, respectively, in the metre named *anushtubh*.[4]

[1] Incidently introduced in this place, showing a computation serviceable in prosody.—SU'R. and *Manó.* To calculate the variations of verse, which also are found by the sum of permutations [§ 113].—GAN'.

[2] *Sanscrit* prosody distinguishes metre in which the four verses of the stanza are alike; or the alternate ones only so; or all four dissimilar. Asiat. Res. vol. 10, Syn. tab. v. vi. & vii.

[3] The number of possible varieties of verse found by the rule of permutation [§ 113] is the same with the continued multiplication of two: this number being taken, because the varieties of syllables are so many; long and short. Accordingly this is assumed for the common multiplier. The product of its continued multiplication is to be found also by this method of squaring and multiplying [§ 127]; assuming for the period a number equal to that of syllables in the verse. The varieties of alternately similar verse, are the same with those of an uniform verse containing twice as many syllables; and the changes in four dissimilar verses are the same with those of one verse comprising four times as many syllables: excepting, however, that these permutations, embracing all the possible varieties, comprehend those of like and half-alike metre. Wherefore the number first found is squared, and this again squared, for twice, or four times, the number of places; and the roots of these squares subtracted, for the permutations of like and alternately like verses. GAN'. &c.

The product of multiplication and squaring is the amount of the last term of the progression, (the first term and common multiplier being equal). RANG.

[4] As. Res. vol. 10, p. 438, (Syn. tab.) p. 469.

Statement: Increase two-fold, 2; Period, 8.

Answer: Variations of like verses, 256; of alternately alike verses, 65280; of dissimilar verses, 4294901760.[1]

[1] Possible varieties of the four verses of a tetrastic containing 32 syllables (8 to a verse) are 4294967296 [2 raised to its 32d power]: of which 4294901760 are dissimilar; and 65536 [2 raised to its 16th power] similar: whereof 65280 alternately alike; and 256 [2 raised to its eighth power] wholly alike.—*Manó.* &c.

CHAPTER VI.

PLANE FIGURE.[1]

133. RULE: A side is assumed.[2] The other side, in the rival direction, is

[1] *Cshétra-vyavahára*, determination of plane figure. *Cshétra*, as expounded by GAŃÉSA, signifies plane surface, bounded by a figure; as triangle, &c. *Vyavahára* is the ascertainment of its dimensions, as diagonal, perpendicular, area, &c.

RANGANÁT'HA distinguishes the sorts of plane figure, precisely as the commentator of BRAHMEGUPTA. See CHAT. on BRAHM. 12, § 21. GAŃÉSA says plane figure is four-fold; triangle, quadrangle, circle and bow. Triangle *(tryasra, tricóna* or *tribhuja)* is a figure containing *(tri)* three *(asra* or *cóna)* angles, and consisting of as many *(bhuja)* sides. Quadrangle or tetragon *(chaturasra, chaturcóna, chaturbhuja)* is a figure comprising *(chatur)* four *(asra*, &c.) angles or sides. The circle and bow (he observes) need no definition. Triangle is either *(játya)* rectangular, as that which is first treated of in the text; or it is *(tribhuja)* trilateral [and oblique] like the fruit of the *Śríngáta* (Trapa natans). This again is distinguished according as the *(lamba)* perpendicular falls within or without the figure: viz. *antar-lamba*, acutangular; *bahirlamba*, obtusangular. Quadrangle also is in the first place twofold: with equal, or with unequal, diagonals. The first of these, or equi-diagonal tetragon *(sama-carńa)* comprises four distinctions: 1st. *sama-chaturbhuja*, equilateral, a square; 2d. *vishama-chaturbhuja*, a trapezium; 3d. *áyata-dírgha-chaturasra*, oblong quadrangle, an oblique parallelogram; 4th, *áyata-sama-lamba*, oblong with equal perpendiculars; that is, a rectangle. The second sort of quadrangle, or the tetragon with unequal diagonals, *(vishama-carńa,)* embraces six sorts: 1st. *sama-chaturbhuja*, equilateral, a rhomb; 2d. *sama-tribhuja*, containing three sides equal; 3d. *sama-dwi-dwi-bhuja*, consisting of two pairs of equal sides, a rhomboid; 4th. *sama-dwi-bhuja*, having two sides equal; 5th. *vishama-chaturbhuja*, composed of four unequal sides, a trapezium; 6th. *sama-lamba*, having equal perpendiculars, a trapezoïd. The several sorts of figures, observes the commentator, are fourteen; the circle and bow being but of one kind each. He adds, that pentagons *(panchásra)*, &c. comprise triangles [and are reducible to them].

[2] *Báhu, dósh, bhuja* and other synonyma of arm are used for the leg of a triangle, or side of a quadrangle or polygon: so called, as resembling the human arm. GAŃ. and SÚR.

called the upright,[1] whether in a triangle or tetragon, by persons conversant with the subject.

134. The square-root of the sum of the squares of those legs is the diagonal.[2] The square-root, extracted from the difference of the squares of the diagonal and side, is the upright: and that, extracted from the difference of the squares of the diagonal and upright, is the side.[3]

135.[4] Twice the product of two quantities, added to the square of their difference, will be the sum of their squares. The product of their sum and difference will be the difference of their squares: as must be every where understood by the intelligent calculator.[5]

136. Example. Where the upright is four and the side three, what is the hypotenuse? Tell me also the upright from the hypotenuse and side; and the side from the upright and hypotenuse.

Statement: 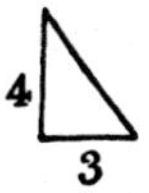 4 ◺ 3 Side 3; Upright 4. Sum of their squares 25. Or

[1] Either leg being selected to retain this appellation, the others are distinguished by different denominations. That, which proceeds in the opposite direction, meaning at right angles, is called *cóti*, *uchch'hraya*, *uchch'hriti*, or any other term signifying upright or elevated. Both are alike sides of the triangle or of the tetragon, differing only in assumed situation and name.—GAŃ. and SU'R. The *cóti* or upright is the cathetus.

[2] A thread or oblique line from the two extremities of the legs, joining them, is the *carńa*, also termed *śruti*, *śravańa*, or by any other words importing ear. It is the diagonal or diameter of a tetragon.—SU'R. RANG. &c. Or, in the case of a triangle, it is the diagonal of the parallelogram, whereof the triangle is the half: and is the hypothenuse of a right-angled triangle.

[3] The rule concerns *(játya)* rectangular triangles. The proof is given both algebraically and geometrically* by GAŃÉŚA; and the first demonstration is exhibited, both with and without algebra, by SU'RYADA'SA. RANGANA'T'HA cites one of those demonstrations from his own brother CAMALA'CARA; and the other from his father NRISINHA, in the *Vártica*, or critical remarks on the *(Vásaná)* annotations of the *Śirómańi*; and censures the *Śríngára-tilaca* for denying any proof of the rule besides experience. BHA'SCARA has himself given a demonstration of the rule in his algebraical work. *Víj. Gań.* § 146.

[4] A stanza of six verses of *anushtubh* metre.

[5] GAŃÉŚA here also gives both an algebraic and a geometrical proof of the latter rule; and an algebraical one only of the first. See *Víj. Gań.* under § 148; whence the latter demonstration is borrowed; and § 147, where the first of the rules is given and demonstrated.

* *Cshétragatópapatti*, geometrical demonstration.

Upapatti avyacta-criyayá, proof by algebra.

product of the sides, doubled, 24; square of the difference 1: added together, 25. The square-root of this is the hypotenuse 5.

5, 3. Difference of the squares [of 5 and 3] 16. Or sum 8, multiplied by the difference 2, makes 16. Its square-root is the upright 4.

4, 5. Difference of squares, found as before, 9. Its square root is the side 3.

137. Example. Where the side measures three and a quarter; and the upright, as much; tell me, quickly, mathematician, what is the length of the hypotenuse?

Statement: $\frac{13}{4}$, $\frac{13}{4}$. Sum of the squares $\frac{338}{16}$ or $\frac{169}{8}$. Since this has no [assignable] root, the hypotenuse is a surd. A method of finding its approximate root [follows:]

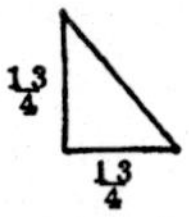

138. Rule: From the product of numerator and denominator,[1] multiplied by any large square number assumed, extract the square-root: that, divided by the denominator taken into the root of the multiplier, will be an approximation.[2]

This irrational hypotenuse $\frac{169}{8}$ [is proposed]. The product of its numerator and denominator is 1352. Multiplied by a myriad (the square of a hundred), the product is 13520000. Its root is 3677 nearly.[3] This divided by the denominator taken into the square-root of the multiplier, viz. 800, gives the approximate root 4 $\frac{477}{800}$. It is the hypotenuse. So in every similar instance.

[1] If the surd be not a fraction, unity may be put for the denominator, and the rule holds good. GAN'.

[2] Here two quantities are assumed: the denominator and the arbitrary square number. The multiplication of the numerator by the denominator is equivalent to the multiplication of the fraction by the denominator twice; that is, by the square of the denominator. The surd, having been thus multiplied by that and the arbitrary number, both squares, the square-root of the product is divided by the denominator and by the root of the arbitrary number. The quotient is the root of the irrational quantity.—GAN'. &c. A like rule occurs in S'RÍD'HARA's compendium.—*Gan'. sár.* § 138.

[3] The remainder being unnoticed.

139. Rule.[1] A side is put. From that multiplied by twice some assumed number, and divided by one less than the square of the assumed number, an upright is obtained. This, being set apart, is multiplied by the arbitrary number, and the side as put is subtracted; the remainder will be the hypotenuse. Such a triangle is termed rectangular.

140. Or a side is put. Its square, divided by an arbitrary number, is set down in two places: and the arbitrary number being added and subtracted, and the sum and difference halved, the results are the hypotenuse and upright. Or, in like manner, the side and hypotenuse may be deduced from the upright. Both results are rational quantities.

141. Example. The side being in both cases twelve, tell quickly, by both methods, several uprights and hypotenuses, which shall be rational numbers.

Statement: Side 12. Assumption 2. The side, multiplied by twice that, viz. 4, is 48. Divide by the square of the arbitrary number less one, viz. 3, the quotient is the upright 16. This upright, multiplied by the assumed number, is 32: from which subtract the given side, the remainder is the hypotenuse 20. See

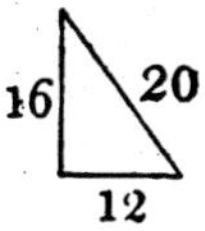

16 20
12

Assume three. The upright is 9; and hypotenuse 15. Or, putting five, the upright is 5, and hypotenuse 13.

[1] Either the side or upright being given, to find the other two sides.—SÚR. To find the upright and hypotenuse from the side; or the side and hypotenuse from the upright.—GAŃ. The problem is an indeterminate one, as is intimated by the author. The second rule is in substance the same with BRAHMEGUPTA's for the upright and diagonal of a rectangle. See BRAHM. 12, § 35.

SÚRYADÁSA demonstrates the first rule thus: 'In some triangle (which should be less than that which has the given side) the upright is taken at double of some assumed number, and the side is taken at one less than the assumed number. Then make proportion, "as this side to this upright, so is the given side to its upright." Whence the given side, multiplied by twice the assumed number, and divided by one less than its square, is the upright. When this upright so found is multiplied by the assumed number, the product is the sum of the side and hypotenuse: when divided by it, the quotient is the difference of the side and hypotenuse: for they increase and decrease by virtue of that assumed number. Thus, subtracting the given side from that sum, the remainder is the hypotenuse: or, adding it to the difference, the sum is the hypotenuse. SÚR.

By the second method: the side, as put, 12. Its square 144. Divide by 2, the arbitrary number being two, the quotient is 72. Add and subtract the arbitrary number, and halve the sum and difference: the hypotenuse and upright are found: viz. upright 35, hypotenuse 37. See

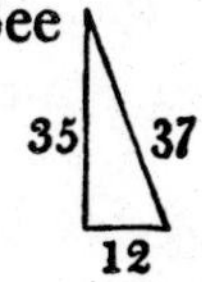

35 37
12

Assume four. The upright is 16, and hypotenuse 20. Assuming six, the upright is 9 and hypotenuse 15.[1]

142. Rule:[2] Twice the hypotenuse taken into an arbitrary number, being divided by the square of the arbitrary number added to one, the quotient is the upright. This taken apart is to be multiplied by the number put: the difference between the product and the hypotenuse is the side.[3]

143. Example: Hypotenuse being measured by eighty-five, say promptly, learned man, what uprights and sides will be rational?

Statement: The hypotenuse 85, being doubled, is 170; and multiplied by an arbitrary number two, is 340. This, divided by the square of the

The demonstration of the second method is given by GAŃÉSÁ, and similarly by SÚRYADÁSA and RANGANÁT'HA. 'Assume any number for the difference between the upright and hypotenuse. The difference of their squares (which is equal to the square of the given side) being divided by that assumed difference, the quotient is the sum of the upright and hypotenuse. For the difference of the squares is equal to the product of the sum and difference of the roots. (§ 135.) The upright and hypotenuse are therefore found by the rule of concurrence' (§ 55). GAŃ. &c.

[1] In like manner, if the upright be given, 16. Its square 256, divided by the arbitrary number 2, is 128. The arbitrary number subtracted and added, makes 126 and 130; which halved gives the side 63 and hypotenuse 65. GANG. and SÚR.

[2] From the hypotenuse given, to find the side and upright in rational numbers.—GAŃ. The problem is an indeterminate one.

[3] Let the upright in a figure be any assumed number doubled; and the hypotenuse be unity added to the square of that arbitrary number. Thence a proportion, as before: If with this hypotenuse, this upright; then with the given hypotenuse, what is the upright? It is consequently found: viz. twice the given hypotenuse multiplied by the arbitrary number, and divided by the square of that number with unity added to it. If that be multiplied by the arbitrary number, the product is the sum of the hypotenuse and side; if divided by it, their difference. Hence, by the rule of concurrence (§ 55), the side and hypotenuse are found. But here, for brevity, the hypotenuse, being already known, is subtracted from the sum of that and the side. SÚR.

arbitrary number added to one, viz. 5, is the upright 68. This upright, multiplied by the arbitrary number, makes 136: and subtracting the hypotenuse, the side comes out 51. See

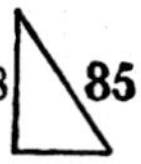

68 85

51

Or putting four, the upright will be 40; and side 75. See 40 85

75

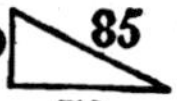

144. Rule: Or else hypotenuse is doubled and divided by the square of an assumed number added to one. Hypotenuse, less that quotient, is the upright. The same quotient, multiplied by the assumed number, is the side.[1]

The same hypotenuse 85. Putting two, the upright and side are 51 and 68. Or, with four, they are 75 and 40.

Here the difference between side and upright is in name only, and not essential.

145. Rule:[2] Let twice the product of two assumed numbers be the upright; and the difference of their squares, the side: the sum of their squares will be the hypotenuse, and a rational number.[3]

[1] The assumed upright in the small triangle was before taken at twice a number put. The assumption is now two, and hypotenuse is put as there stated. Then proportion being made as before, the quotient is multiplied by the arbitrary number, because, in comparison with the preceding, it was just so much less. The quotient, as it comes out, is the difference between the hypotenuse and side: and, that being subtracted from the hypotenuse, the residue is the side.—Sú'r. This and the preceding rule are founded on the same principle; differing only in the order of the operation and names of the sides: the same numbers come out for the side and upright in one mode, which were found for the upright and side by the other.

[2] Having taught the mode of finding a third side, from any two, of hypotenuse, upright and side; and in like manner from one, the other two; the author now shows a method of finding all three rational [none being given.]—Gań'. The problem is an indeterminate one.

[3] The demonstration is by resolution of a quadratic equation involving several unknown: Let the length of the side be *ya* 1, and that of the upright *ca* 1. The sum of their squares is *ya v* 1 *ca v* 1. It is a square quantity. Putting it equal to *ni v* 1, the root of this side of the equation is *ni* 1; and those of the other side are to be found by the rule of the affected square.* Assuming either term for the affected square, the other will be the additive. Let *ya v* 1 be the proposed square, and *ca v* 1 the additive. Then the coefficient being a square, the roots are to be found by the rule (*Víj. gán.* § 95). Here a fraction of *ca* is put; an arbitrary number for the numerator, and another arbitrary

* *Víj. gań*. ch. 3.

146. Example. Tell quickly, friend, three numbers, none being given, with which as upright, side and hypotenuse, a rectangular triangle may be [constructed.]

Statement. Let two numbers be put, 1 and 2. From these the side, upright and hypotenuse are found, 4, 3, 5. Or, putting 2 and 3, the side, upright and hypotenuse deduced from them, are 12, 5, 13. Or let the assumed numbers be 2 and 4: from which will result 16, 12, 20. In like manner, manifold [answers are obtained].

147. Rule[1]. The square of the ground intercepted between the root and tip, is divided by the [length of the] bambu; and the quotient severally added to, and subtracted from, the bambu: the moieties [of the sum and difference] will be the two portions of it representing hypotenuse and upright.[2]

148. Example.[3] If a bambu, measuring thirty-two cubits and standing upon level ground, be broken in one place, by the force of the wind, and

one for the denominator. For instance *ca* $\frac{2}{3}$. Then by the method taught (*Víj. gaṅ.* § 95) the least and greatest roots come out *ca* $\frac{5}{12}$, *ca* $\frac{13}{12}$ Here, in the place of the numerator of the least root, is the difference of the squares of the assumed numbers; and, in that of the denominator, twice their product. So, in place of the numerator of the greatest root, is the sum of the squares; and, in that of the denominator, twice the product. The least root is the value of *ya*, the fraction *ca* $\frac{5}{12}$. Then, by the pulverizer,* the multiplier and quotient come out 5 and 12. The multiplier is the value of *ya* and is the side. The quotient is the value of *ca* and is the upright 12. Substituting with it for *ca* in the greatest root, this is found 13. It is the value of *ni* and is the hypotenuse. Thus the side, upright and hypotenuse are obtained 5, 12, 13. This is the operation directed by the rule, § 145. GAṄ.

[1] The sum of hypotenuse and upright being known, as also the side, to discriminate the hypotenuse and upright.—GAṄ. The rule bears reference to the example which follows.

[2] The height from the root to the fracture is the upright. The remaining portion of the bambu is hypotenuse. The whole bambu, therefore, is the sum of hypotenuse and upright. The ground intercepted between the root and tip is the side: it is equal to the square root of the difference between the squares of the hypotenuse and upright. Hence the square of the side, divided by the sum of the hypotenuse and side, is their difference [§ 135]. With these (sum and difference) the upright and hypotenuse are found by the rule of concurrence (§ 55). GAṄ.

[3] See Arithm. of BRAHMEGUPTA under § 41; and *Víj.-gaṅ.* § 124; where the same example occurs.

* *Víja-gaṅita*, ch. 2.

the tip of it meet the ground at sixteen cubits: say, mathematician, at how many cubits from the root is it broken?

Statement. Bambu 32. Interval between the root and tip of the bambu 16. It is the side of the triangle. Proceeding as directed, the upper and lower portions of the bambu are found 20 and 12. See figure

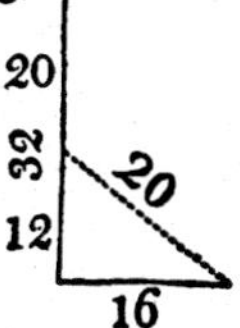

149. Rule.[1] The square [of the height] of the pillar is divided by the distance of the snake from his hole; the quotient is to be subtracted from that distance. The meeting of the snake and peacock is from the snake's hole half the remainder, in cubits.[2]

150. Example.[3] A snake's hole is at the foot of a pillar, and a peacock is perched on its summit. Seeing a snake, at the distance of thrice the pillar, gliding towards his hole, he pounces obliquely upon him. Say quickly at how many cubits from the snake's hole do they meet, both proceeding an equal distance?

Statement. Pillar 9. It is the upright. Distance of the snake from his hole 27. It is the sum of hypotenuse and side. Proceeding as directed, the meeting is found in cubits; viz. 12.[4] See figure 9 12 15 15 27

[1] The sum of the side and hypotenuse being known, as also the upright, to discriminate the hypotenuse and side.
GAṄ.

[2] The rule bears reference to the example which follows. The principle is the same with that of the preceding rule.

[3] This occurs also in some copies of the *Víja-gańita*, after § 139; as appears from the commentary of SÚRYADÁSA, giving an interpretation of it in that place. It is borrowed from the Arithm. of BRAHMEGUPTA under § 41, with a change of a snake and a peacock substituted for a rat and a cat.

[4] Subtracted from the sum of hypotenuse and side, this leaves 15 for the hypotenuse. The snake had proceeded the same distance of 15 cubits towards his hole, as the peacock in pouncing upon him. Their progress is therefore equal.
SÚR.

151. Rule.[1] The quotient of the square of the side divided by the difference between the hypotenuse and upright is twice set down · and the difference is subtracted from the quotient [in one place] and added to it [in the other]. The moieties [of the remainder and sum] are in their order the upright and hypotenuse.[2]

This[3] is to be generally applied by the intelligent mathematician.

152. Friend, the space, between the lotus [as it stood] and the spot where it is submerged, is the side. The lotus as seen [above water] is the difference between the hypotenuse and upright. The stalk is the upright:[4] for the depth of water is measured by it. Say, what is the depth of water?

153. Example.[5] In a certain lake swarming with ruddy geese[6] and cranes, the tip of a bud of lotus was seen a span above the surface of the water. Forced by the wind, it gradually advanced, and was submerged at the distance of two cubits. Compute quickly, mathematician, the depth of water.

Statement: Diff. of hypotenuse and upright $\frac{1}{2}$ cubit. Side 2 cubits. Proceeding as directed, the upright and hypotenuse are found, viz. upright $\frac{15}{4}$. It is the depth of water. Adding to it the height of the bud, the hypotenuse comes out $\frac{17}{4}$. See

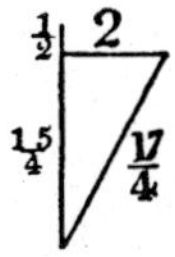

$\frac{1}{2}$ 2

$\frac{15}{4}$ $\frac{17}{4}$

154. Rule.[7] The height of the tree, multiplied by its distance from the

[1] The difference between the hypotenuse and upright being known, as also the side, to find the upright and hypotenuse. GAN'.

[2] The demonstration, distinctly set forth under a preceding rule, is applicable to this. GAN'.

[3] Beginning from the instance of the broken bambu (§ 147) and including what follows. GAN'.

[4] The sides, constituting the figure in the example which follows, are here set forth, to assist the apprehension of the student. SU'R. and GAN'.

[5] See Arithm. of BRAHM. under § 41; and *Vij.-gan'.* § 125: where the same example is inserted.

[6] Anas Casarca.

[7] The sum of the hypotenuse and upper portion of the upright being given, and the lower portion being known; as also the side: to discriminate the portion of the upright from the hypotenuse.—GAN'. As in several preceding instances, a reference to the example is requisite to the understanding of the rule. The same problem occurs in BRAHMEGUPTA's Arithmetic, § 39; and is repeated in the *Vija-gańita*, § 126.

pond, is divided by twice the height of the tree added to the space between the tree and pond: the quotient will be the measure of the leap.

155. Example. From a tree a hundred cubits high, an ape descended and went to a pond two hundred cubits distant: while another ape, vaulting to some height off the tree, proceeded with velocity diagonally to the same spot. If the space travelled by them be equal, tell me, quickly, learned man, the height of the leap, if thou have diligently studied calculation.

Statement: Tree 100 cubits. Distance of it from the pond 200. Proceeding as directed, the height of the leap comes out 50.* See

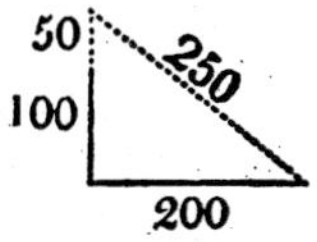

50 250 100 200

156. Rule.[2] From twice the square of the hypotenuse subtract the sum of the upright and side multiplied by itself, and extract the square-root of the remainder. Set down the sum twice, and let the root be subtracted in one place and added in the other. The moieties will be measures of the side and upright.[3]

157. Example. Where the hypotenuse is seven above ten; and the sum of the side and upright, three above twenty; tell them to me, my friend.

Statement: Hypotenuse 17. Sum of side and upright 23. Proceeding as directed, the side and upright are found 8 and 15. See

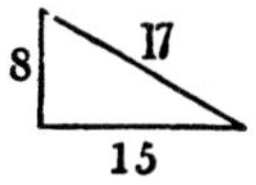

8 17 15

* The hypotenuse is 250: and the entire upright 150.

[2] Hypotenuse being known, as also the sum of the side and upright, or their difference; to discriminate those sides. GAŃ.

[3] In like manner, the difference of the side and upright being given, the same rule is applicable.—GAŃ. Using the difference instead of the sum.

The principle of the rule is this: the square of the hypotenuse is the sum of the squares of the sides. But the sum of the squares, with twice the product of the sides added to it, is the square of the sum; and, with the same subtracted, is the square of the difference. Hence, cancelling equal quantities affirmative and negative, twice the square of the hypotenuse will be the sum of the squares of the sum and difference. Therefore, subtracting from twice the square of hypotenuse the square of the sum, the remainder is the square of the difference; or conversely, subtracting the square of the difference, the residue is the square of the sum. The square-root is the sum or the difference. With these, the sides are found by the rule of concurrence. GAŃ. and SÚR.

158. Example. Where the difference of the side and upright is seven and hypotenuse is thirteen, say quickly, eminent mathematician, what are the side and upright?[1]

Statement. Difference of side and upright 7. Hypotenuse 13. Proceeding as directed, the side and upright come out 5 and 12. See

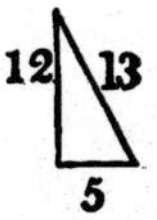

12 13
5

159. Rule.[2] The product of two erect bambus being divided by their sum, the quotient is the perpendicular[3] from the junction [intersection] of threads passing reciprocally from the root [of one] to the tip [of the other.]. The two bambus, multiplied by an assumed base, and divided by their sum, are the portions of the base on the respective sides of the perpendicular.

160. Example.[4] Tell the perpendicular drawn from the intersection of strings stretched mutually from the roots to the summits of two bambus fifteen and ten cubits high standing upon ground of unknown extent.

Statement: Bambus 15, 10. The perpendicular is found 6.

Next to find the segments of the base: let the ground be assumed 5; the segments come out 3 and 2. Or putting 10, they are 6 and 4. Or taking 15, they are 9 and 6. See the figures

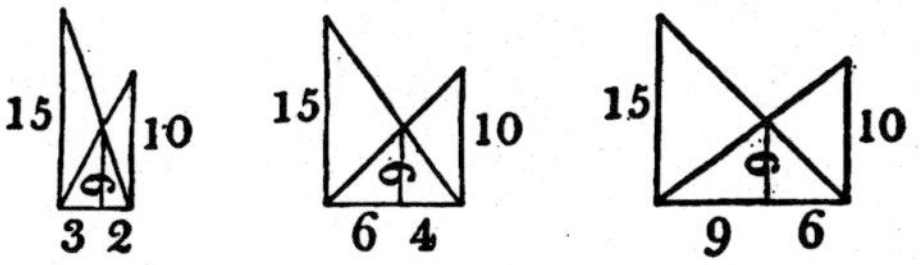

15 10 6
3 2

15 10 6
6 4

15 10 6
9 6

In every instance the perpendicular is the same: viz. 6.*

The proof is in every case by the rule of three: if with a side equal to the

[1] This example of a case where the difference of the sides is given, is omitted by SU'RYADA'SA, but noticed by GAN'E'SA. Copies of the text vary; some containing, and others omitting, the instance.

[2] Having taught fully the method of finding the sides in a right-angled triangle, the author next propounds a special problem.—GAN'. To find the perpendicular, the base being unknown.—SU'R.

[3] *Lamba, Avalamba, Valamba, Ad'hólamba*, the perpendicular.

[4] See *Víja-gańita*, § 127.

* However the base may vary by assuming a greater or less quantity for it, the perpendicular will still be the same. GAN'.

base, the bambu be the upright, then with the segment of the base what will be the upright?[1]

161. Aphorism.[2] That figure, though rectilinear, of which sides are proposed by some presumptuous person, wherein one side[3] exceeds or equals the sum of the other sides, may be known to be no figure.

162. Example: Where sides are proposed two, three, six and twelve in a quadrilateral, or three, six and nine in a triangle, by some presumptuous dunce, know it to be no figure.

Statement: The figures are both incongruous. Let strait rods exactly of the length of the proposed sides be placed on the ground, the incongruity will be apparent.[4]

163—164. Rule[5] in two couplets: In a triangle, the sum of two sides, being multiplied by their difference, is divided by their base:[6] the quotient

[1] On each side of the perpendicular, are segments of the base relative to the greater and smaller bambus, and larger or less analogously to them. Hence this proportion. "If with the sum of the bambus, this sum of the segments equal to the entire base be obtained, then, with the smaller bambu, what is had?" The answer gives the segment, which is relative to the least bambu Again: "if with a side equal to the whole base, the higher bambu be the upright, then with a side equal to the segment found as above, what is had?" The answer gives the perpendicular let fall from the intersection of the threads. Here a multiplicator and a divisor equal to the entire base are both cancelled as equal and contrary: and there remain the product of the two bambus for numerator and their sum for denominator. Hence the rule. GAŃ.

[2] The aphorism explains the nature of impossible figures proposed by dunces.—SU'R. It serves as a definition of plane figure *(cshétra)*.—GAŃ. In a triangle or other plane rectilinear figure, one side is always less than the sum of the rest. If equal, the perpendicular is nought, and there is no complete figure. If greater, the sides do not meet.—SU'R. Containing no area, it is no figure.—*Caum.* RANG.

[3] The principal or greatest side.—GAŃ. *Caum.* RANG.

[4] The rods will not meet.—SU'R.

[5] In any triangle to find the perpendicular, segments and area. This is introductory to a fuller consideration of areas.—GAŃ. and SU'R. It is taken from BRAHMEGUPTA, 12, § 22.

[6] *Bhúmi, bhú, cu, mahí*, or any other term signifying earth; the ground or base of a triangle or other plane figure. Any one of the sides is taken for the base; and the rest are termed simply sides. GAŃE'SA restricts the term to the greatest side. See note § 168.

Lamba, &c. the perpendicular. See note on § 159.

is subtracted from, and added to, the base which is twice set down: and being halved, the results are segments corresponding to those sides.[1]

164. The square-root of the difference of squares of the side and its own segment of the base becomes the perpendicular. Half the base, multiplied by the perpendicular,[2] is in a triangle the exact[3] area.[4]

Ábád'há, abad'há, avabad'há, segment of the base. These are terms introduced by earlier writers. From the point, where a perpendicular falling from the apex *(mastaca)* meets the base, the two portions or divisions of the ground on their respective sides [or, if the perpendicular fall without the figure in an obtuse-angled one, on the same side] are distinguished by this name.

Phala, Gańita, Cshétra-phala, Sama-cósht́a-miti; the measure of like compartments, or number of equal squares of the same denomination (as cubit, fathom, finger, &c.) in which the dimension of the side is given: the area or superficial content. It is the product of multiplication of length by breadth. GAŃ. and SU'R.

[1] The relative, dependent, or corresponding segments. The smaller segment answers to the less side; the larger segment to the greater side. GAŃ.

[2] Or half the perpendicular taken into the base. GAŃ.

[3] *Sphut́a-phala* distinct or precise area; opposed to *asphut́a*—or *st'hula-phala* indistinct or gross area. See § 167—and Arithm. of BRAHM. § 21.

[4] Demonstration: In both the right-angled triangles formed in the proposed triangular figure, one on each side of the perpendicular, this line is the upright; the side is hypotenuse, and the correspondent segment is side. Hence, subtracting the square of the perpendicular from the square of the side, the remainder is square of the segment. So, subtracting the square of the other side, there remains the square of the segment answering to it. Their difference is the difference of the squares of the segments and is equal to the difference of the squares of the sides; since an equal quantity has been taken from each: for any two quantities, less an equal quantity, have the same difference. It is equal to the product of the sum and difference of the simple quantities. Therefore the sum of the sides, multiplied by their difference, is the difference of the squares of the segments. But the base is the sum of the segments. The difference of the squares, divided by that, is the difference of the segments. From which, by the rule of concurrence (§ 55) the segments are found.

The square-root of the difference between the squares of the side and segment (taken as hypotenuse and side) is the upright. It is the perpendicular.

Dividing the triangle by a line across the middle, and placing the two halves [or parts] of the upper portion disjoined by the perpendicular, on the two sides of the lower portion, an oblong is formed; in which the half of the perpendicular is one side, and the base is the other. See Wherefore half the perpendicular, multiplied by the base, is the area or number of equal compartments. Or half the base, multiplied by the perpendicular, is just so much.—GAŃ.

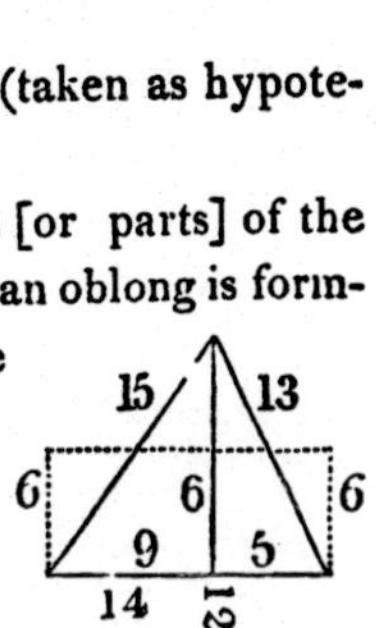

If with the sum of the sides, this difference be had, then with this sum of the segments, that is,

165. Example. In a triangular figure, in which the base is fourteen and its sides thirteen and fifteen, tell quickly the length of the perpendicular, the segments, and the dimensions by like compartments termed area.

Statement: Base 14. Sides 13 and 15. Proceeding as directed, the segments are found, 5 and 9; and the perpendicular, 12: the area, 84. See

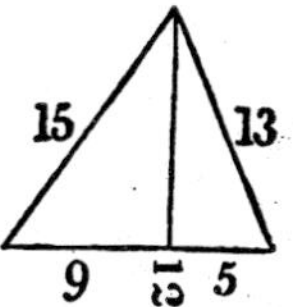

166. Example. In a triangle, wherein the sides measure ten and seventeen, and the base nine, tell me promptly, expert mathematician, the segments, perpendicular and area.

Statement: Sides 10 and 17. Base 9. By the rule § 163, the quotient found is 21. This cannot be subtracted from the base. Wherefore the base is subtracted from it. Half the remainder is the segment, 6; and is negative: that is to say, is in the contrary direction.[1] Thus the two segments

with the base which is their sum, what is obtained? Here, as the demand increases, the fruit decreases: wherefore, by the inverse rule of three § 74, the difference of the sides, multiplied by their sum, and divided by the base, gives the difference of the segments. With that and the base, which is their sum, the segments are found by the rule of concurrence § 55.

In an acute-angled triangle, two right-angled triangles are formed by the perpendicular within it. The side becomes an hypotenuse, the segment a side, and the unknown perpendicular an upright alike in both. Hence (§ 134) the square-root of the difference of the squares of the side and segment is the perpendicular.

The perpendicular is the breadth; and the base is the length. It is exactly so in the lower part; but not so in the upper part: for there the figure terminates in a sharp point. Wherefore half the length is the length to be multiplied. If two triangles be placed within a quadrilateral, it is readily perceived, that the triangle is half the quadrilateral. Or if an acute-angled triangle be figured, two right-angled triangles are formed by the perpendicular; and their bases are the segments. The moieties of the segments, multiplied by the perpendicular, are the areas of the two rectangular triangles. Their sum is the area of the proposed triangle.—SÚR.

In an obtuse-angled triangle also, the base multiplied by half the perpendicular is the area.
GAŃ.

[1] When the perpendicular falls without the base, as overpassing the angle in consequence of the side exceeding the base, the quotient found by the rule § 163 cannot be taken from the base: for both origins of sides are situated in the same quarter from the fall of the perpendicular. There-

are found 6 and 15. From which, both ways too, the perpendicular comes out 8. The area, 36. See

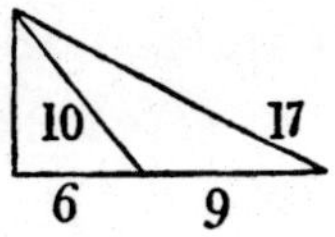

10 17
6 9

167. Rule.[1] Half the sum of all the sides is set down in four places; and the sides are severally subtracted. The remainders being multiplied together, the square-root of the product is the area, inexact in the quadrilateral, but pronounced exact in the triangle.[2]

168. Example. In a quadrilateral figure, of which the base[3] is fourteen, the summit[4] nine, the flanks thirteen and twelve, and the perpendicular twelve, tell the area as it was taught by the ancients.

fore subtracting the base from the quotient, half the residue is the segment and situated on the contrary side, being negative. Wherefore, as both segments stand on the same side, the smaller is comprehended in the greater; and, in respect of it, is negative. Thus all is congruous and unexceptionable.—GAŃ. When the sum of the segments is to be taken, as they have contrary signs, affirmative and negative, the difference of the quantities is that sum.—SU'R. See *Vij.-gań.* § 5.

[1] For finding the gross area of a quadrilateral; and, by extension of the rule, the exact area of a triangle.—GAŃ. For finding the area by a method delivered by ŚRÍD'HARA, as a general one common to all figures.—RANG. Excepting an equidiagonal quadrilateral.—*Caum.* ŚRÍD'HARA's rule, which is here censured, occurs in his compendium of Arithmetic.—*Gań. sár.* § 126. See likewise Arithm. of BRAHMEGUPTA, § 21.

[2] In the case of a triangle, half the sum of the three sides is four times set down; the three sides are subtracted severally in three instances: in the fourth, it remains unchanged. The square-root of the product of such four quantities is the exact area.—GAŃ.

If the three remainders be added together, their sum is equal to half the sum of all the sides. The product of the continual multiplication of the three remainders being taken into the sum of those remainders, the product so obtained is equal to the product of the square of the perpendicular taken into the square of half the base. It is a square quantity: for a square, multiplied by a square, gives a square. The square-root being extracted, the product of the perpendicular by half the base is the result: and that is the area of the triangle. Therefore the true area is thus found. In a quadrilateral, the product of the multiplication does not give a square quantity: but an irrational one. Its approximate root is the area of the figure; not, however, the true one: for, when divided by the perpendicular, it should give half the sum of the base and summit.—SU'R.

[3] The greatest of the four sides is called the base.—GAŃ. This definition is, however, too restricted. See § 185 and 178. The notion of it is taken from BRAHMEGUPTA. Arithm. § 38.

[4] *Muc'ha, vadana,* or other term expressing mouth: the side opposite to the base; the summit.

Statement: Base 14. Summit 9. Sides 13 and 12. Perp. 12.

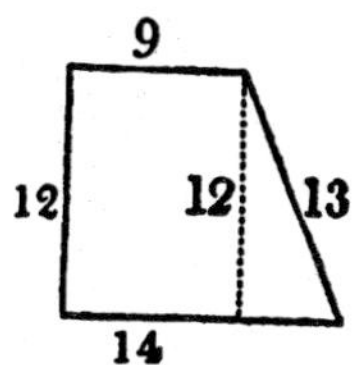

By the method directed, the result obtained is the surd 19800, of which the approximated root is somewhat less than a hundred and forty-one: 141. That, however, is not in this figure the true area. But, found by the method which will be set forth (§ 175), the true area is 138.

Statement of the triangle before instanced

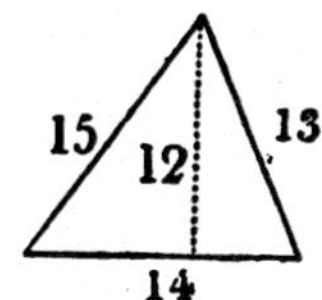

By this method the area comes out the same: viz. 84.

169—170. Aphorism comprised in a stanza and a half: Since the diagonals of the quadrilateral are indeterminate, how should the area be in this case determinate? The diagonals, found as assumed by the ancients,[1] do not answer in another case. With the same sides, there are other diagonals; and the area of the figure is accordingly manifold.

For, in a quadrilateral, opposite angles, being made to approach, contract their diagonal as they advance inwards: while the other angles, receding outwards, lengthen their diagonal. Therefore it is said, " with the same sides, there are other diagonals."

171. How can a person, neither specifying one of the perpendiculars, nor either of the diagonals, ask the rest?[2] or how can he demand a determinate area, while they are indefinite?

172. Such a questioner is a blundering devil.[3] Still more so is he, who answers the question. For he considers not the indefinite nature of the lines[4] in a quadrilateral figure.

[1] By Śríd'hara and the rest. GAṆ.

[2] The perpendiculars, diagonals, &c. GAṆ.

[3] *Piśácha* (a demon or vampire). So termed, because he blunders. SÚR.

[4] Of the diagonal and perpendicular lines. SÚR.

173—175. Rule[1] in two and a half stanzas: Let one diagonal of an equilateral tetragon be put as it is given. Then subtract its square from four times the square of the side. The square-root of the remainder is the measure of the second diagonal.

174. The product of unequal diagonals multiplied together, being divided by two, will be the precise area in an equilateral tetragon. But in a regular one with equal diagonals, as also in an oblong,[2] the product of the side and upright will be so.

175. In any other quadrilateral with equal perpendiculars,[3] the moiety of the sum of the base and summit, multiplied by the perpendicular, [is the area.]

176. Mathematician, tell both diagonals and the area of an equilateral quadrangular figure, whose side is the square of five: and the area of it, the diagonals being equal: also [the area] of an oblong, the breadth of which is six and the length eight.

Statement of first figure 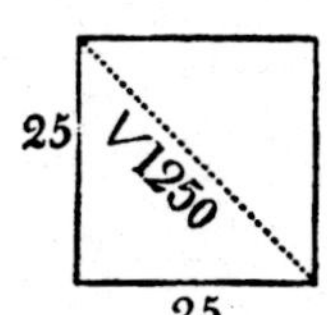

Here, taking the square-root of the sum of the squares (§ 134), the diagonal comes out the surd 1250, alike both ways. The area 625.

Assume one diagonal thirty; the other is found 40; and the area 600. See

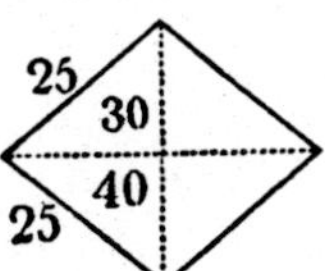

Put one diagonal fourteen: the other is found 48; and area 336. See

[1] In an equilateral tetragon, one diagonal being given, to find the second diagonal and the area: also in an equi-perpendicular tetragon [trapezoid] to find the area.—GAN'. Equilateral tetragons are twofold: with equal, and with unequal, diagonals. The first rule regards the equilateral tetragon with unequal diagonals [the rhomb.] SU'R.

[2] *Áyata:* a long quadrilateral which has pairs of equal sides. GAN'.

[3] In an unequal quadrilateral figure, to find the area.—SUR. In any quadrilateral with two, or with three, equal sides, or with all unequal, but having equal perpendiculars. RANG

Statement of the oblong 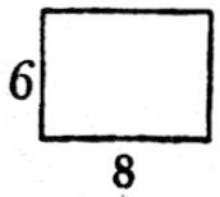 6 8 Area 48.

177. Example. Where the summit is eleven; the base twice as much as the summit; and the flanks thirteen and twenty; and the perpendicular twelve; say what will be the area?

Statement: 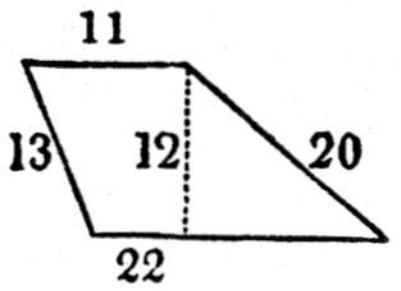 11 13 12 20 22 The gross area (§ 167) is 250. The true area (§ 175) is 198.

Or making three portions of the figure, and severally finding their areas, and summing them, the principle may be shown. 11 5 6 13 12 20 6 16 22

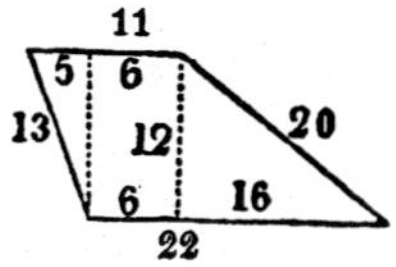

178. Example. Declare the diagonal, perpendicular and dimensions of the area, in a figure of which the summit is fifty-one, the base seventy-five, the left side sixty-eight, and the other side twice twenty.

179. Aphorism showing the connexion of area, perpendicular and diagonal:

If the perpendicular be known, the diagonal is so: if the diagonal be known, the perpendicular is so: if they be definite, the area is determinate.

For, if the diagonal be indefinite, so is the perpendicular. Such is the meaning.

179 continued. Rule for finding the perpendicular:[1] In the triangle within the quadrilateral, the perpendicular is found as before taught:[2] the diagonal and side being sides, and the base a base.[3]

Here, to find the perpendicular, a diagonal, proceeding from the extre-

[1] The diagonal being either given or assumed. GAN'.

[2] See § 163 and 164.

[3] The summit becomes base of the second triangle; the diagonal is one leg; and the remaining side of the quadrilateral, the other. RANG.

mity of the left side to the origin of the right one, is assumed, put at seventy-seven. See

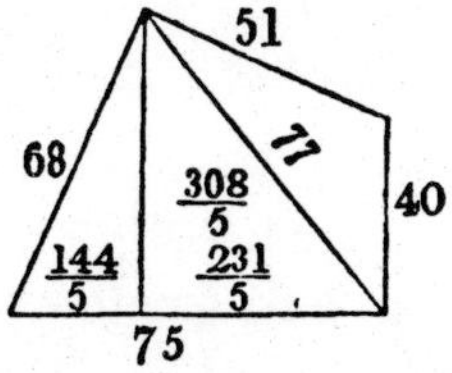

68 51 77 40 $\frac{308}{5}$ $\frac{144}{5}$ $\frac{231}{5}$ 75

By this a triangle is constituted within the quadrilateral. In it that diagonal is one side, 77; the left side is another, 68; the base continues such, 75. Then, proceeding by the rule (§ 163—164), the segments are found $\frac{144}{5}$ and $\frac{231}{5}$; and the perpendicular, $\frac{308}{5}$. See figure.

180. Rule to find the diagonal, when the perpendicular is known:

The square-root of the difference of the squares of the perpendicular and its adjoining side is pronounced the segment. The square of the base less that segment being added to the square of the perpendicular, the square-root of the sum is the diagonal.

In that quadrilateral, the perpendicular from the extremity of the left side is put $\frac{308}{5}$. Hence the segment is found $\frac{144}{5}$; and by the rule (§ 180) the diagonal comes out 77.

181—182. Rule to find the second diagonal [two stanzas]:

In this figure, first a diagonal is assumed.[1] In the two triangles situated one on either side of the diagonal, this diagonal is made the base of each; and the other sides are given: the perpendiculars and segments[2] must be found. Then the square of the difference of two segments on the same side[3] being added to the square of the sum of the perpendiculars, the square-root of the sum of those squares will be the second diagonal in all tetragons.[4]

In the same quadrilateral, the length of the diagonal passing from the extremity of the left side to the origin of the right one, is put 77. Within the figure cut by that diagonal line, two triangles are formed, one on each

[1] Either arbitrarily [see § 183] or as given by the conditions of the question. GAN'.

[2] The two perpendiculars and the four segments. GANG.

[3] Square of the interval of two segments measured from the same extremity.

[4] In the figure, which is divided by the diagonal line, two triangles are contained: one on each side of that line; and their perpendiculars, which fall one on each side of the diagonal, are thence found. The difference between two segments on the same side will be the interval between the perpendiculars. It is taken as the upright of a triangle. Producing one perpendicular by the ad-

side of the diagonal. Taking the diagonal for the base of each, and the two other sides as given, the two perpendiculars and the several segments must be found by the method before taught. See figure

viz. Perpendiculars 24 and 60. Segment of the base of the one part 45 and 32; of the other 32 and 45. Difference of the segments on the same side (that is, so much of the base as is intercepted between the perpendiculars) 13. Its square 169. Sum of the perpendiculars 84. Its square 7056. Sum of the squares 7225. Square-root of the sum 85. It is the length of the second diagonal. So in every like instance.

183—184. Rule restricting the arbitrary assumption of a diagonal [a stanza and a half:] The sum of the shortest pair of sides containing the diagonal being taken as a base, and the remaining two as the legs [of a triangle,] the perpendicular is to be found: and, in like manner, with the other diagonal. The diagonal cannot by any means be longer than the corresponding base, nor shorter than the perpendicular answering to the other. Adverting to these limits an intelligent person may assume a diagonal.

For a quadrilateral, contracting as the opposite angles approach, becomes a triangle; wherein the sum of the least pair of sides about one angle is the

dition of the other, the sum is made the side of the triangle. The second diagonal is hypotenuse.

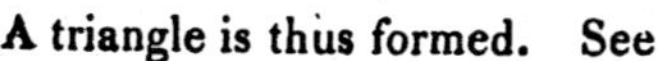

A triangle is thus formed. See

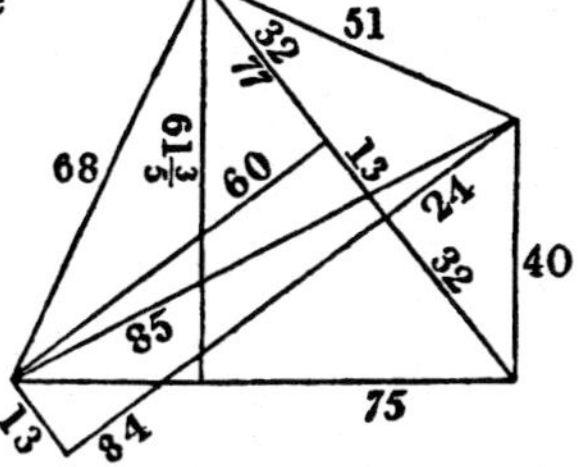

From this is deduced, that the square-root of the sum of the squares of the upright and side will be the second diagonal: and the rule is demonstrated. GAN.

In an equilateral tetragon, and in a trapezium of which the greatest side is the base and the least is the summit, there is no interval between the perpendiculars; and the second diagonal is the sum of the perpendiculars. *Ibid.*

base; and the other two are taken as the legs. The perpendicular is found in the manner before taught. Hence the shrinking diagonal cannot by any means be less than the perpendicular; nor the other be greater than the base. It is so both ways. This, even though it were not mentioned, would be readily perceived by the intelligent student.

184. Rule to find the area [half a stanza:] The sum of the areas of the two triangles on either side of the diagonal is assuredly[1] the area in this figure.

In the figure last specified, the areas of the two triangles are 924 and 2310. The sum of which is 3234; the area of the tetragon.

185—186. Rule[2] [two stanzas:] Making the difference between the base and summit of a [trapezoid, or] quadrilateral that has equal perpendiculars, the base [of a triangle], and the sides [its] legs, the segments of it and the length of the perpendicular are to be found as for a triangle. From the base of the trapezoid subtracting the segment; and adding the square of the remainder to the square of the perpendicular, the square-root of the sum will be the diagonal.[3]

In a [trapezoid, or] quadrangle that has equal perpendiculars, the sum of the base and least flank is greater than the aggregate of the summit and other flank.

[1] It is the true and correct area, contrasted with the gross or inexact area of former writers. GAN'. and SU'R.

[2] To find definite diagonals, when neither is given; nor the perpendicular; but the condition that the perpendiculars be equal; which is a sufficient limitation of the problem.

[3] In a quadrilateral figure having equal perpendiculars, the intermediate portion between the extreme perpendiculars being taken away, there remain two rectangular triangles on the outer side. Uniting them together, a triangle is formed, in which the flanks are legs, and the base less the summit is the base. Hence the perpendicular in this triangle, found by the rule before taught (§ 164), is precisely the perpendicular of the tetragon; and the segments, which are found (§ 163), lie between the perpendicular and the corresponding sides. The base of the tetragon, less either of the segments, is the side of a rectangular triangle within the same tetragon; and the perpendicular is its upright: wherefore the square-root of the sum of their squares is the correspondent diagonal: and, in like manner, with the other segment, the diagonal resting on the other perpendicular is found. GAN'.

187—189. The sides measuring fifty-two and one less than forty; the summit equal to twenty-five, and the base sixty: this was given as an example by former writers for a figure having unequal perpendiculars; and definite measures of the diagonals were stated, fifty-six and sixty-three. Assign to it other diagonals; and those particularly which appertain to it as a figure with equal perpendiculars.

Statement:

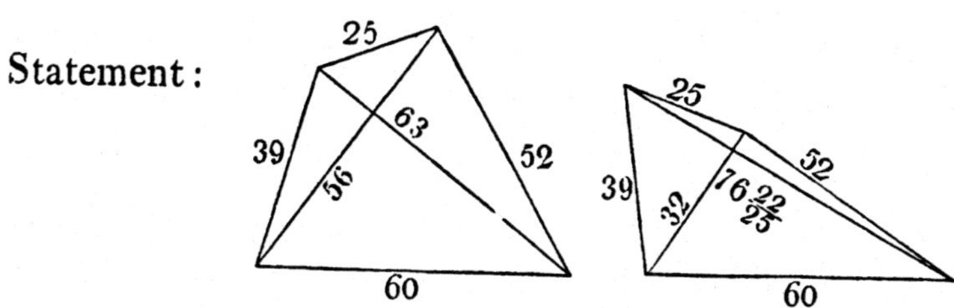

Here assuming one diagonal sixty-three, 63, the other is found as before, 56. Or, putting thirty-two instead of fifty-six for a diagonal, the other, found by the process before shown, comes out in two portions, both surds, 621 and 2700. The sum of the roots [as extracted by approximation] is the second diagonal 76 $\frac{22}{25}$. See figure of a triangle put to find the perpendicular:

Here the segments are found $\frac{3}{5}$ and $\frac{172}{5}$; and the perpendicular, the surd $\frac{38016}{25}$; of which the root found by approximation is 38 $\frac{622}{625}$. It is the equal perpendicular of that tetragon.

Next the sum of the squares of the perpendicular and difference between base and segment: Base of the tetragon, 60: least segment $\frac{3}{5}$; difference $\frac{297}{5}$. Square of the difference $\frac{88209}{25}$. Square of the perpendicular, which was a surd root, $\frac{38016}{25}$. Sum $\frac{126225}{25}$; or, dividing by the denominator, 5049. It is the square of one diagonal. So base 60; greater segment $\frac{172}{5}$; difference $\frac{128}{5}$. Its square $\frac{16384}{25}$. Square of the surd perpendicular $\frac{38016}{25}$. Sum $\frac{54400}{25}$; or, dividing by the denominator, 2176. It is the sum of the squares of the perpendicular and difference between base and greater segment; and is the square of the second diagonal. Extracting the roots of these squares by approximation, the two diagonals come out $71\frac{1}{25}$ and $46\frac{16}{25}$.

See

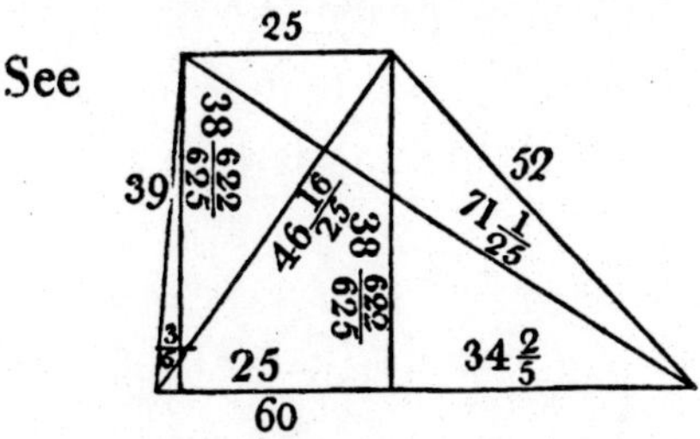

In this tetragon with equal perpendiculars the short side 39 added to the base 60, makes 99: which is greater than the aggregate of the summit and other flank, 77. Such is the limitation.

Thus, with the same sides, may be many various diagonals in the tetragon. Yet though indeterminate, diagonals have been sought as determinate, by BRA'HMEGUPTA and others. Their rule is as follows.

190. Rule:[1] The sums of the products of the sides about both the diagonals being divided by each other, multiply the quotients by the sum of the products of opposite sides; the square-roots of the results are the diagonals in a trápezium.

The objection to this mode of finding the diagonals is its operoseness, as I shall show by proposing a shorter method.

191—192. Rule [two stanzas]: The uprights and sides of two assumed rectangular triangles,[2] being multiplied by the reciprocal hypotenuses, become sides [of a quadrilateral]: and in this manner is constituted a trapezium, in which the diagonals are deducible from the two triangles.[3] The product of the uprights, added to the product of the sides, is one diagonal; the sum of the products of uprights and sides reciprocally multiplied, is the

[1] A couplet cited from BRAHMEGUPTA. 12. § 28.

[2] Assumed conformably with the rule contained in § 145. An objection, to which the commentator GAN'ÉSA adverts, and which he endeavours to obviate; is that this shorter method requires sagacity in the selection of assumed triangles; and that the longer method is adapted to all capacities.

[3] This method of constructing a trapezium is taken from BRAHMEGUPTA. 12. § 38.

other.[1] When this short method presented, why an operose one was practised by former writers, we know not.[2]

[1] A trapezium is divided into four triangles by its intersecting diagonals; and conversely, by the junction of four triangles, a trapezium is constituted. For that purpose, four triangles are assumed in this manner. Two triangles are first put in the mode directed (§ 145), the sides of which are all rational. Such sides, multiplied by any assumed number, will constitute other rectangular triangles, of which also the sides will be rational. By the twofold multiplication of hypotenuse, upright and side of one assumed triangle by the upright and side of the other, four triangles are formed, such that turning and adapting them and placing the multiples of the hypotenuses for sides, this trapezium is composed.

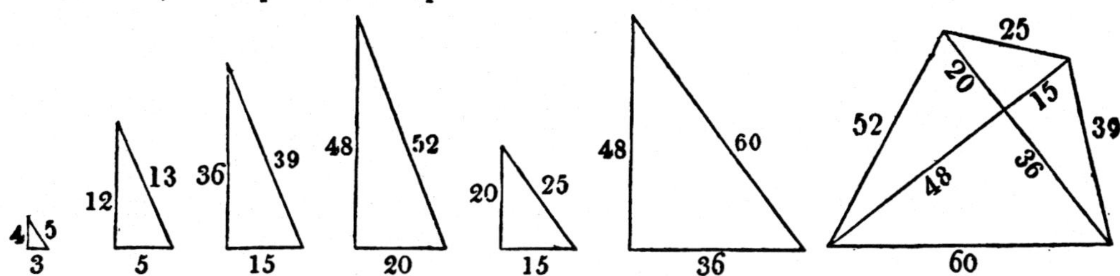

Here the uprights and sides of the arbitrary triangles, reciprocally multiplied by the hypotenuses, become sides of the quadrilateral: and hence the directions of the rule (§ 191).

In a trapezium so constituted, it is apparent, that the one diagonal is composed of two parts; one the product of the uprights, the other the product of the sides of the arbitrary triangles. The other diagonal consists of two parts, the products of the reciprocal multiplication of uprights and sides. These two portions are the perpendiculars: for there is no interval between the points of intersection. This holds, provided the shortest side be the summit; the longest, the base; and the rest, the flanks. But, if the component triangles be otherwise adapted, the summit and a flank change places. See

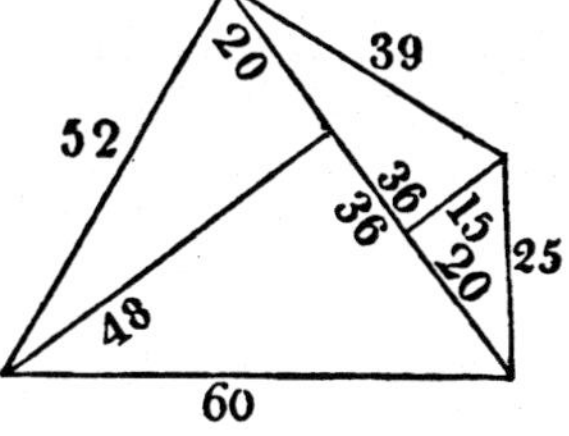

Here the two portions of the first diagonal, as above found (viz. 48 and 13) do not face; but are separated by an interval, which is equal to the difference between the two portions of the other diagonal (36 and 20) viz. 16. It is the difference of two segments on the same side, found by a preceding rule (§181—182); and is the interval between the intersections of the perpendiculars; and is taken for the upright of a triangle, as already explained (§ 181, note): the sum of the two portions of diagonal equal to the two perpendiculars is made the side. The square-root of the sum of the squares of such upright and side is equal to the product of the hypotenuses (13 and 5): wherefore the author adds " if the summit and flank change places, the first* diagonal will be the product of the hypotenuses."

From the demonstration of BRAHMEGUPTA's rule (Arithm. of BRAHM. § 28) may be deduced

* So the MSS. But BHÁSCARA's text exhibits second.

Assuming two rectangular triangles, 4 5 3 12 13 5 Multiply the upright and side of one by the hypotenuse of the other: the greatest of the products is taken for the base; the least for the summit; and the other two for the flanks. See

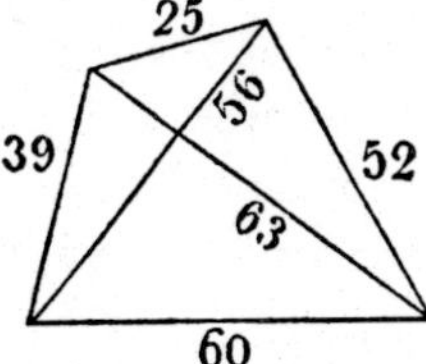

Here, with much labor [by the former method] the diagonals are found 63 and 56.

With the same pair of rectangular triangles, the products of uprights and sides reciprocally multiplied are 36 and 20: the sum of which is one dia-

grounds of a succinct proof, that the diagonal is found by multiplication of the hypotenuses, when the summit is not the least side. For, if the two derivative triangles be fitted together by bringing the hypotenuses in contact, the trapezium is such as is produced by the transposition of the summit and a flank, and the diagonal is the product of the hypotenuses of the generating triangles. GAN'.

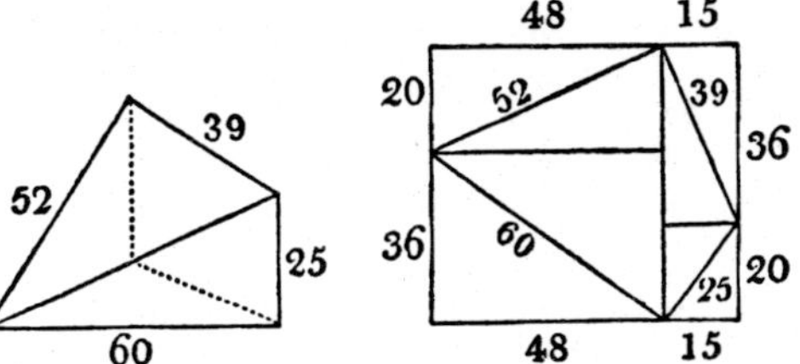

[2] In like manner, for the tetragon before instanced (§ 178), to find the diagonals, a pair of rectangular triangles is put 4 5 3 15 17 8 Proceeding as directed, the diagonals come out 77 and 84. In the figure instanced, a transposition of the flank and summit takes place

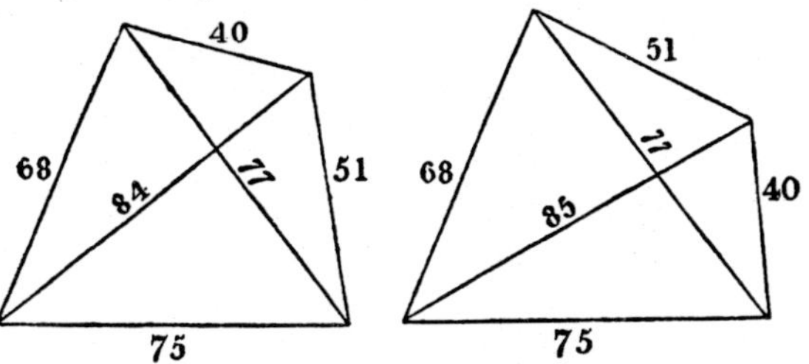

wherefore the product of the hypotenuse of the two rectangular triangles will be the second diagonal: and they thus come out 77 and 85. GAN'.

gonal, 56. The products of uprights multiplied together, and sides taken into each other, are 48 and 15: their sum is the other diagonal, 63. Thus they are found with ease.

But if the summit and flank change places, and the figure be stated accordingly, the second diagonal will be the product of the hypotenuses of the two rectangular triangles: viz. 65. See

193—194. Example.[1] In a figure, in which the base is three hundred, the summit a hundred and twenty-five, the flanks two hundred and sixty and one hundred and ninety-five, one diagonal two hundred and eighty and the other three hundred and fifteen, and the perpendiculars a hundred and eighty-nine and two hundred and twenty-four; what are the portions of the perpendiculars and diagonals below the intersections of them? and the perpendicular let fall from the intersection of the diagonals; with the segments answering to it? and the perpendicular of the needle formed by the prolongation of the flanks until they meet? as well as the segments corresponding to it; and the measure of both the needle's sides? All this declare, mathematician, if thou be thoroughly skilled in this [science of][2] plane figure.

Statement:

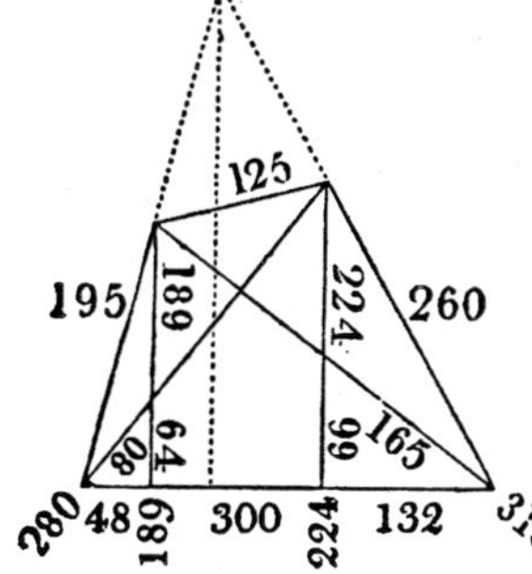

Length of the base 300. Summit 125. Flanks 260 and 195. Diagonals 280 and 315. Perpendiculars 189 and 224.

[1] Having thus, from § 173 to this place, shown the method of finding the area, &c. in the fourteen sorts of quadrilaterals, the author now exhibits another trapezium, proposing questions concerning segments produced by intersections.—GAŃ. The author proposes a question in the form of an example.—GANG. For the instruction of the pupil, he exhibits the figure called *(suchí)* a needle. *Manór.*

The problem is taken from BRAHMEGUPTA with a slight variation; and this example differs from his only in the scale, his numbers being here reduced to fifths. Arithm. of BRAHM. § 32.

[2] *Manóranjana.*

195—196. Rule (two stanzas): The interval between the perpendicular and its correspondent flank is termed the *sand'hi*[1] or link of that perpendicular. The base, less the link or segment, is called the *pít'ha* or complement of the same. The link or segment contiguous to that portion [of perpendicular or diagonal] which is sought, is twice set down. Multiplied by the other perpendicular in one instance, and by the diagonal in the other, and divided [in both instances] by the complement belonging to the other [perpendicular], the quotients will be the lower portions of the perpendicular and diagonal below the intersection.

Statement: Perpendicular 189. Flank contiguous to it 195. Segment intercepted between them (found by § 134) 48. It is the link. The second segment (found by § 195) is 252, and is called the complement.

In like manner the second perpendicular 224. The flank contiguous to it 260. Interval between them, being the segment called link, 132. Complement 168.

Now to find the lower portion of the first perpendicular 189. Its link, separately multiplied by the other perpendicular 224 and by the diagonal 280, and divided by the other complement 168, gives quotients 64 the lower portion of the perpendicular, and 80 the lower portion of the diagonal.

So for the second perpendicular 224, its link 132, severally multiplied by the other perpendicular 189 and by the diagonal 315, and divided by the other complement 252, gives 99 for the lower portion of the perpendicular and 165 for that of the diagonal.

197. Rule to find the perpendicular below the intersection of the diagonals: The perpendiculars, multiplied by the base and divided by the respective complements, are the erect poles: from which the perpendicular let fall from the intersection of the diagonals, as also the segments of the base, are to be found as before.[2]

Statement: Proceeding as directed, the erect poles are found 225 and 400. Whence, by a former rule (§ 159), the perpendicular below the intersection

[1] *Sand'hi* union, alliance; intervention, connecting link.
Pít'ha lit. stool. Here the complement of the segment.

[2] By the rule § 159.

of the diagonals is deduced, 144; and the segments of the base 108 and 192.

See figure

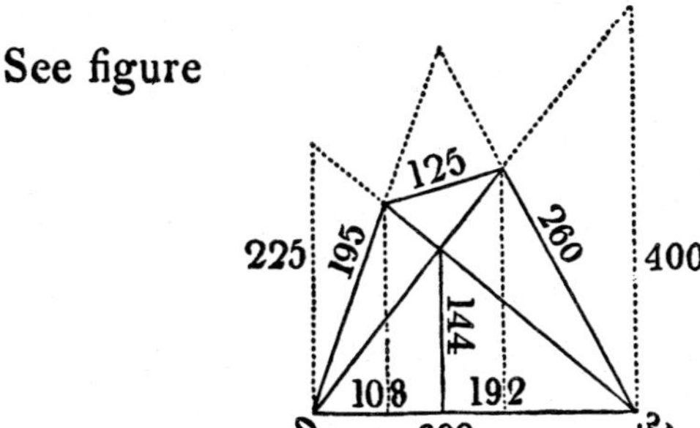

198—200. Rule to find the perpendicular of the needle[1], its legs and the segments of its base [three stanzas]: The proper link, multiplied by the other perpendicular and divided by its own, is termed the mean;[1] and the sum of this and the opposite link is called the divisor. Those two quantities, namely, the mean and the opposite link, being multiplied by the base and divided by that divisor, will be the respective segments of the needle's base. The other perpendicular, multiplied by the base and divided by the divisor, will be the perpendicular of the needle. The flanks, multiplied by the perpendicular of the needle and divided by their respective perpendiculars, will be the legs of the needle.[2] Thus may the subdivision of a

[1] *Súchí*, needle; the triangle formed by the flanks of the trapezium produced until they meet.

Sand'hi. See preceding note.

Sama, mean; a fourth proportional to the two perpendiculars and the link or segment.

Hara, divisor; the sum of such fourth proportional and the other link or segment.

[2] The needle, or figure resulting from the prolongation of the flanks of the trapezium, is a triangle, of which the sides are those prolonged flanks; and the base, the same with the base of the trapezium; and the perpendicular, the perpendicular of the needle: to find which, another similar and interior triangle is formed, in which the flank of the trapezium is one side, and a line drawn from its extremity parallel to the other leg of the needle is the second side: the perpendicular [of the trapezium] is perpendicular [of this interior triangle]; the link is one segment of the base; and the mean, as it is called, is the other. See

195 125 260 48 132 189 300 224 or this 195 125 260 48 132 189 300 224

Here to find the segment denominated the mean. In proportion as the opposite perpendicular is less or greater than the proper perpendicular, so is the segment termed the mean less or greater

plane figure be conducted by the intelligent, by means of the rule of three.[1]

Here the perpendicular being 224, its link is 132. This, multiplied by the other perpendicular, viz. 189, and divided by its own, viz. 224, gives the mean as it is named; $\frac{891}{8}$. The sum of this and the other link 48 is the divisor, as it is called, $\frac{1275}{8}$. The mean and other link taken into the base, being divided by this divisor, give the segments of the needle's base $\frac{1536}{17}$ and $\frac{3564}{17}$. The other perpendicular 189, multiplied by the base and divided by the same divisor, yields the perpendicular of the needle $\frac{6048}{17}$. The sides 195 and 260, multiplied by the needle's perpendicular and divided by their own perpendiculars respectively, viz. 189 and 224, give the legs of the needle, which are the sides of the trapezium produced: $\frac{6240}{17}$ and $\frac{7020}{17}$. See

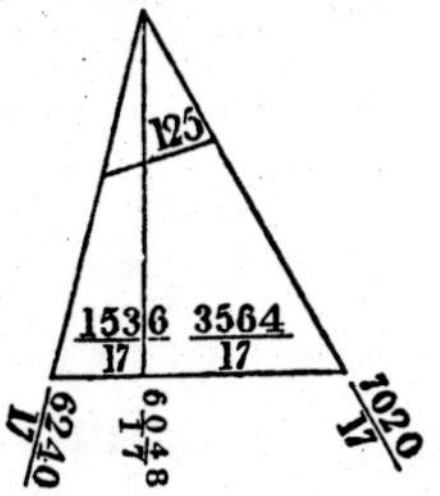

Thus, in all instances, under this head, taking the divisor for the argument, and making the multiplicand or multiplicator, as the case may be, the fruit or requisition, the rule of three is to be inferred by the intelligent mathematician.

than that called link: for, according as the side contiguous to the perpendicular is greater or less, so is the parallel side also greater or less; and so likewise is the segment contiguous thereto. Hence this proportion with the opposite perpendicular: 'If the proper perpendicular have this its segment, what has the opposite perpendicular?' The proportional resulting is the other segment termed the mean in the constructed triangle: and the sum of that and of the other segment called the opposite link will be the base of the constituted triangle. It is denominated the divisor. To find the perpendicular of the needle and the corresponding segments of its base accordingly, the proportion is this: 'If for this base these be the segments, what are they for the needle's base, which is equal to the entire base?' And, 'for that base, if this be the perpendicular, what is the perpendicular for the needle's base, which is equal to the whole base?' and to find the legs of the needle, 'if the hypotenuse answering to an upright equal to the perpendicular be the side contiguous to it, what is the hypotenuse answering to an upright equal to the perpendicular of the needle?' In like manner, the other leg is deduced from the other perpendicular. GAN'.

[1] From one part of a figure given, another member of it is deduced by the intelligent, through the rule of proportion. SU'R.

201. Rule:[1] When the diameter of a circle[2] is multiplied by three thousand nine hundred and twenty-seven and divided by twelve hundred and fifty, the quotient is the near[3] circumference: or multiplied by twenty-two and divided by seven, it is the gross circumference adapted to practice.[4]

202. Example. Where the measure of the diameter is seven, friend, tell the measure of the circumference: and where the circumference is twenty-two, find the diameter.

[1] To deduce the circumference of a circle from its diameter, and the diameter from the circumference. GAŃ.

[2] *Vritta, vartula*, a circle.

Vyása, vishcambha, vistriti, vistára, the breadth or diameter of a circle.

Parid'hi, pariñáha, vritti, némi (and other synonyma of the felloe of a wheel), the circumference or compass of a circle.

[3] *Súcshma*, delicate or fine; nearly precise; contrasted with *st'húla*, gross, or somewhat less exact, but sufficient for common purposes.—GANG. SU'R.

BRAHMEGUPTA puts the ratio of the circumference to the diameter as three to one for the gross value, and takes the root of ten times the square of the diameter for the neat value of the circumference. See Arithm. of BRAHM. §40. Also ŚRÍD'HARA's *Gań. sár.*

[4] As the diameter increases or diminishes, so does the circumference increase or diminish: therefore to find the one from the other, make proportion, as the diameter of a known circle to the known circumference, so is the given diameter to the circumference sought: and conversely, as the circumference to the diameter, so is the given circumference to the diameter sought.

Further: the semidiameter is equal to the side of an equilateral hexagon within the circle: as will be shown. From this the side of an equilateral dodecagon may be found in this manner: the semidiameter being hypotenuse, and half the side of the hexagon, the side; the square-root of the difference of their squares is the upright: subtracting which from the semidiameter the remainder is the arrow [or versed sine]. Again, this arrow being the upright, and the half side of the hexagon, a side; the square-root of the sum of their squares is the side of the dodecagon. See

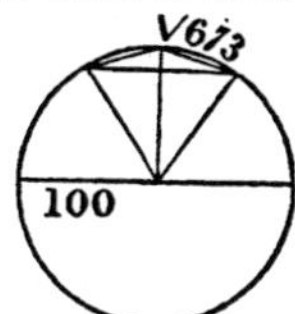

From which, in like manner, may be found the side of a polygon with twenty-four sides: and so on, doubling the number of sides in the polygon, until the side be near to the arc. The sum of such sides will be the circumference of the circle nearly. Thus, the diameter being a hundred, the side of the dodecagon is the surd 673; and that of a polygon of three hundred and eighty-four sides is nearly equal to the arc. By computation it comes out the surd 98683. Now the proportion, if to the square of the diameter put at a hundred, viz. 10000, this be the circumference, viz. the surd, 98683, then to the square of the assumed diameter twelve hundred and fifty, viz. 1562500, what will be the circumference? Answer: the root 3927 without remainder. GAŃ.

Statement: (circle, diameter 7)

Answer: Circumference 21 $\frac{1239}{1250}$, or gross circumference 22.

Statement: (circle, circumference 22)

Reversing multiplier and divisor, the diameter comes out 7 $\frac{11}{3927}$; or gross diameter 7.

203. Rule:. In a circle, a quarter of the diameter multiplied by the circumference is the area. That multiplied by four[1] is the net all around the ball.[2] This content of the surface of the sphere, multiplied by the diameter and divided by six is the precise solid, termed cubic, content within the sphere.[3]

204. Example. Intelligent friend, if thou know well the spotless *Lílávatí*, say what is the area of a circle, the diameter of which is measured by seven? and the surface of a globe, or area like a net upon a ball, the diameter being seven? and the solid content within the same sphere?

[1] Or rejecting equal multiplier and divisor, the circumference multiplied by the diameter is the surface. GAN'.

[2] *Prishta-phala*, superficial content: compared to the net formed by the string, with which cloth is tied to make a playing ball.

G'hana-phala, solid content: compared to a cube, and denominated from it cubic.

[3] Dividing the circle into two equal parts, cut the content of each into any number of equal angular spaces, and expand it so that the circumference become a straight line. See

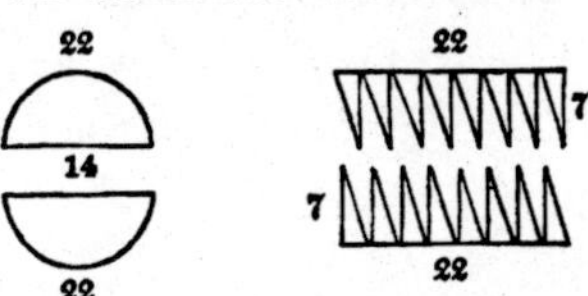

Then let the two portions approach so as the sharp angular spaces of the one may enter into the similar intermediate vacant spaces of the other: thus constituting an oblong, of which the semi-diameter is one side and half the circumference the other. See 7 [oblong figure] 22 or 21 $\frac{1239}{1250}$

The product of their multiplication is the area. Half by half is a quarter. Therefore a quarter of the diameter by the circumference is equal to the area.

See in the *Gólád'hyáya* (spherics) of the *Sidd'hánta-śirómani*, a demonstration of the rule, that the surface of the sphere is four times the area of the great circle, or equal to the circumference multiplied by the diameter.

Statement: 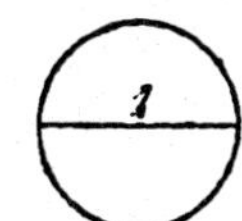

Answer: Area of the circle 38 $\frac{2423}{5000}$. Superficial content of the sphere 153 $\frac{1173}{1250}$. Solid content of the sphere 179 $\frac{1487}{2500}$.

205—206. Rule: a stanza and a half. The square of the diameter being multiplied by three thousand nine hundred and twenty-seven, and divided by five thousand, the quotient is the nearly precise area; or multiplied by eleven and divided by fourteen,[1] it is the gross area adapted to common practice. Half the cube of the diameter, with its twenty-first part added to it,[2] is the solid content of the sphere.

The area of the circle, nearly precise, comes out as before 38 $\frac{2423}{50}$, or gross area 38 $\frac{1}{2}$. Gross solid content 179 $\frac{2}{3}$.

206—207. Rule:[3] a stanza and a half. The sum and difference of the chord and diameter being multiplied together, and the square-root of the product being subtracted from the diameter, half the remainder is the arrow.[4]

To demonstrate the rule for the solid content of the sphere: suppose the sphere divided into as many little pyramids, or long needles with an acute tip and square base,* as is the number by which the surface is measured; and in length [height] equal to half the diameter of the sphere: the base of each pyramid is an unit of the scale by which the dimensions of the surface are reckoned: and, the altitude being a semidiameter, one-third of the product of their multiplication is the content: for a needle-shaped excavation is one-third of a regular equilateral excavation, as will be shown [§ 221]. Therefore [unit taken into] a sixth part of the diameter is the content of one such pyramidical portion: and that multiplied by the surface gives the solid content of the sphere.
GAŃ.

1 Multiplied by 22, and divided by (7 × 4) 28; or abridged by reduction to least terms, $\frac{11}{14}$. See GAŃ. &c.

2 Multiplied by 22, and divided by (7 × 6) 42; or multiplied by 11 and divided by 21. Then $21:11::2:\frac{22}{21}$ or $1+\frac{1}{21}$. See GAŃ. &c.

3 In a circle cut by a right line, to find the chord, arrow, &c. That is, either the chord, the arrow, or the diameter, being unknown, and the other two given, to find the one from the others.
GAŃ. SÚR.

4 A portion of the circumference is a bow. The right line between its extremities, like the string of a bow, is its chord. The line between them is the arrow, as resembling one set on a bow.
GAŃ. SÚR.

Dhanush, *chápa* and other synonyma of bow; an arc or portion of the circumference of a circle. *Jívá*, *jyá*, *jyacá*, *guńa*, *maurví* and other synonyma of bow-string; the chord of an arc. *Śara*, *ishu* and other synonyma of arrow; the versed sine.

* *Múrd'han*, base: lit. head or skull. *Agra*, tip or point: sharp summit.

The diameter, less the arrow, being multiplied by the arrow, twice the square-root of the product is the chord. The square of half the chord being divided by the arrow, the quotient added to the arrow is pronounced[1] to be the diameter of the circle.[2]

208. Example. In a circle, of which the diameter is ten, the chord being measured by six, say friend what is the arrow: and from the arrow tell the chord; and from chord and arrow, the diameter.

1 By teachers: that is, it has been so declared by the ancients. GAŃ.

BRAHMEGUPTA divides the square of the chord by four times the versed sine. See *Arithm.* of BRAHM. § 41.

2 On plane ground, with an arbitrary radius,* describe a circle; and through the centre draw a vertical diameter: then, on the circumference, at an arbitrary distance, make two marks; and the line between them, within the circle, across the diameter, is the chord; the portion of the circumference below the chord is the arc: and the portion of the diameter between the chord and arc is the arrow. Statement of a circle to exhibit these lines

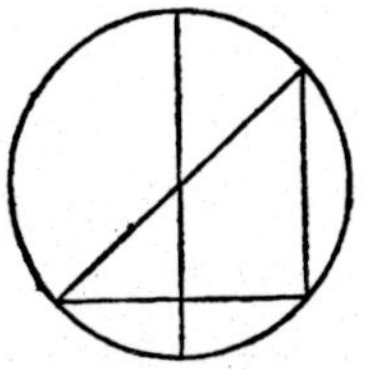

Thus, if the arrow be unknown, to find it, a triangle is constituted within the circle; where the chord is side, a thread stretched from the tip of the chord over the diameter to the circumference is [hypotenuse; and a line uniting their extremities is] the upright. That is to be first found. The square-root of the difference of the squares of the diameter and chord, which are hypotenuse and side, is the upright. But the product of their sum and difference is the difference of their squares: the root of which is the upright, and is measured on the vertical diameter. Thus the sum of the two portions of the diameter is equal to the diameter. Now, under the rule of concurrence (§ 55), the less portion only being required, the difference is subtracted from the diameter; and the remainder being halved is the arrow.

To deduce the chord from the arrow: another triangle is constituted within the circle; wherein the semidiameter less the arrow is the upright, the semidiameter is hypotenuse; and the square-root of the differences of these squares will be the half of the chord; and this doubled is the chord.

Now to find the diameter. The root of the difference of the squares of hypotenuse and upright before gave half the chord: now the square of this will be the difference of the squares of hypotenuse and upright. That being divided by the arrow and added to it, the result is the diameter. SU'R.

The following rule for finding the arc is cited by GAŃE'SA from ARYABHAT'T'A: "Six times the square of the arrow being added to the square of the chord, the square-root of the sum is the arc."

* *Carcáta;* compass; lit. a crab; meaning the radius.

Céndra; centre.

Statement: Diameter 10. When the chord is 6, the length of the arrow comes out 1. See

Or, arrow 1. The chord is found 6. Or from the chord and arrow the diameter is deduced 10.

209—211. Rule:[1] three stanzas. By 103923, 84853, 70534, 60000, 52055, 45922, or 41031, multiply the diameter of the circle, and divide the respective products by 120000; the quotients a.e severally, in their order, the sides of polygons, from the triangle to the enneagon, [inscribed] within the circle.[2]

[1] To find the sides of regular inscribed polygons. SU'R. GAN'.

[2] Divide the circumference by the number of sides of the polygon, and find the chord of the arc which is the quotient. For this purpose one commentator (SU'R.) refers to the subsequent rule (§ 213) in this treatise; and another (GAN'.) to the rule, in the author's astronomical work, (*Sidd'hánta*) for finding chords.

Or the demonstration, says GAN'ÉS'A, may be otherwise given. Describe a circle with any radius* at pleasure, divide it into three equal parts and mark the points; and from those points, with the same radius, describe three circles, which will be equal in circumference to the first circle; and it is thus manifest, that the side of the regular hexagon within the circle is half a diameter. See

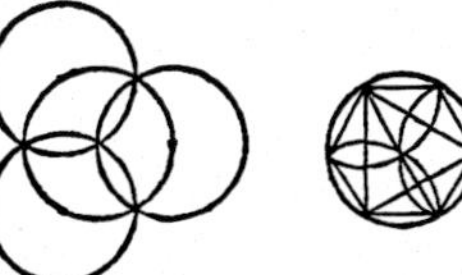

The side of a triangle [inscribed] within a circle is the upright; the diameter is hypotenuse and the side of the hexagon is side of the rectangular triangle. See the same figure. Therefore the square-root of the difference of the squares of the semidiameter and diameter is the side of the equilateral [inscribed] triangle: viz. for the proposed diameter (120000) 103923.

The side of a regular tetragon is hypotenuse, the semidiameter is upright, and side. See

Wherefore the square-root of twice the square of the semidiameter is the side of the [inscribed] tetragon: viz. for the diameter assumed, 84853.

The side of the regular octagon is hypotenuse, half the side of the tetragon is upright, and the difference between that and the semidiameter is the side. See

Wherefore the square-root of the sum of the squares of the half the side of the tetragon and the semidiameter less the half side of the tetragon is the side of the regular [inscribed] octagon: viz. for the diameter as put, 45922.

* *Carcata.* See above.

212. Example. Within a circle, of which the diameter is two thousand, tell me severally the sides of the inscribed equilateral triangle and other polygons.

The proof of the sides of the regular pentagon, heptagon and enneagon cannot be shown in a similar manner. GAN'.

First to find the side of a triangle inscribed within the circle, describe with a radius* equal to the proposed semidiameter, a circle; and draw a vertical diameter line through its centre. Then dividing the circumference into three equal parts, draw a triangular figure and another opposite to it; let two other diameters join the summits [angles] of those triangles. Thus there are six angles within the circle; and the interval between each pair of angles is equal to a semidiameter: for the diameter, which is in contact with two sides of a quadrilateral within the hexagon, is, from the centre of the circle to the side, a quarter of a diameter above the centre and just so much below it; and the sum of two quarters is half a diameter.

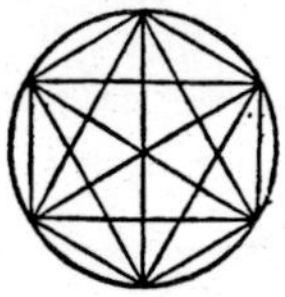

Now the length of a chord between an extremity of a diameter and an extremity of a side of the triangle, is equal to a semidiameter. It is a side of a rectangular triangle, of which the diameter is hypotenuse, and the square-root of the difference of their squares is the upright and is the measure of the side of the inscribed triangle. Ex. Diameter 120000. Side 60000. Difference of their squares 10800000000. Its square-root 103923.

Or a hexagon being described within the circle as before, and three diameters being drawn through the centre to the six angles, three equilateral quadrangular figures are constituted, wherein the four sides are equal to semidiameters: the short diagonal too is equal to a semidiameter; and the long diagonal is equal to the side of the inscribed triangle: and that is unknown. To find it, put the less diagonal equal to the semidiameter and proceeding as directed by the rule (§ 181-2), the greater diagonal is found, and is the side of the inscribed triangle. Example: assumed diagonal 60000. Its square 3600000000, subtracted from four times the square of the side 14400000000, leaves 10800000000. Its square-root is the greater diagonal 103923; and is the side of the inscribed triangle.

The method of finding the side of the triangle and of the hexagon has been thus shown. That of the inscribed tetragon is next propounded. Describing a circle as before, draw through the centre a diameter east and west and one north and south: and four lines are to be then drawn in the manner of chords, uniting their extremities. Thus a tetragon is inscribed in the circle. In each quadrant, another rectangular triangle is formed; wherein a semidiameter is side, and a semidiameter also upright; and the square-root of the sum of their squares will be the side of the tetragon. For example, in the proposed instance, side 60000, upright 60000. Sum of their squares 7200000000. Its square-root 84853.

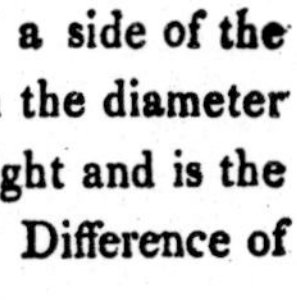

The side of the pentagon is the square-root of five times the square of the radius less the radius.†

* *Carcata;* opening of the compasses. See above.

† *Trijyá;* sine of three signs.

Statement: 2000

Answer: Side of the triangle $1732\frac{1}{20}$; of the tetragon $1414\frac{13}{60}$; of the

Or describing a circle as before, and dividing it into five equal parts, construct a pentagon within the circle. Draw a line between the extremities of two sides at pleasure; and two figures are thus formed; one of which is a trapezium, and the other a triangle: and the line drawn is the common base of both. Assume that chord arbitrarily: its arrow. found as directed, will be the perpendicular. Thus, in the same triangle, two rectangular triangles are constituted; in which half the base is the side, the perpendicular is upright, and the square-root of the sum of their squares is hypotenuse, and is the side of the pentagon. Example: putting the length of the arbitrary chord which is the base of the two figures, at a value near to the diameter, viz. 114140, the arrow comes out by the rule (§ 206) 41435; and the side of the pentagon is thence deduced 70534.

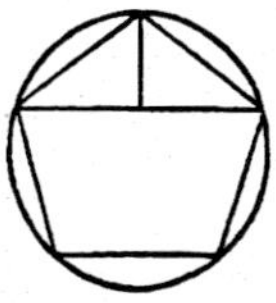

The side of the hexagon is half the diameter, as before shown.

For the heptagon, describe a circle as before, and within it a heptagon; draw a line between the extremities of two sides at pleasure, and three lines through the centre to the angles indicated by those sides: an unequal quadrilateral is thus formed: of which the two greater sides, as well as the least diagonal, are equal to a semidiameter. Assume the value of the greater diagonal arbitrarily: it is the chord of the arc encompassing two sides. Hence finding the arrow in the manner directed, it is the side of a small rectangular triangle, in which half the base or chord is the upright; whence the hypotenuse or side of the heptagon is deducible. Ex. Putting 93804 for the chord; the arrow inferred from it is 22579; and the side of the heptagon 52055.

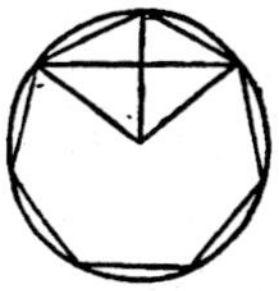

Or by a preceding rule (§ 181) the short diagonal, equal to a semidiameter, is the base of the two triangles on either side of it. The perpendicular thence deduced (§ 163—164) being doubled is the greater diagonal.

To demonstrate the side of the octagon: describe a circle as before and two diameter lines, dividing the circle into four parts. Then draw two sides in each of those parts; and eight angles are thus delineated. The line between the extremities of two sides, in form of a chord, is the side of an inscribed tetragon. The line from the center of the circle to the corner of the side is equal to half the diameter. Thus an unequal quadrilateral is constituted; which is divided by the line across it, forming two triangles, in which one side is a semidiameter and the base also is equal to a semidiameter; and half the side of the inscribed tetragon is the perpendicular: whence the other side is to be inferred. It comes out 45922.

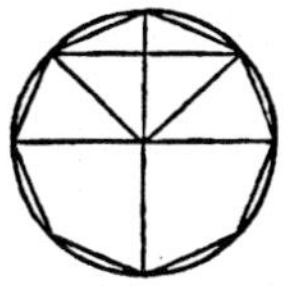

Next the proof of the side of the nonagon is shown. A circle being described as before, in-

pentagon $1175\frac{17}{30}$; of the hexagon 1000; of the heptagon $867\frac{7}{12}$; of the octagon $765\frac{11}{30}$; of the nonagon $683\frac{17}{20}$. See

From variously assumed diameters, other chords are deducible; as will be shown by us under the head of construction of sines, in the treatise on spherics.

The following rule teaches a short method of finding the gross chords.

213. Rule: The circumference less the arc being multiplied by the arc, the product is termed first.[1] From the quarter of the square of the circumference multiplied by five, subtract that first product: by the remainder divide the first product taken into four times the diameter: the quotient will be the chord.[2]

scribe a triangle in it. Thus the circle is divided into three parts. Three equal chords being drawn in each of those portions, an enneagon is thus inscribed in the circle: and three oblongs are formed within the same; of which the base is equal to the side of the inscribed triangle. Two perpendiculars being drawn in the oblong, it is divided into three portions, the first and last of which are triangles; and the intermediate one is a tetragon. The base in each of them is a third part of the side of the inscribed triangle. It is the upright of a rectangular triangle; the perpendicular is its side; and the square-root of the sum of their squares is hypotenuse, and is the side of the enneagon.

To find the perpendicular, put an assumed chord equal to half the chord of the [inscribed] tetragon; find its arrow in the manner directed; and subtract that from the arrow of the chord of the [inscribed] triangle: the remainder is the perpendicular. Thus the perpendicular comes out 21989: it is the side of a rectangular triangle. The third part of the inscribed triangle is 34641: it is the upright. The square-root of the sum of their squares is 41031: and is the side of the inscribed enneagon. Thus all is congruous. SU'R.

1 *Prat'hama, ádya*, first [product].

2 This, according to the remark of the commentators, is merely a rough mode of calculation, giving the gross, not the near, nor precise, chords. The rule appears from their explanations of the principle of it, to be grounded on considering the circle as converted into a rectangular triangle, in which the proposed arc is a side, its complement to the semicircle is the upright, and the other semicircle is hypotenuse. The difference between the squares of such upright and hypotenuse is the square of the arc and is the *first* product in the rule. When the pro-

214. Example. Where the semidiameter is a hundred and twenty, and the arc of the circle is measured by an eighteenth multiplied by one and so forth [up to nine,[1]] tell quickly the chords of those arcs.

Statement: Diameter 240. Here the circumference is 754 [nearly].

Arcs being taken, multiples of an eighteenth thereof, the chords are to be sought.

Or for the sake of facility, abridging both circumference and arcs by the eighteenth part of the circumference, the same chords are found. Thus, circumference 18. Arcs 1. 2. 3. 4. 5. 6. 7. 8. 9. Proceeding as directed, the chords come out 42. 82. 120. 154. 184. 208. 226. 236. 240. See

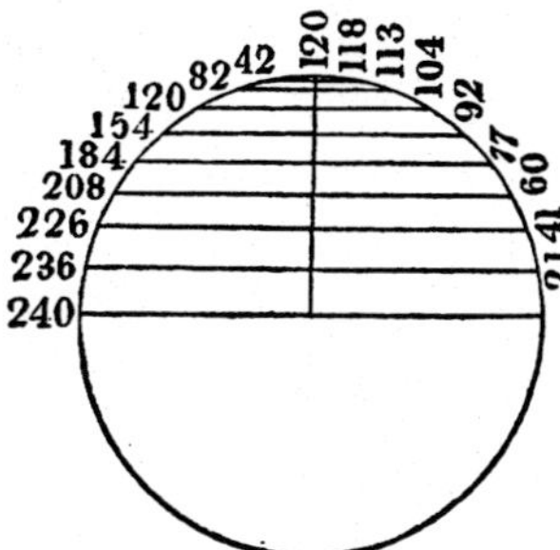

In like manner, with other diameters [chords of assigned arcs may be found.][2]

215. Rule:[3] The square of the circumference is multiplied by a quarter of the chord and by five, and divided by the chord added to four times the diameter; the quotient being subtracted from a quarter of the square of the circumference, the square-root of the remainder, taken from the half of the circumference, will leave the arc.[4]

posed arc is a semicircle, the chord is a maximum: and so is the first product; and this is equal to the square of the semicircumference or quarter of the square of the circumference. Then, as this maximum is to the greatest chord, or four times the one to four times the other, so is the first product for the proposed arc to the chord of that arc. This proportion, however, is modified, by adding to the first term of it the square of the complement of the proposed arc to the semicircle.

[1] Up to nine, or half the number of arcs: for the chords of the eighth and tenth will be the same; and so will those of the seventh and eleventh: and so forth. GAṄ.

[2] GANG. &c.

[3] To find the arc from the chord given.

[4] This is analogous to the preceding rule. The complement of the arc is found by a rough approximation.

216. Example. From the chords, which have been here found, now tell the length of the arcs, if, mathematician, thou have skill in computing the relation of arc and chord.[1]

Statement: Chords 42. 82. 120. 154. 184. 208. 226. 236. 240.

Circumference abridged 18. Arcs thence found 1. 2. 3. 4. 5. 6. 7. 8. 9. They must be multiplied by the eighteenth part of the circumference.[2]

[1] To find the area of the bow or segment of a circle, the following rule is given in VISHN'U's *Gan'ita-sára*, as cited by GANGA'DHARA; and the like rule is taught by CE'S'AVA quoted by his son GAN'E'S'A: 'The arrow being multiplied by half the sum of the chord and arrow, and a twentieth part of the product being added, the sum is the area of the segment.' S'RI'D'HARA's rule, as cited by GAN'E'S'A, is 'the square of the arrow multiplied by half the sum of the chord and arrow, being multiplied by ten and divided by nine, the square-root of the product is the area of the bow.' GAN'E'S'A adds: 'the chord and arrow being given, find the diameter; and from this the circumference; and thence the arc. Then from the extremities of the arc draw lines to the centre of the circle. Find the area of the sector* by multiplying half the arc by the semidiameter; and the area of the triangle by taking half the chord into the semidiameter less the arrow. Subtracting the area of the triangle from the area of the sector, the difference is the area of the segment.' The *Manóranjana* gives a similar rule: but finds the area of the sector by the proportion 'as the whole circumference is to the whole area, so is the proposed arc to the area of the sector.'

[2] The commentator SU'RYADA'SA notices other figures omitted, as he thinks, by the author; and GANGA'DHARA quotes from the *Gan'ita-sára* of VISHN'U an enumeration of them; the most material of which are specified in S'RI'D'HARA's *Gan'ita-sára*. They are reducible, however, according to these authors, to the simple figures which have been treated of: and the principal ones are, the *Gaja-danta* or elephant's tusk, which may be treated as a triangle.—S'RI'. *Bálendu*, or crescent, [a lunule or meniscus,] which may be considered as composed of two triangles.—S'RI'. *Yava* or barley-corn, [a convex lens,] treated as consisting either of two triangles or two bows.—GANG. *Némi* or felloe, considered as a quadrilateral.—S'RI'. and SU'R. *Vajra* or thunderbolt, treated as comprising two triangles.—SU'R. Or a quadrilateral with two bows or two trapezia.—GANG. Or two quadrilaterals.—S'RI'. *Panchacóna* or pentagon, composed of a triangle and a trapezium.—GANG. *Shad'bhuja* or hexagon, a quadrilateral and two triangles, or two quadrilaterals.—GANG. *Saptásra* or heptagon, five triangles.—GANG. Besides *Sanc'ha* or conch; *Mrìdanga* or great drum; and several others.

* *Vrìtta-c'han'da*, portion of a circle.

CHAPTER VII.

EXCAVATIONS[1] *and CONTENT of SOLIDS.*

217—218. RULE:[2] a couplet and a half. Taking the breadth in several places,[3] let the sum of the measures be divided by the number of places: the quotient is the mean measure.[4] So likewise with the length and depth.[5] The area of the plane figure, multiplied by the depth, will be the number of solid cubits contained in the excavation.

219—220. Example: two stanzas. Where the length of the cavity, owing to the slant of the sides, is measured by ten, eleven and twelve cubits in three several places, its breadth by six, five and seven, and its depth

[1] *C'háta-vyavahára.* The author treats first of excavations; secondly of stacks of bricks and the like; thirdly of sawing of timber and cutting of stones; and fourthly of stores of grain; in as many distinct chapters.

[2] For measuring an excavation, the sides of which are trapezia. GAŃ.

[3] *Vistára,* breadth.

Dairghya, length.

Béd'ha, béd'hana, depth.

C'háta, an excavation, or a cavity *(garta),* as a pond, well, or fountain, &c.

Sama-c'háta, a cavity having the figure of a regular solid with equal sides: a parallelipipedon, cylinder, &c.

Vishama-c'háta, one, the sides of which are unequal: an irregular solid.

Súchí-c'háta, an acute one: a pyramid or cone.

Sama-miti, mean measure.

G'hana-phala, g'hana-hasta-sanc'hyá, c'háta-sanc'hyá, the content of the excavation; or of a solid alike in figure.

[4] The greater the number of the places, the nearer will the mean measure be to to the truth, and the more exact will be the consequent computation. GAŃ.

[5] The irregular solid is reduced to a regular one, to find its content. SÚR.

by two, four and three; tell me, friend, how many solid cubits are contained in that excavation?

Statement:	12	11	10 L.
	7	5	6 B.
	3	4	2 D.

Here finding the mean measure, the breadth is 6 cubits, the length 11, and the depth 3. See

Answer. The number of solid cubits is found 198.

221. Rule:[1] a couplet and a half. The aggregate of the areas at the top and at the bottom, and of that resulting from the sum [of the sides of the summit and base], being divided by six, the quotient is the mean area: that, multiplied by the depth, is the neat[2] content.[3] A third part of the content of the regular equal solid is the content of the acute one.[4]

222. Example. Tell the quantity of the excavation in a well, of which the length and breadth are equal to twelve and ten cubits at its mouth, and half as much at the bottom, and of which the depth, friend, is seven cubits.

Statement: Length 12. Breadth 10. Depth 7. Sum of the sides 18 and 15.

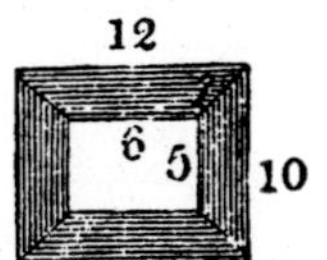

Area at the mouth 120; at the bottom 30; reckoned by the sum of the sides 270. Total 420. Mean area 70. Solid content 490.

[1] To find the content of a prism, pyramid, cylinder and cone.

[2] Contrasted with the result of the preceding rule, which gave a gross or approximated measure.

[3] Half the sum of the breadth at the mouth and bottom is the mean breadth; and half the sum of the length at the mouth and bottom is the mean length: their product is the area at the middle of the parallelipipedon. [Four times that is the product of the sums of the length and breadth.] This, added to once the area at the mouth and once the area at the bottom, is six times the mean area. GAN'.

[4] As the bottom of the acute excavation is deep, by finding an area for it in the manner before directed, the regular equal solid is produced: wherefore proportion is made; if such be the content, assuming three places, what is the content taking one? Thus the content of the regular equal solid, divided by three, is that of the acute one. SU'R.

223. Example. In a quadrangular excavation equal to twelve cubits, what is the content, if the depth be measured by nine? and in a round one, of which the diameter is ten and depth five? and tell me separately, friend, the content of both acute solids.

Statement:

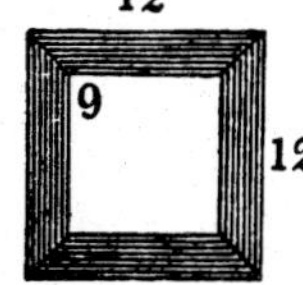

Product of the side and upright 144; multiplied by the depth, is the exact content 1296. Content of the acute solid 432.

Statement

Content nearly exact $\frac{3927}{10}$; of the acute solid $\frac{1309}{10}$. Or gross content [of the cylinder] $\frac{2750}{7}$ [of the cone] $\frac{2750}{21}$.

CHAPTER VIII.

STACKS.[1]

224—225. Rule:[2] a stanza and a half. The area of the plane figure [or base] of the stack, multiplied by the height,[3] will be the solid content. The content of the whole pile, being divided by that of one brick, the number of bricks is found. The height of the stack, divided by that of one brick, gives the number of layers. So likewise with piles of stones.[4]

226—227. Example: two stanzas. The bricks of the pile being eighteen fingers long, twelve broad and three high, and the stack being five cubits broad, eight long, and three high, say what is the solid content of that pile? and what the number of bricks? and how many the layers?

Statement: 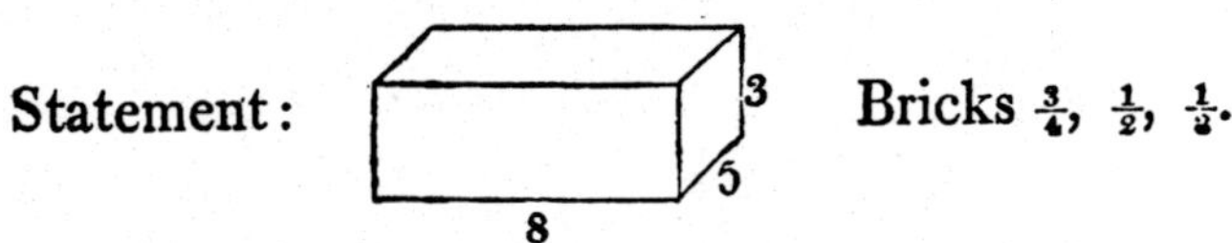

Bricks $\frac{3}{4}$, $\frac{1}{2}$, $\frac{1}{8}$.

Answer: Solid content of the brick $\frac{3}{64}$; of the stack 120. Number of bricks 2560. Number of layers 24.

So likewise in the case of a pile of stones.

[1] *Chiti-vyavahára.*

[2] To find the solid content of a stack or pile of bricks, or of stones or other things of uniform dimensions: also the number of bricks and of strata contained in the stack.

[3] *Chiti;* a pile or stack: an oblong with quadrangular sides.
Uchch'hraya, Uchch'hriti, Auchchya, height.
Stara, layer or stratum.

[4] The principle of the rule is obvious: being the extension of the preceding rule concerning the content of excavations, to a solid pile; and the application of the rule of proportion. Gan.

CHAPTER IX.

SAW.[1]

228. RULE: two half stanzas.[2] Half the sum of the thickness at both extremities, multiplied by the length in fingers; and the product again multiplied by the number of sections of the timber, and divided by five hundred and seventy-six,[3] will be the measure in cubits.

229. Example. Tell me quickly, friend, what will be the reckoning in cubits, for a timber the thickness of which is twenty fingers at the root; and sixteen fingers at the tip, and the length a hundred fingers, and which is cut by four sections.

Statement:

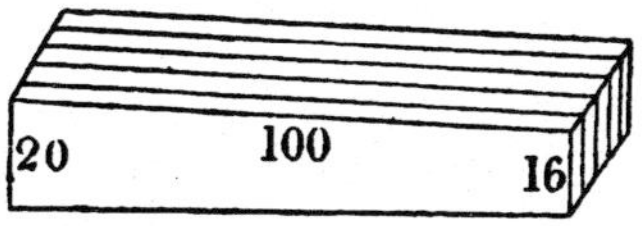

20 100 16

Half the sum of the thickness at the two extremities 18, multiplied by the length, makes 1800; and by the sections, 7200; divided by 576, gives the quotient in cubits $\frac{25}{2}$.

230. Rule: half a stanza. But when the wood is cut across, the superficial measure is found by the multiplication of the thickness and breadth, in the mode above mentioned.[4]

[1] *Cracacha-vyavahára:* determination of the reckoning concerning the saw (*cracacha*) or iron instrument with a jagged edge for cutting wood. SÚR.

[2] The concluding half of one stanza begun in the preceding rule (225), and the first half of another stanza of like metre completed in the following rule (230).

[3] To reduce superficial fingers to superficial cubits.

[4] If the breadth be unequal, the mean breadth must be taken; and so must the mean thickness, as before directed, if that be unequal. GAŃ. SÚR.

231. Maxim. The price for the stack of bricks or the pile of stones, or for excavation and sawing, is settled by the agreement of the workman, according to the softness or hardness of the materials.[1]

232. Example. Tell me what will be the superficial measure in cubits, for nine cross sections of a timber, of which the breadth is thirty-two fingers, and thickness sixteen.

Statement: 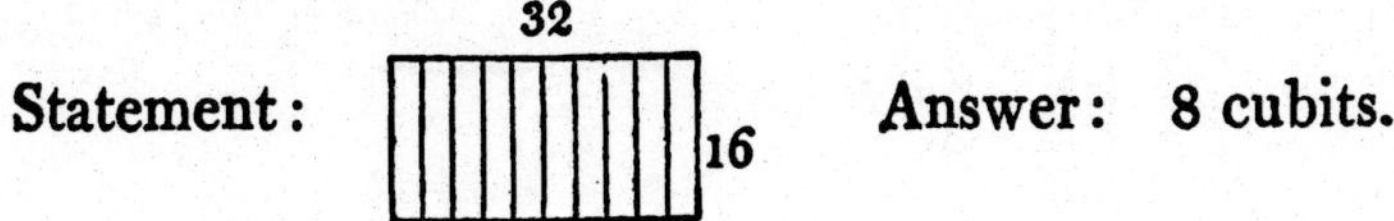

Answer: 8 cubits.

[1] This is levelled at certain preceding writers, who have given rules for computing specific prices or wages, as A'RYA-BHAT́T́A quoted by GAŃE'S'A, and as BRAHMEGUPTA (see Arithm. of BRAHM. § 49); particularly in the instance of sawyers' work, by varying the divisors according to the difference of the timber.

CHAPTER X.

MOUND[1] *of GRAIN.*

233. Rule. The tenth part of the circumference is equal to the depth [height[2]] in the case of coarse grain; the eleventh part, in that of fine; and the ninth, in the instance of bearded corn.[3] A sixth of the circumference being squared and multiplied by the depth [height], the product will be the solid cubits:[4] and they are *c'háris* of *Magad'ha.*[5]

234. Example. Mathematician, tell me quickly how many *c'háris* are

[1] *Rási-vyavahára* determination of a mound, meaning of grain.

[2] *Bédha* depth. See § 217. Here it is the perpendicular from the top of the mound of corn to the ground.—GAN'. It is the height in the middle from the ground to the summit of the mound of grain.—SU'R.

[3] *Anu, sucshma-d'hánya,* fine grain, as mustard seed, &c.—GAN'. As Paspalum Kora, &c.—*Manór.* As wheat, &c.—SU'R.

Ananu, st'húla-d'hánya, coarse grain, as chiches (Cicer arietinum).—GAN'. and SU'R. As wheat, &c.—*Manóranjana.* Barley, &c.—CH. on BRAHM.

Súcin, súca-d'hánya, bearded corn, as rice, &c.

The coarser the grain, the higher the mound. The rule is founded on trial and experience; and, for other sorts of grain, other proportions may be taken, as $9\frac{1}{2}$ or $10\frac{1}{2}$, or 12 times the height, equal to the circumference.—GAN'. and SU'R. The rule, as it is given in the text, is taken from BRAHMEGUPTA.—Arithm. of BRAHM. § 50.

[4] This is a rough calculation, in which the diameter is taken at one-third of the circumference. The content may be found with greater precision by taking a more nearly correct proportion between the circumference and diameter. GAN'.

[5] See § 7. The proportion of the *c'hári* or other dry measure of any province to the solid cubit being determined, a rule may be readily formed for computing the number of such measures in a conical mound of grain. GAN'ÉSA accordingly delivers rules by him devised for the *c'hári* of *Nandigráma* and for that of *Dévagiri:* ' the circumference measured by the human cubit, squared and

contained in a mound of coarse grain standing on even ground, the circumference of which (mound) measures sixty cubits? and separately say how many in a like mound of fine grain and in one of bearded corn?

Statement: 60 ○ Circumference 60. Height 6.

Answer: 600 *c'háris* of coarse grain. But of fine grain, height $\frac{60}{11}$, and quantity thence deduced $\frac{6000}{11}$. So, of bearded corn, height $\frac{60}{9}$, and quantity $\frac{6000}{9}$ *c'háris.*

235. Rule: In the case of a mound piled against the side of a wall, or against the inside or outside of a corner of it, the product is to be sought with the circumference multiplied by two, four, and one and a third; and is to be divided by its own multiplier.[1]

236—237. Example: two stanzas. Tell me promptly, friend, the number of solid cubits contained in a mound of grain, which rests against the side of a wall, and the circumference of which measures thirty cubits; and that contained in one piled in the inner corner and measuring fifteen cubits; as also in one raised against the outer corner and measuring nine times five cubits.

divided by sixteen, gives the *chári* of *Nandigráma;* and by sixty, that of *Dévagiri.*'* He further observes, that a vessel measuring a span every way contains a *maṅa*; that one measured by a cubit every way, taking the natural human cubit, contains eight *maṅas*; and that the cubit, intended by the text, is a measure in use with artisans, called in vulgar speech *gaj* [or *gaz*]; and a *c'hárí*, equal to such a solid cubit, will contain twenty-five *maṅas* and three quarters.

[1] Against the wall, the mound is half a cone; in the inner corner, a quarter of one; and against the outer corner, three quarters. The circumference intended is a like portion of a circular base; and the rule finds the content of a complete cone, and then divides it in the proportion of the part. See GAṄ. &c.

* In the vernacular dialect, *Nandigaon* and *Déógir:* the latter is better known by the name of *Dauletabad*, which the Emperor MUHAMMED conferred on it in the 14th century. The Hindus, however, have continued to it its ancient name of *Dévagiri*, mountain of the gods. *Nandigráma*, the town or village of *Nandi* (SIVA's bull and vehicle), retains the antique name; and is situated about 65 miles west of *Dévagiri;* and is accordingly said by this commentator in the colophon of his work to be near that remarkable place.

Statement: [diagram: rectangle with semicircles marked 30, 15, 45]

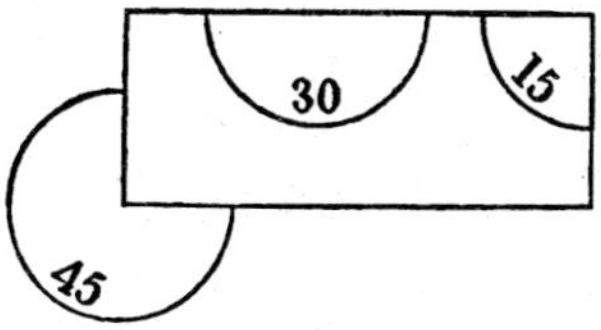

Twice the first-mentioned circumference is 60. Four times the next is 60. The last multiplied by one and a third is likewise 60. With these the product is alike 600. This, being divided by the respective multipliers, gives the several answers, 300, 150 and 450.[1]

[1] For coarse grain: but the product is $\frac{6000}{11}$ for fine; and $\frac{6000}{9}$ for bearded corn: and the answers are $\frac{3000}{11}$, $\frac{1500}{11}$, $\frac{4500}{11}$; and $\frac{3000}{9}$, $\frac{1500}{9}$, $\frac{4500}{9}$. GAN'. &c.

CHAPTER XI.

SHADOW[1] *of a GNOMON.*

238. Rule.[2] The number five hundred and seventy-six being divided by the difference of the squares of the differences of both shadows and of the two hypotenuses,[3] and the quotient being added to one, the difference of the hypotenuses is multiplied by the square-root of that sum; and the product being added to, and subtracted from, the difference of the shadows, the moieties of the sum and difference are the shadows.[4]

[1] *Ch'háyá-vyavahára:* determination of shadow; that is, measurement by means of a gnomon.

[2] The difference of the shadows, and difference of the hypotenuses being given, to find the length of the shadows and hypotenuses. SÚR.

This rule is the first in the chapter, according to all the commentators except SÚRYADÁSA, who begins with the next, § 240; and places this after § 244.

[3] *Ch'háyá*, *bhá*, *prabhá*, and other synonyma: shadow.

Śancu, *nara*, *nri*; a gnomon. It is usually twelve fingers long.

Carńa, hypotenuse of the triangle, of which the gnomon is the perpendicular, and the shadow the base.

[4] The rule, as the author hints in the example, which follows, is founded on an algebraic solution. It is given at length in the commentary of GAŃÉSA. The gnomon and shadow, with the line which joins their extremities, constitute a rectangular triangle, in which the gnomon is the upright, the shadow is the side, and the line joining their extremities the hypotenuse. In like manner another such is constituted; and joining their flanks, a triangle is formed. See

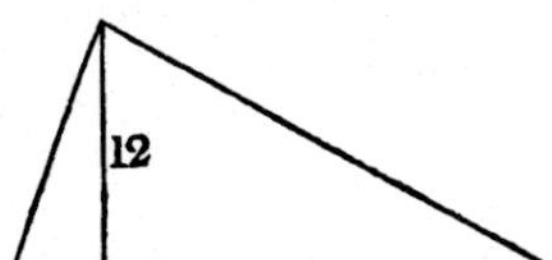

Herein the gnomon is the perpendicular; the two hypotenuses are the sides; the two shadows are the segments; and the sum of these is the base. Put this equal to *ya* 1; and to pursue the investigation, let the difference of the shadows be given 11; and the difference of the hypotenuses 7.

239. Example. The ingenious man, who tells the shadows, of which the difference is measured by nineteen, and the difference of hypotenuses by thirteen, I take to be thoroughly acquainted with the whole of algebra as well as arithmetic.

Statement: Difference of the shadows 19. Difference of hypotenuses 13. [Gnomon 12.]

Difference of their squares 192. By this divide 576: quotient 3. Add one. Sum 4. Square-root 2. By which multiply the difference of hypotenuses 13: product 26. Add it to, and subtract it from, the difference of the shadows 19; and halve the sum and difference: the shadows are found $\frac{45}{2}$ and $\frac{7}{2}$.

Then, by the rule of concurrence (§ 55), the segments are *ya* $\frac{1}{2}$ *ru* $\frac{11}{2}$ and *ya* $\frac{1}{2}$ *ru* $\frac{\dot{1}1}{2}$. The square of the greater segment, added to the square of the perpendicular twelve, is the square of the greater side: *ya v* $\frac{1}{4}$ *ya* $\frac{22}{4}$ *ru* $\frac{697}{4}$. This is one side of an equation. The difference of the squares of the segments is equal to the difference of the squares of the sides; as has been before shown (§ 164, note). But the difference of squares is equal to the product of the sum and difference. Sum *ya* 1; diff. *ru* 11; product *ya* 11. It is the difference of the squares of the sides. Divided by the difference of the simple quantities, the quotient is the sum *ya* $\frac{11}{7}$. The sum and difference added together and halved give the greater side *ya* $\frac{11}{14}$ *ru* $\frac{49}{14}$. Its square is *ya v* $\frac{121}{196}$ *ya* $\frac{1078}{196}$ *ru* $\frac{2401}{196}$. It is the second side of the equation. Reducing both to the same denomination and dropping the denominator, the equation becomes

ya v 49 *ya* 1078 *ru* 34153.
ya v 121 *ya* 1078 *ru* 2401

Now, when equal subtraction is made, the residue [or remaining coefficient] of the square of the unknown is the difference of the squares of the differences of the shadows and hypotenuses. The residue of the simple unknown term is nought. The absolute numbers on both sides being abridged by the square of the difference of hypotenuses as common divisor, there remains on one side of the equation the square of the difference of the hypotenuses, and on the other side the square of the difference of the shadows added to five hundred and seventy-six. Subtraction of like quantities being made, the residue of the absolute number is the difference of the squares of the differences of the hypotenuses and shadows added to five hundred and seventy-six. The remaining term involving the square of the unknown, being divided by the coefficient of the same, gives unity. The remainder of the absolute number being abridged by the common divisor, there results the number five hundred and seventy-six divided by the difference of the squares of the differences of the shadows and hypotenuses together with one. Its square-root is the root of the absolute number. But the absolute number was previously abridged by the square of the difference of the hypotenuses: wherefore the root must be multiplied by the difference of the hypotenuses. Hence the rule § 238. It is the value of *yávat-távat* as found by the equation: and is the base. It is the sum of the shadows. The difference of the shadows being added and subtracted, the moieties will be the shadows, by the rule of concurrence (§ 55). GAŃ.

Under the rule § 134, the gnomon being the upright, and the shadow the side, the square-root of the sum of their squares is the hypotenuse; viz. $\frac{51}{2}$ and $\frac{25}{2}$.

240. Rule:[1] half a stanza. The gnomon, multiplied by the distance of its foot from the foot of the light, and divided by the height of the torch's flame less the gnomon, will be the shadow.[2]

241. Example. If the base between the gnomon and torch be three cubits, and the elevation of the light, three cubits and a half, say quickly, friend, how much will be the shadow of a gnomon, which measures twelve fingers?

Statement: $\frac{7}{2}$; $\frac{1}{2}$; 3 $\frac{1}{2}$ Answer: Shadow 12 fingers.

242. Rule:[3] half a stanza. The gnomon being multiplied by the distance between the light and it, and divided by the shadow; and the quotient being added to the gnomon; the sum is the elevation of the torch.[4]

[1] The elevation of the light and [horizontal] distance of its foot from the foot of the gnomon being given, to find the shadow. GAN'.

[2] As the height of the light increases, the shadow of the gnomon decreases; and as the light is lowered, the length of the shadow augments. Now a line drawn strait, in the direction of the diagonal, from the light, meets the extremity of the gnomon's shadow. In like manner, taking off from the tip of the torch's flame a height equal to the gnomon's, and placing the light there, a diagonal line drawn as before meets the base of the gnomon. Thus the base between the foot of the gnomon and that of the light is the side of the triangle, and the height of the light less the gnomon is the upright. Hence the proportion: 'as the height of the torch less the gnomon, is to the distance of its foot from that of the gnomon, so is the gnomon to the shadow.' Whence the rule.

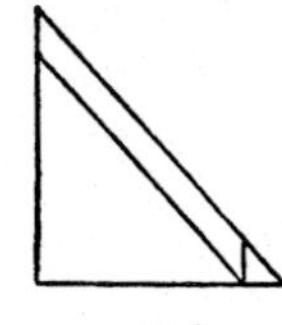

SU'R.

[3] To find the elevation of the torch; the length of the shadow being given, and the [horizontal] distance. SU'R.

[4] The demonstration proceeds on the proportion 'as the side measured by the shadow of the gnomon is to an upright equal to the gnomon, so is a base equal to the distance of the gnomon from the light, to a proportional,' which is the elevation of the torch less the height of the gnomon. SU'R.

243. Example. If the base between the torch and gnomon be three cubits, and the shadow be equal to sixteen fingers, how much will be the elevation of the torch? And tell me what is the distance between the torch and gnomon [if the elevation be given]?

Statement: Answer: Height of the torch $\frac{11}{4}$.

244. Rule:[1] half a stanza. The shadow, multiplied by the elevation of the light less the gnomon and divided by the gnomon, will be the interval between the gnomon and light.

Example, as before proposed (§ 243).

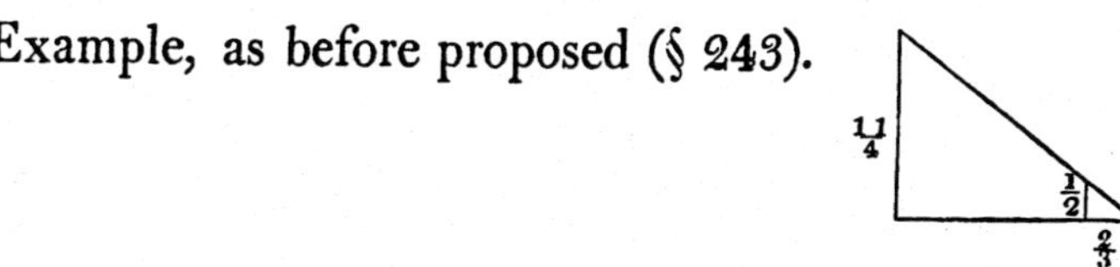

Answer: Distance 3 cubits.

245. Rule:[2] a stanza and a half. The length of a shadow multiplied by the distance between the terminations of the shadows and divided by the difference of the length of the shadows, will be the base. The product of the base and the gnomon, divided by the length of the shadow, gives the elevation of the torch's flame.[3]

In like manner is all this, which has been before declared, pervaded by the rule of three with its variations,[4] as the universe is by the Deity.

[1] To find the [horizontal] distance; the elevation of the torch and length of the shadow being given. SŪR. and GAṄ.

[2] The gnomon being set up successively in two places, the distance between which is known, and the length of the two shadows being given, to find the elevation of the light, and the base. SŪR. and GAṄ.

[3] The rule is borrowed from BRAHMEGUPTA. See Arithm. of BRAHM. § 54.

[4] The double rule of proportion, or rule of five or more quantities, &c.—GAṄ. The author intimates, that the whole preceding system of computation, as well the rules contained under the present head, as those before delivered, is founded on the rule of proportion. GAṄ.

246. Example. The shadow of a gnomon measuring twelve fingers being found to be eight, and that of the same placed on a spot two cubits further in the same direction being measured twelve fingers, say, intelligent mathematician, how much is the distance of the shadow[1] from the torch, and the height of the light, if thou be conversant with computation, as it is termed, of shadow?

Statement:

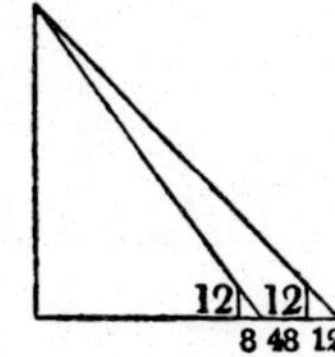

Here the interval between the termination of the shadows is in fingers 52; and the shadows are 8 and 12. The first of these, viz. 8, multiplied by the interval 52, and divided by the difference of the length of the shadows 4, gives the length of the base 104. It is the distance between the foot of the torch and the tip of the first shadow. So the length of the base to the tip of the second shadow is 156.

The product of the base and gnomon, divided by the shadow, gives both ways the same elevation of the light: viz. $6\frac{1}{2}$ cubits.

In like manner.][2] As under the present head of measurement of shadow, the solution is obtained by putting a proportion: viz. 'if so much of the shadow, as is the excess of the second above the first, give the base intercepted between the tips of the shadows, what will the first give?' The distances of the terminations of the shadows from the foot of the torch are in this manner severally found. Then a second proportion is put: 'if, the shadow being the side, the gnomon be the upright; then, the base being the side, what will be the upright?' The elevation of the torch is thus found: and is both ways [that is, computed with either shadow,] alike.

So the whole sets of five or more terms are explained by twice putting three terms and so forth.

As the being, who relieves the minds of his worshippers from suffering, and who is the sole cause of the production of this universe, pervades the

[1] All the commentators appear to have read gnomon in this place; but one copy of the text exhibits shadow as the reading: and this seems to be correct.

[2] Reference to the text: § 245.

whole, and does so with his various manifestations, as worlds, paradises,[1] mountains, rivers, gods, demons, men, trees,[2] and cities; so is all this collection of instructions for computations pervaded by the rule of three terms.

Then why has it been set forth by so many different [writers,[3] with much labour, and at great length]? The answer is

247. Whatever is computed either in algebra or in this [arithmetic] by means of a multiplier and a divisor, may be comprehended by the sagacious learned as the rule of three terms. Yet has it been composed by wise instructors in miscellaneous and other manifold rules, teaching its easy variations, thinking thereby to increase the intelligence of such dull comprehensions as ours.

[1] *Bhuvana*, worlds; heaven, earth, and the intermediate region. *Bhavana*, paradises, the several abodes of BRAHMA' and the rest of the gods.

[2] *Naga*, either tree or mountain. The term, however, is read in the text by none of the commentators besides GAN'ÉŚA.

[3] As ŚRÍD'HARA and the rest.—*Manó.* As BRAHMEGUPTA and others.—GANG.

CHAPTER XII.

PULVERIZER.[1]

248—252. RULE. In the first place, as preparatory to the investigation

[1] *Cuttaca-vyavahára* or *cuttacád'hyáya* determination of a grinding or pulverizing multiplier, or quantity such, that a given number being multiplied by it, and the product added to a given quantity, the sum (or, if the additive be negative, the difference) may be divisible by a given divisor without remainder.

See *Víja-Gańita*, chapter 2, from which this is borrowed, the contents being copied, (with some variation of the order,) nearly word for word. For this, as well as the following chapter 13, on Combination, belongs to algebra rather than arithmetic ; according to the remark of the commentator GAŃE'SA BHATTA : and they are here introduced, as he observes, and treated without employing algebraic forms, to gratify such as are unacquainted with analysis.

The commentator begins by asking 'why this subject has been admitted into a treatise of arithmetic, while a passage of A'RYA-BHATTA expressly distinguishes it from both arithmetic and algebra: " the multifarious doctrine of the planets, arithmetic, the pulverizer, *(cuttaca)* and analysis *(víja)*, and the rest of the science treating of seen* objects ;" and BRAHMEGUPTA, at the beginning of his chapter on Arithmetic, excludes it from this head; when describing the complete mathematician (see Arithm. of BRAHM. § 1)? The commentator proceeds to answer,—' Mathematics consist of two branches treating of known and of unknown quantity; as expressly declared: " The science of computation *(gańita)* is pronounced two-fold, denominated *vyacta* and *avyacta* (distinct and indistinct)." The investigation of the pulverizer, like the problem of the affected square, *(varga-pracriti*. See *Víja-gańita*, ch. 3), is comprehended in algebra, being subservient to its solutions; as hinted by the author. (See *Víja*. § 99). The separate mention of the head of investigation of the pulverizer, in passages of A'RYA-BHATTA and other ancient authors, as well as in those of BHA'SCARA and the rest (" By arithmetic, by algebra, by investigation of the pulverizer, and by resolution of the affected square, answers are found") is designed as an intimation of the difficulty and importance of the matter; not to indicate it as the subject of a separate treatise: and this, no less then the head of combination treated in the next chapter (chapter 13), with other

* Seen, or physical; as opposed to astrology, which is considered to be conversant with matters of an unseen and spiritual nature, the invisible influence which connects effects with causes.

of a pulverizer,[1] the dividend, divisor and additive quantity[2] are, if practicable, to be reduced by some number.[3] If the number, by which the dividend and divisor are both measured, do not also measure the additive quantity, the question is an ill put [or impossible] one.

249—251. The last remainder, when the dividend and divisor are mutually divided, is their common measure.[3] Being divided by that common measure, they are termed reduced quantities.[4] Divide mutually the

topics (all exclusive of arithmetic, which comprises logistics and the rest of the enumerated heads terminating with measurement by shadow,) falls within algebra, as the precepts of the rules concur with exercise of sagacity to effect the solution. (See *Víja*, § 224). It is then true, concludes this commentator, that mathematics consist but of two branches. Nevertheless the subjects of this and of the following chapter are here introduced, to be treated without reference to algebraic solutions, as the *Bhadra-gańita* and other problems* have found place in the arithmetical treatises of NÁRÁYAŃA and other writers, to be there wrought without algebra; and for the same purpose of gratifying such as are not conversant with this branch. GAŃ.

In BRAHMEGUPTA's work the whole of algebra is comprised under this title of *Cuttacád'hyáya*, chapter on the pulverizer. See BRAHM. ch. 18, and CHATURVÉDA on BRAHMEGUPTA, ch. 12, § 66.

[1] *Cuttaca* or *Cutta*, from *cutt*, to grind or pulverize; (to multiply: all verbs importing tendency to destruction also signifying multiplication.—GAŃ.)

The term is here employed in a sense independent of its etymology to signify a multiplier such, that a given dividend being multiplied by it, and a given quantity added to (or subtracted from) the product, the sum (or difference) may be measured by a given divisor. SÚR. on *Víj.-gań.* and *Líl.* RANG. on *Vás.* GAŃ. on *Líl.*

The derivative import is, however, retained in the present version to distinguish this from multiplier in general; *cuttaca* being restricted to the particular multiplier of the problem in question.

[2] *Cshépa*, or *cshépaca*, or *yuti*, additive. From *cship* to cast or throw in, and from *yu* to mix. A quantity superinduced, being either affirmative or negative, and consequently in some examples an additive, in others a subtractive, term.

Viśudd'hi, subtractive quantity, contradistinguished from *cshépa* additive, when this is restricted to an affirmative one. See § 263.

[3] *Apavártana*, abridgment; abbreviation.—GAŃ. Depression or reduction to least terms; division without remainder: also the number which serves to divide without residue; the common measure, or common divisor of equal division. SÚR.

[4] *Dríd'ha*, firm: reduced by the common divisor to the least term. The word is applicable to the reduced additive, as well as to the dividend and divisor. BRAHMEGUPTA uses *nich'héda* and *nirapavarta* in this sense.—*Brahm.* 18, § 9.

* *Bhadra-gańita*, on the construction of magical squares, &c. is the 13th head termed *vyavahára*, as *Anca-pása* on combination of numerals, is the 14th, in NÁRÁYAŃA's treatise of arithmetic entitled *Caumudí*.

Q

reduced dividend and divisor, until unity be the remainder in the dividend. Place the quotients one under the other; and the additive quantity beneath them, and cipher at the bottom.[1] By the penult multiply the number next above it and add the lowest term. Then reject the last and repeat the operation until a pair of numbers be left. The uppermost of these being abraded[2] by the reduced dividend, the remainder is the quotient. The other [or lowermost] being in like manner abraded by the reduced divisor, the remainder is the multiplier.

252. Thus precisely is the operation when the quotients are an even number.[3] But, if they be odd, the numbers as found must be subtracted from their respective abraders, the residues will be the true quotient and multiplier.

253. Example. Say quickly, mathematician, what is that multiplier, by which two hundred and twenty-one being multiplied, and sixty-five added to the product, the sum divided by a hundred and ninety-five becomes exhausted?

Statement: Dividend 221 / Divisor 195 Additive 65.

Here the dividend and divisor being mutually divided, the last of the remainders (or divisors) is 13. By this common measure, the dividend, divisor and additive, being reduced to their least terms, are Divd. 17 / Divr. 15 Addve. 5.

The reduced dividend and divisor being divided reciprocally, and the quotients put one under the other, the additive under them, and cipher at the bottom, the series which results is 1 / 7 / 5 / 0 Then multiplying by the penult the number above it and proceeding as directed, the two quantities are

[1] *Tashta*, abraded; from *tacsh*, to pare or abrade: divided, but the residue taken, disregarding the quotient: reduced to a residue.—SU'R. As it were a residue after repeated subtractions. GANG.

Tacshaṅa, the abrader; the divisor employed in such operation.

[2] *P'hala-valli*, the series of quotients; to be reduced by the operation forthwith directed to only two terms.

[3] Even, as 2, 4, 6, &c.—*Manór.*

obtained 40/35 These being abraded by the reduced dividend and divisor 17 and 15, the quotient and multiplier are obtained 6 and 5. Or, by the subsequent rule (§ 262), adding them to their abraders multiplied by an assumed number, the quotient and multiplier [putting 1] are 23 and 20; or, putting 2, they are 40 and 35: and so forth.[1]

254. Rule: The multiplier is also found by the method of the pulverizer, the additive quantity and dividend, being either reduced by a common measure [or used unreduced.][2] But if the additive and divisor be so reduced, the multiplier found, being multiplied by the common measure, is the true one.

255. Example. If thou be expert in the investigation of such questions, tell me the precise multiplier by which a hundred being multiplied, with ninety added to the product, or subtracted from it,[3] the sum, or the difference, may be divisible by sixty-three without a remainder.

Statement: Dividend 100 Additive
Divisor 63 or subtractive 90.

The quotient and multiplier are found by proceeding as before, 30 and 18.

Or, the dividend and additive being reduced by the common measure ten,

Dividend 10
Divisor 63 Additive 9.

Placing the quotients of reciprocal division, the additive quantity and cipher, one under the other, the series is 0, 6, 3, 9, 0. And the multiplier is found by the former process 45. The quotient (3) is here not to be taken; and the quotients [of the series] being an odd number, the multiplier 45 is to be subtracted from its own abrader 63; the true multiplier is thus found 18. The dividend being multiplied by that multiplier, and the additive quantity being added, and the sum divided by the divisor, the quotient is found 30.

[1] Putting 3, they are 57 and 50. SU'R. and GANG.

[2] GAN'.

[3] An example of the subsequent rule § 256.

Or, the divisor and additive quantity are reduced by the common measure nine. Dividend 100 Divisor 7 Additive 10.

Here the quotients, the additive and cipher make the series 14 3 10 0 The multiplier is found 2, which, multiplied by the common measure 9, gives the true multiplier 18.

Or, the dividend and additive are reduced, and further the divisor and additive, by common measures. Dividend 10 Divisor 7 Additive 1.

Proceeding as before, the series is 1 2 1 0 The multiplier hence deduced is 2; which, taken into the common measure 9, gives 18: and hence, by multiplication and division, the quotient comes out 30.

Or, adding the quotient and multiplier as found, to [multiples of] their divisors, the quotient and multiplier are 130 and 81; or 230 and 144; and so forth.[1]

256. Rule: half a stanza. The multiplier and quotient, as found for an additive quantity, being subtracted from their respective abraders, answer for the same as a subtractive quantity.[2]

Here the quotient and multiplier, as found for the additive quantity ninety in the preceding example, namely 30 and 18, being subtracted from their respective abraders, namely 100 and 63; the remainders are the quotient and multiplier, which answer when ninety is subtractive: viz. 70 and 45.

Or, these being added to arbitrary multiples of their respective abraders, the quotient and multiplier are 170 and 108; or 270 and 131; &c.

257. Example.[3] Tell me, mathematician, the multipliers severally, by

[1] As 330 and 201; &c. GANG.

[2] The rule serves when the additive quantity is negative.—GAN'. SU'R. It is followed in the *Vija-gańita* by half a stanza relating to the change induced by reversing the sign, affirmative or negative, of the dividend. See *Vij.-gań.* § 59.

[3] This additional example is unnoticed by GAŃE'SA; but expounded by the rest of the commentators, and found in all copies of the text that have been collated. See a corresponding one with an essential variation however in the reading; *Vij.-gań.* § 67.

which sixty being multiplied, and sixteen being added to the product, or subtracted from it, the sum or difference may be divisible by thirteen without a remainder.

Statement: Dividend 60 Divisor 13 Additive 16.

The series of quotients, found as before, is 4
1
1
1
1
16
0

Hence the multiplier and quotient are deduced 2 and 8. But the quotients [of the series] are here uneven: wherefore the multiplier and quotient must be subtracted from their abraders 13 and 60: and the multiplier and quotient, answering to the additive quantity sixteen, are 11 and 52. These being subtracted from the abraders, the multiplier and quotient, corresponding to the subtractive quantity sixteen, are 2 and 8.

258. Rule:[1] a stanza and a half. The intelligent calculator should take a like quotient [of both divisions] in the abrading of the numbers for the multiplier and quotient [sought]. But the multiplier and quotient may be found as before, the additive quantity being [first] abraded by the divisor; the quotient, however, must have added to it the quotient obtained in the abrading of the additive. But, in the case of a subtractive quantity, it is subtracted.

259. Example. What is the multiplier, by which five being multiplied and twenty-three added to the product, or subtracted from it, the sum or difference may be divided by three without remainder.

Statement: Dividend 5 Divisor 3 Additive 23.

Here the series is 1 1 23 0 and the pair of numbers found as before 46 23 They

[1] Applicable when the additive quantity exceeds the dividend and divisor. GAN.

are abraded by the dividend and divisor 5/3. The lower number being abraded by 3, the quotient is 7 [and residue 2]. The upper number being abraded by 5, the quotient would be 9 [and residue 1]: nine, however, is not taken; but, under the rule for taking like quotients, seven only, [and the residue consequently is eleven]. Thus the multiplier and quotient come out 2 and 11.

And by the former rule (§ 256) the multiplier and quotient answering to the same as a negative quantity are found, 1 and 6.* Added to arbitrary multiples of their abraders, double for example so as the quotient may be affirmative, the multiplier and quotient are 7 and 4.† So in every [similar] case.

Or, statement for the second rule: Dividend 5 Additive
Divisor 3 abraded 2.‡

The multiplier and quotient hence found as before are 2 and 4. These subtracted from their respective divisors, give 1 and 1; as answering to the subtractive quantity. The quotient obtained in the abrading of the additive, [viz. 7] being added in one instance and subtracted in the other,[4] the results are 2 and 11 answering to the additive quantity; and 1 and 6 answering to the subtractive: or, to obtain an affirmative quotient, add to the latter twice their divisors,[5] and the result is 7 and 4.

260. Rule:[6] If there be no additive quantity; or if the additive be measured by the divisor; the multiplier may be considered as cipher, and the quotient as the additive divided by the divisor.[7]

261. Example. Tell me promptly, mathematician, the multiplier by which five being multiplied and added to cipher, or added to sixty-five, the division by thirteen shall in both cases be without remainder.

Statement: Dividend 5
Divisor 13 Additive 0.

* The difference between 5 and 11, viz. $5-11=-6$. The quotient therefore is negative.

† Thus 10 $(5\times2)-6=4$.

‡ 23, abraded by the divisor 3, gives the quotient 7 and residue 2.

4 $4+7=11$ and $1-7=-6$.

5 $1+(3\times2)=7$ and $-6+(5\times2)=4$.

6 Applicable if there be no additive; or if it be divisible by the divisor without remainder.

7 It is so in the latter case: but in the former (where the additive is null) the quotient is cipher. —SÚR. &c. See *Víj.-gań.* § 63.

There being no additive, the multiplier and quotient are 0 and 0; or 13 and 5; or 26 and 10; and so forth.

Statement: Dividend 5 / Divisor 13 — Additive 65.

By the rule (§ 260) the multiplier and quotient come out 0 and 5; or 13 and 10; [or 26 and 15;] and so forth.

Rule:[1] Or, the dividend and additive being abraded by the divisor, the multiplier may thence be found as before; and the quotient from it, by multiplying the dividend, adding the additive, and dividing by the divisor.

In the former example (§ 253) the reduced dividend, divisor and additive furnish this statement: Dividend 17 / Divisor 15 — Additive 5.

Abraded by the divisor (15) the additive and dividend become 5 and 2; and the statement now is Dividend 2 / Divisor 15 — Additive 5.

Proceeding as before the two terms found are 5 / 35. The lower one, abraded by the divisor (15), gives the multiplier 5. Whence, by multiplying with it the dividend (17) and adding (the additive) and dividing (by the divisor), the quotient comes out 6.

262. Rule for finding divers multipliers and quotients in every case: half a stanza. The multiplier and quotient, being added to their respective [abrading] divisors multiplied by assumed numbers, become manifold.[2]

The influence and operation of this rule have been already shown in various instances.

263. Rule for a constant pulverizer:[3] two half stanzas. Unity being taken for the additive quantity, or for the subtractive, the multiplier and quotient, which may be thence deduced, being severally multiplied by an

[1] This is found in one copy of the text; and is expounded by a single commentator GANGÁD'HARA; but unnoticed by the rest. It occurs, however, in the similar chapter of the *Víja-gańita*, § 62.

[2] See *Víj.-gań.* § 64.

[3] *St'hira-cuttaca,* steady pulverizer. See explanation of the term in the commentary on BRAHMEGUPTA's algebra.—*Brahm.* ch. 18, § 9—11. *Drĭdha-cuttaca* is there used as a synonymous term.

arbitrary additive or subtractive, and abraded by the respective divisors, will be the multiplier and quotient for such assumed quantity.[1]

In the first example (§ 253) the reduced dividend and divisor with additive unity furnish this statement: Dividend 17 / Divisor 15 Additive 1.

Here the multiplier and quotient (found in the usual manner) are 7 and 8. These, multiplied by an assumed additive five, and abraded by the respective divisors (15 and 17), give the multiplier and quotient 5 and 6, for that additive.

Next unity being the subtractive quantity, the multiplier and quotient thence deduced are 8 and 9. These, multiplied by five and abraded by the respective divisors, give 10 and 11.

So in every [similar] case.[2]

Of this method of investigation great use is made in the computation of planets.[3] On that account something is here said [by way of instance.]

264. A stanza and a half. Let the remainder of seconds be made the subtractive quantity,[4] sixty the dividend, and terrestrial days[5] the divisor. The quotient deduced therefrom will be the seconds; and the multiplier will be the remainder of minutes. From this again the minutes and remainder of degrees are found: and so on upwards.[6] In like manner, from the remainder of exceeding months and deficient days,[7] may be found the solar and lunar days.

The finding of [the place of] the planet and the elapsed days, from the remainder of seconds in the planet's place, is thus shown. It is as follows. Sixty is there made the dividend; terrestrial days, the divisor; and the re-

[1] See *Vij.-gan.* § 71.

[2] See *Gólád'hyaya.*

[3] See BRAHMEGUPTA, ch. 18, § 9—12.

[4] The present rule is for finding a planet's place and the elapsed time, when the fraction above seconds is alone given. GAŃ.

[5] The number of terrestrial days (nycthemera) in a *calpa* is stated at 1577916450000. *Śirómani*, computation of planets, ch. 1, § 20—21.

[6] The dividend varies, when the question ascends above the sexagesimal scale, to signs, revolutions, &c.

[7] *Ád'hi-mása*, additive months; and *Avama* (or *Cshaya*) *dina*, subtractive days. See *Śirómani* on planets, ch. 1, § 42. The exceeding months, or *more* lunar than solar months, in a *calpa*, are 1593300000. The deficient days or *fewer* terrestrial days than lunar, in a *calpa*, are 25082550000.

mainder of seconds, the subtractive quantity: with which the multiplier and quotient are to be found. The quotient will be seconds; and the multiplier the remainder of minutes. From this remainder of minutes taken [as the subtractive quantity] the quotient deduced will be minutes; and the multiplier, the remainder of degrees. The residue of degrees is next the subtractive quantity; terrestrial days, the divisor; and thirty, the dividend: the quotient will be degrees; and the multiplier, the remainder of signs. Then twelve is made the dividend; terrestrial days, the divisor; and the remainder of signs, the subtractive quantity: the quotient will be signs; and the multiplier, the remainder of revolutions. Lastly, the revolutions in a *calpa* become the dividend; terrestrial days, the divisor; and the remainder of revolutions, the subtractive quantity: the quotient will be the elapsed revolutions; and the multiplier, the number of elapsed days.[1] Examples of this occur [in the *Śirómańi*] in the chapter of the problems.[2]

In like manner the exceeding months in a *calpa* are made the dividend; solar days, the divisor;[3] and the remainder of exceeding months, the subtractive quantity: the quotient will be the elapsed additional months; and the multiplier, the elapsed solar days. So the deficient days in a *calpa* are made the dividend; lunar days, the divisor;[4] and the remainder of deficient days, the subtractive quantity: the quotient will be the elapsed fewer days; and the multiplier, the elapsed lunar days.[5]

[1] The elapsed days of the *calpa* to the time for which the planet's place is found. The method of computing elapsed days to any given time is taught in the *Śirómańi* on planets, ch. 1, § 47—49.

[2] *Tri-praśná'd'hyáya.* Also in the *Góla'd'hyáya*, and *Mad'hyagrahá'd'hyáya.* GANG.

[3] The solar days, each equal to the sun's passage through one degree of its annual revolution, are 1555200000000 in the *calpa*. See *Śirómańi* 1, § 40.

[4] The lunar days, reckoning thirty to the month or synodical revolution, are 1602999000000 in the *calpa*. See *Śirómańi* 1, § 40.

[5] These may be illustrated, as the preceding astronomical example is, and rendered distinctly intelligible, by instances given by the commentators GANGÁ'D'HARA and GAŃÉŚA, and the *Manóranjana*, in arbitrary numbers. Put the terrestrial days in a *calpa* 19, the revolutions of the planet 10, the elapsed days 12. Then, by the proportion 19 | 10 | 12 | the planet's place comes out in revolutions, signs, &c. $6^r\ 3^s\ 23^\circ\ 41'\ 3''\ \frac{3}{19}$. In bringing out the seconds, the remainder of seconds is 3. From this, by an inverse process, the planet's place is to be found. Here the remainder of seconds is the subtractive quantity 3; the dividend is 60; and the divisor, 19. Proceeding as directed (§ 256) the multiplier and quotient are found 1 and 3. The quotient is the number of seconds 3; and the multiplier is the remainder of minutes 1. Let this be the subtractive quantity, 1; the dividend 60; and the divisor, 19. Proceeding as directed, the multiplier and quotient

265. Rule for a conjunct pulverizer.[1] If the divisor be the same and the multipliers various;[2] then, making the sum of those multipliers the dividend, and the sum of the remainders a single remainder, and applying the foregoing method of investigation, the precise multiplier so found is denominated a conjunct one.

266. Example. What quantity is it, which multiplied by five, and divided by sixty-three; gives a residue of seven; and the same multiplied by ten and divided by sixty-three, a remainder of fourteen? declare the number.[3]

Here the sum of the multipliers is made the dividend; and the sum of the residues, a subtractive quantity; and the statement is Dividend 15 / Divisor 63 Subtractive 21. Or reduced to least terms Dividend 5 / Divisor 21 Subtractive 7.

Proceeding as before, the multiplier is found 14.*

are found 13 and 41. The minutes are therefore 41; and the remainder of degrees 13. This again being the subtractive quantity, 13; the dividend, 30; and the divisor, 19; the multiplier and quotient are 15 and 23. The degrees then are 23; and the remainder of signs 15. The subtractive quantity then being 15; the dividend 12; and the divisor 19; the multiplier and quotient are 6 and 3. Thus the signs are 3; and the remainder of revolutions 6. This becomes the subtractive quantity, 6; the dividend, 10; and divisor, 19; whence the multiplier and quotient come out 12 and 6. The revolutions therefore are 6; and the elapsed time is 12. GANG. GAN'. &c.

[1] *Sanślisht́a-cut́t́aca* or *sanślisht́a-sphut́a-cut́t́aca*, a distinct pulverizing multiplier belonging to conjunct residues.—GAN'. A multiplier (*cut́t́aca*) consequent on conjunction; one deduced from the sum of multipliers and that of remainders. SU'R.

[2] Whether two, three or more.—GAN'. on *Líl.* and CRI'SHN'. on *Víj.*

[3] See another example in the *Gólá'd'hyáya* or spherics of the astronomical portion of the *Śirómańi.* GAN'. and CRI'SHN'.

* The quotient as it comes out in this operation is not to be taken: but it is to be separately sought with the several original multipliers applied to this quantity and divided by the divisor as given. GAN'.

CHAPTER XIII.

COMBINATION.[1]

267. Rule The product of multiplication of the arithmetical series beginning and increasing by unity and continued to the number of places, will be the variations of number with specific figures: that, divided by the number of digits and multiplied by the sum of the figures, being repeated in the places of figures and added together, will be the sum of the permutations.

268. Example. How many variations of number can there be with two and eight? or with three, nine and eight? or with the continued series from two to nine? and tell promptly the several sums of their numbers.

Statement 1st Example: 2. 8.

Here the number of places is 2. The product of the series from 1 to the number of places and increasing by unity, (1, 2.) will be 2. Thus the permutations of number are found 2.

That product 2, multiplied by the sum of the figures, 10 [2 and 8] is 20; and divided by the number of digits 2, is 10. This, repeated in the places of figures [10 and added together, is 110; the sum of the numbers.
10]

Statement 2d Example: 3. 9. 8.

The arithmetical series is 1. 2. 3; of which the product is 6; and so many

[1] *Anca-pása-vyavahára* or *Anca-pásá'd'hyáya*, concatenation of digits: a mutual mixing of the numbers, as it were a rope or halter of numerals: their variations being likened to a coil. See GAŃ. and SÚR.

The subject is more fully treated in the *Gańita-caumudí* of NÁRÁYAŃA PAŃDITA.

[2] To find the number of the permutations and the sum or amount of them, with specific numbers. GAŃ. and SÚR.

are the variations of number. That, multiplied by the sum 20, is 120; which, divided by the number of digits 3, gives 40; and this repeated in the places of figures [40 and summed, makes 4440 the sum of the
40
40]
numbers.

Statement 3d Example: 2. 3. 4. 5. 6. 7. 8. 9.

The arithmetical series beginning and increasing by unity is 1. 2. 3. 4. 5. 6. 7. 8. The product gives the permutation of numbers 40320. This, multiplied by the sum of the figures 44, is 1774080; and divided by the number of terms 8, is 221760; and the quotient being repeated in the eight places of figures and summed, the total is the sum of the numbers 24639999935360.

269. Example. How many are the variations of form of the god SAMBHU by the exchange of his ten attributes held reciprocally in his several hands: namely the rope, the elephant's hook, the serpent, the tabor, the skull, the trident, the bedstead, the dagger, the arrow, and the bow:[1] as those of HARI by the exchange of the mace, the discus, the lotus and the conch?

Statement: Number of places 10.

In the same mode, as above shown, the variations of form are found 3628800.

So the variations of form of HARI are 24.

[1] SAMBHU or S'IVA is represented with ten arms, and holding in his ten hands the ten weapons or symbols here specified; and, by changing the several attributes from one hand to another, a variation may be effected in the representation of the idol: in the same manner as the image of HARI or VISHN'U is varied by the exchange of his four symbols in his four hands. The twenty-four different representations of VISHN'U, arising from this diversity in the manner of placing the weapons or attributes in his four hands, are distinguished by as many discriminative titles of the god allotted to those figures in the theogonies or *Purán'as*. It does not appear that distinct titles have been in like manner assigned to any part of the more than three millions of varied representations of S'IVA.

The ten attributes of *S'iva* are, 1st, *pás'a*, a rope or chain for binding an elephant; 2d, *ancus'a*, a hook for guiding an elephant; 3d, a serpent; 4th, *'damaru*, a tabor; 5th, a human skull; 6th, a trident; 7th, *c'hatwánga*, a bedstead, or a club in form of the foot of one; 8th, a dagger; 9th, an arrow; 10th, a bow.

270. Rule:[1] The permutations found as before, being divided by the combinations separately computed for as many places as are filled by like digits, will be the variations of number; from which the sum of the numbers will be found as before.

271. Example. How many are the numbers with two, two, one and one? and tell me quickly, mathematician their sum: also with four, eight, five, five and five; if thou be conversant with the rule of permutation of numbers.

Statement 1st Example: 2.2.1.1.

Here the permutations found as before (§ 267) are 24: First, two places are filled by like digits (2.2.); and the combinations for that number of places are 2. Next two other places are filled by like digits (1.1.); and the combinations for these places are also 2. Total 4. The permutations as before 24, divided by (4) the twofold combinations for two pairs of like digits, give 6 for the variations of number: viz. 2211, 2121, 2112, 1212, 1221, 1122.[2] The sum of the numbers is found as before 9999.[3]

Statement 2d Example: 4.8.5.5.5.

Here the permutations found as before are 120; which, divided by the combinations for three places 6, give the variations 20: viz. 48555, 84555, 54855, 58455, 55485, 55845, 55548, 55584, 45855, 45585, 45558, 85455, 85545, 85554, 54585, 58545, 55458, 55854, 54558, 58554.

The sum of the numbers comes out 1199988.[4]

272. Rule:[5] half a stanza. The series of the numbers decreasing by unity from the last[6] to the number of places, being multiplied together, will be the variations of number, with dissimilar digits.

[1] Special; being applicable when two or more of the digits are alike.

[2] The enumeration of the possible combinations is termed *prastára*.

[3] The variations 6, multiplied by the sum of the figures 6, and divided by the number of digits 4, give 9; which being repeated in four places of figures and summed makes 9999.

[4] Variations 20, multiplied by the sum of the figures 27, give 540; which, divided by the number of digits 5, makes 108: and this being repeated in five places of figures and summed, yields 1199988.

[5] To find the variations for a definite number of places with indeterminate digits. GAN'.

[6] That is, from nine. GAN'. &c.

273. Example. How many are the variations of number with any digits except cipher exchanged in six places of figures? If thou know, declare them.

The last number is nine. Decreasing by unity, for as many as are the places of figures, the statement of the series is 9. 8. 7. 6. 5. 4. The product of these is 60480.[1]

274. Rule:[2] a stanza and a half. If the sum of the digits be determinate, the arithmetical series of numbers from one less than the sum of the digits, decreasing by unity, and continued to one less than the places, being divided by one and so forth, and the quotients being multiplied together, the product will be equal to the variations of the number.

This rule must be understood to hold good, provided the sum of the digits be less than the number of places added to nine.

A compendium only has been here delivered for fear of prolixity: since the ocean of calculation has no bounds.

275. Example. How many various numbers are there, with digits standing in five places, the sum of which is thirteen? If thou know, declare them.

Here the sum of the digits less one is 12. The decreasing series from this to one less than the number of digits, divided by unity, &c. being exhibited, the statement is $\frac{12}{1}$ $\frac{11}{2}$ $\frac{10}{3}$ $\frac{9}{4}$. The product of their multiplication [$\frac{11880}{24}$] is equal to the variations of the number, 495.[3]

276. Though neither multiplier, nor divisor, be asked, nor square, nor cube, still presumptuous inexpert scholars in arithmetic will assuredly fail in [problems on] this combination of numbers.

[1] The combinations of two dissimilar digits, excluding cipher, are 72; with three, 504; with four, 3024; with five, 15120; with six 60480.

[2] To find the combinations with indeterminate digits for a definite sum and a specific number of places. GAN'.

[3] 91111, 52222, 13333; each five ways. 55111, 22333; each ten ways. 82111, 73111, 64111, 43222, 61222; each twenty ways. 72211, 53311, 44221, 44311; each thirty ways. 63211, 54211, 53221, 43321; each sixty ways. Total four hundred and ninety-five.

277. Joy and happiness is indeed ever increasing in this world for those who have *Lílávatí* clasped to their throats,[1] decorated as the members are with neat reduction of fractions, multiplication and involution, pure and perfect as are the solutions, and tasteful as is the speech which is exemplified.

[1] By constant repetition of the text. This stanza, ambiguously expressed and bearing a double import, implies a simile: as a charming woman closely embraced, whose person is embellished by an assemblage of elegant qualities, who is pure and perfect in her conduct, and who utters agreeable discourse. See GAN.

VÍJA-GAŃITA,

OR

AVYACTA-GAŃITA;

ELEMENTAL ARITHMETIC OR ALGEBRA.

CHAPTER I.

ALGORITHM or LOGISTICS.[1]

SECTION I.

INVOCATION and INTRODUCTION.

I REVERE the unapparent primary matter, which the *Sánc'hyas*[2] declare to be productive of the intelligent principle, being directed to that production by the sentient being: for it is the sole element of all which is apparent. I adore the ruling power, which sages conversant with the nature

[1] *Paricarma-trinśati;* thirty operations or modes of process. *Lílá.* c. 2, § 2.

[2] Not the followers of CAPILA, but those of PÁTANJALI. The same term *Sánc'hya,* as relating to another member of the period, signifies sages conversant with theology and the nature of soul; and, corresponding again to another member of it, the same word intends calculators and mathematicians, whose business is with *Sanc'hyá* number. Throughout the stanza the same words are employed in threefold acceptations: and, in translating it, the distinct meanings are repeated in separate members of a period: because the ambiguity of the original could not be preserved by a version of it as of a single sentence.

of soul pronounce to be the cause of knowledge, being so explained by a holy person: for it is the one element of all which is apparent. I venerate that unapparent computation, which calculators affirm to be the means of comprehension, being expounded by a fit person: for it is the single element of all which is apparent.

2. Since the arithmetic of apparent [or known] quantity, which has been already propounded in a former treatise, is founded on that of unapparent [or unknown] quantity; and since questions to be solved can hardly be understood by any, and not at all by such as have dull apprehensions, without the application of unapparent quantity; therefore I now propound the operations of analysis.[1]

[1] *Víja* cause, origin; primary cause *(ádi-cárańa)*.—SU'R. Hence signifying in mathematics, analysis, algebra.

Víja-criyá: operation of analysis; elemental or algebraic solution. See explanation of the title of *Víja-gańita*, causal or elemental arithmetic, ch. 7, § 174.

SECTION II.

Logistics of Negative and Affirmative Quantities.

ADDITION.

3. Rule for addition of affirmative and negative quantities: half a stanza. In the addition of two negative or two affirmative[1] quantities, the sum must be taken: but the difference of an affirmative and a negative quantity is their addition.[2]

4. Example. Tell quickly the result of the numbers three and four, negative or affirmative, taken together: that is, affirmative and negative, or both negative or both affirmative, as separate instances: if thou know the addition of affirmative and negative quantities.

The characters, denoting the quantities known and unknown,[3] should be first written to indicate them generally; and those, which become negative, should be then marked with a dot over them.

Statement: 3. 4. Adding them, the sum is found 7.

Statement: $\dot{3}$. $\dot{4}$. Adding them, the sum is $\dot{7}$.

Statement: 3. $\dot{4}$. Taking the difference, the result of addition comes out $\dot{1}$.

[1] *Rĭńa* or *cshaya*, minus; literally debt or loss: negative quantity.

D'hana or *swa*, plus; literally wealth or property: affirmative or positive quantity.

[2] For a demonstration of the rule, the commentators, SU'RYADA'SA and CRĬSHŃA, exhibit familiar examples of the comparison of debts and assets.

[3] *Rási*, quantity, is either *vyacta*, absolute, specifically known, (which is termed *rúpa*, form, species;) or it is *avyacta*, indistinct, unapparent, unknown (*ajnyáta*). It may either be a multiple of the arithmetical unit, or a part of it, or the unit itself. See CRĬSHŃA.

Statement: $\dot{3}$. 4. Taking the difference, the result of addition is 1.

So in other instances,[1] and in fractions[2] likewise.

SUBTRACTION.

5. Rule for subtraction of positive and negative quantities: half a stanza. The quantity to be subtracted being affirmative, becomes negative; or, being negative, becomes affirmative: and the addition of the quantities is then made as above directed.[3]

6. Example: half a stanza. Subtracting two from three, affirmative from affirmative, and negative from negative, or the contrary, tell me quickly the result.

[1] For the addition of unknown and compound quantities and surds, see § 18—30.

[2] Whether known or unknown quantities having divisors. Of such as have like denominators, the sum or difference is taken. Else, other previous operations take place for the reduction of them to a common denominator. The same must be understood in subtraction. CRĬSHN'.

[3] In demonstrating this rule, the commentator CRĬSHN'A BHAT'T'A observes, that 'here negation is of three sorts, according to place, time, and things. It is, in short, contrariety. Wherefore the *Lílávatí*, § 166, expresses "The segment is negative, that is to say, is in the contrary direction." As the west is the contrary of east; and the south the converse of north. Thus, of two countries, east and west, if one be taken as positive, the other is relatively negative. So when motion to the east is assumed to be positive, if a planet's motion be westward, then the number of degrees equivalent to the planet's motion is negative. In like manner, if a revolution westward be affirmative, so much as a planet moves eastward, is in respect of a western revolution negative. The same may be understood in regard to south and north, &c. That prior and subsequent times are relatively to each other negative, is familiarly understood in reckoning of days. So in respect of chattels, that, to which a man bears the relation of owner, is considered as positive in regard to him: and the converse [or negative quantity] is that to which another person has the relation of owner. Hence so much as belongs to *Yajnyadatta* in the wealth possessed by *Dévadatta*, is negative in respect of *Dévadatta*. The commentator gives as an example the situation of *Pattana* (Patna) and *Prayága* (Allahábád) relatively to *Ananda-vana* (Benares). *Pattana* on the Ganges bears east of *Váránaśí*, distant fifteen *yójanas*; and *Prayága* on the confluence of the *Gangá* and *Yamuná*, bears west of the same, distant eight *yójanas*. The interval or difference is twenty-three *yójanas*; and is not obtained but by addition of the numbers. Therefore, if the difference between two contrary quantities be required, their sum must be taken. CRĬSHN'.

Statement: 3. 2. The subtrahend, being affirmative, becomes negative; and the result is 1.

Statement: $\dot{3}$. $\dot{2}$. The negative subtrahend becomes affirmative; and the result is $\dot{1}$.

Statement: 3. $\dot{2}$. The negative subtrahend becomes affirmative; and the result is 5.

Statement: $\dot{3}$. 2. The affirmative subtrahend becomes negative; and the result is $\dot{5}$.

MULTIPLICATION.

7. Rule for multiplication [and division] of positive and negative quantities: half a stanza. The product of two quantities both affirmative, is positive.[1] When a positive quantity and a negative one are multiplied together, the product is negative.[2] The same is the case in division.

[1] The sign only of the product is taught. All the operations upon the numbers are the same which were shown in simple arithmetic (*Lílá.* § 14—16). CRĬSHN.

[2] Multiplication, as explained by the commentators,* is a sort of addition resting on repetition of the multiplicand as many times as is the number of the multiplicator. Now a multiplicator is of two sorts, positive or negative. If it be positive, the repetition of the multiplicand, which is affirmative or negative, will give correspondently an affirmative or negative product. The multiplication then of positive quantities is positive; and that of a negative multiplicand by a positive multiplier is negative: as is plain. The question for disquisition concerns a negative multiplier. It has been before observed that negation is contrariety. A negative multiplier, therefore, is a contrary one: that is, it makes a contrary repetition of the multiplicand. Such being the case, if the multiplicand be positive, (the multiplier being negative), the product will be negative; if the multiplicand be negative, the product will be affirmative. In the latter case the multiplication of two negative quantities gives an affirmative product. In the middle instances, either of the two (multiplicator or multiplicand) being positive, and the other negative, the product is negative: as is taught in the text.

Or the proof may be deduced from the process of computation. There is no dispute respecting the multiplication of affirmative quantities: but the discussion arises on that of negative quantity. Now so much at least is known and admitted, that, the multiplicand being separately multiplied by component parts of the multiplier, and the products added together, the sum is the product of the proposed multiplication. Let the multiplicand be 135, and the multiplicatcr 12; and its two parts, (the one arbitrarily assumed, the other equal to the given number less the assumed one,)

* SÚRYADÁSA on *Lílávatí.* GAŃÉŚA on the same. CRĬSHNA-BHAṬṬA on *Vija-ganita.*

8. Example: half a stanza (completing § 6). What is the product of two multiplied by three, positive by positive; and negative by negative; or positive by negative?

Statement: 2. 3. Affirmative multiplied by affirmative is affirmative. Product 6.

Statement: $\dot{2}$. $\dot{3}$. Negative multiplied by negative is positive. Product 6.

Statement: 2. $\dot{3}$. [or $\dot{2}$. 3.] Positive multiplied by negative [or, negative by positive] is negative. Product $\dot{6}$.

The result is the same, if the multiplicator be multiplied by the multiplicand.[1]

DIVISION.

Rule. The same is the case in division (§ 7).[2]

9. Example. The number eight being divided by four, affirmative by affirmative, negative by negative, positive by negative, or negative by positive, tell me quickly, what is the quotient, if thou well know the method.

4 and 8. Then the multiplicand being separately multiplied by those component parts of the multiplicator, give 540 and 1080: which, added together, make the product 1620. In like manner let the assumed portion be $\dot{4}$. The other, (or given number less that) will be 16. Here also, if the multiplicand be separately multiplied by those parts, and the products added together, the same aggregate product should be obtained. But the multiplicand, multiplied severally by those parts, gives 540 and 2160. The sum of these numbers [with the same signs] does not agree with the product of multiplication. It follows therefore, since the right product is not otherwise obtained, that the multiplication of a positive and a negative quantity together give a negative result. For so the addition of 540 and 2160 [with contrary signs] makes the product right: viz. 1620. CRĬSHN'.

[1] It is thus intimated, that either quantity may at pleasure be treated as multiplicator, and the other as multiplicand: and conversely. CRĬSHN'.

[2] If both the dividend and the divisor be affirmative, or both negative, the quotient is affirmative: but, if one be positive and the other negative, the quotient is negative. CRĬSHN'.

Statement: 8. 4. Affirmative divided by affirmative gives an affirmative quotient 2.

Statement: $\dot{8}$. $\dot{4}$. Negative by negative gives an affirmative quotient 2.

Statement: 8. $\dot{4}$. Positive by negative gives a negative quotient $\dot{2}$.

Statement: $\dot{8}$. 4. Negative by positive gives a negative quotient $\dot{2}$.

SQUARE AND SQUARE-ROOT.

10. Rule: half a stanza. The square of an affirmative or of a negative quantity is affirmative; and the root of an affirmative quantity is two-fold, positive and negative. There is no square-root of a negative quantity: for it is not a square.[1]

11. Example. Tell me quickly, friend, what is the square of the number three positive; and of the same negative? Say promptly likewise what is the root of nine affirmative and negative, respectively?

Statement: 3. $\dot{3}$. Answer: the squares come out 9 and 9.

Statement: 9. Answer: the root is 3 or $\dot{3}$.

Statement: $\dot{9}$. Answer: there is no root, since it is no square.

[1] For, if it be maintained, that a negative quantity may be a square, it must be shown what it can be a square of. Now it cannot be the square of an affirmative quantity: for a square is the product of the multiplication of two like quantities; and, if an affirmative one be multiplied by an affirmative, the product is affirmative. Nor can it be the square of a negative quantity: for a negative quantity also, multiplied by a negative one, is positive. Therefore we do not perceive any quantity such, as that its square can be negative. CRISHN.

SECTION III.

CIPHER.

12. Rule for addition and subtraction of cipher: part of a stanza. In the addition of cipher, or subtraction of it, the quantity,[1] positive or negative, remains the same. But, subtracted from cipher, it is reversed.[2]

13. Example: half a stanza. Say what is the number three, positive, or [the same number] negative, or cipher, added to cipher, or subtracted from it?[3]

Statement: 3. 3̇. 0. These, having cipher added to, or subtracted from, them, remain unchanged: 3. 3̇. 0.[4]

Statement: 3. 3̇. 0. Subtracted from cipher, they become 3̇. 3. 0.[5]

[1] Whether absolute, expressed by digits, or unknown, denoted by letter, colour, &c. or an irrational and surd root. Crĭshń.

[2] In both cases of addition, and in the first of subtraction, the absolute number, unknown quantity, or surd, retains its sign, whether positive or negative. In the other case of subtraction, the sign is reversed. Crĭshń.

[3] Or having cipher added to, or subtracted from, it. Crĭshń.

[4] In addition, if either of the quantities be increased or diminished, the result of the addition is just so much greater or less. If then either be reduced to nothing, the other remains unchanged. But subtraction diminishes the proposed quantity by so much as is the amount of the subtrahend; and, if the subtrahend be reduced, the result is augmented: if it be reduced to nought, the result rises to its maximum; the amount of the proposed quantity. Or, if the proposed quantity be itself reduced, the result of the subtraction is diminished accordingly: if reduced to nought, the result is diminished to its greatest degree; the amount of the subtrahend with the subtractive sign. See Crĭshń.

[5] Cipher is neither positive nor negative: and it is therefore exhibited with no distinction of sign. No difference arises from the reversing of it; and none is here shown. Crĭshń.

14. Rule: (completing the stanza, § 12.) In the multiplication and the rest of the operations[1] of cipher, the product is cipher; and so it is in multiplication by cipher: but a quantity, divided by cipher, becomes a fraction the denominator of which is cipher.[2]

15. Example: half a stanza. Tell me the product of cipher multiplied by two;[3] and the quotient of it divided by three, and of three divided by cipher; and the square of nought; and its root.

Statement: Multiplicator 2. Multiplicand 0. Product 0.

[Statement: Multiplicator 0. Multiplicand 2. Product 0[4].]

Statement: Dividend 0. Divisor 3. Quotient 0.

Statement: Dividend 3. Divisor 0. Quotient the fraction $\frac{3}{0}$.

This fraction, of which the denominator is cipher, is termed an infinite quantity.[5]

[1] Multiplication, division, square and square-root. SÚR. and CRĬSHṆ́.

Multiplication and division are each two-fold: viz. multiplication of nought by a quantity; or the multiplication of this by nought: so division of cipher by a quantity; and the division of this by cipher. But square and square-root are each single. CRĬSHṆ́.

[2] The more the multiplicand is diminished, the smaller is the product; and, if it be reduced in the utmost degree, the product is so likewise: now the utmost diminution of a quantity is the same with the reduction of it to nothing: therefore, if the multiplicand be nought, the product is cipher. In like manner, as the multiplier decreases, so does the product; and, if the multiplier be nought, the product is so too. In fact multiplication is repetition: and, if there be nothing to be repeated, what should the multiplicator repeat, however great it be?

So, if the dividend be diminished, the quotient is reduced: and, if the dividend be reduced to nought, the quotient becomes cipher.

As much as the divisor is diminished, so much is the quotient raised. If the divisor be reduced to the utmost, the quotient is to the utmost increased. But, if it can be specified, that the amount of the quotient is so much, it has not been raised to the utmost: for a quantity greater than that can be assigned. The quotient therefore is indefinitely great, and is rightly termed infinite. CRISHṆ́.

[3] Or else multiplying two. CRĬSHṆ́.

[4] CRĬSHṆ́.

[5] *Ananta-rási*, infinite quantity. *C'ha-hara*, fraction having cipher for its denominator.

This fraction, indicating an infinite quantity, is unaltered by addition or subtraction of a finite quantity. For, in reducing the quantities to a common denominator, both the numerator and

16. In this quantity consisting of that which has cipher for its divisor, there is no alteration, though many be inserted or extracted; as no change takes place in the infinite and immutable GOD, at the period of the destruction or creation of worlds, though numerous orders of beings are absorbed or put forth.

Statement: 0. Its square 0. Its root 0.

denominator of the finite quantity, being multiplied by cipher, become nought: and a quantity is unaltered by the addition or subtraction of nought. The numerator of the infinite fraction may indeed be varied by the addition or subtraction of a finite quantity, and so it may by that of another infinite fraction: but whether the finite numerator of a fraction, whose denominator is cipher, be more or less, the quotient of its division by cipher is alike infinite. CRĬSHN'.

This is illustrated by the same commentator through the instance of the shadow of a gnomon, which at sunrise and sunset is infinite; and is equally so, whatever height be given to the gnomon, and whatever number be taken for radius, though the expression will be varied. Thus, if radius be put 120; and the gnomon be 1, 2, 3, or 4; the expression deduced from the proportion, as sine of sun's altitude is to sine of zenith distance; so is gnomon to shadow; becomes $\frac{120}{0}$, $\frac{240}{0}$, $\frac{360}{0}$ or $\frac{480}{0}$. Or, if the gnomon be, as it is usually framed, 12 fingers, and radius be taken at 3438, 120, 100, or 90; the expression will be $\frac{41256}{0}$, $\frac{1440}{0}$, $\frac{1200}{0}$ or $\frac{1080}{0}$; which are all alike infinite. See CRĬSHN'.

SECTION IV.

Arithmetical Operations on Unknown Quantities.

17. "So much as" and the colours "black, blue, yellow and red,"[1] and others besides these, have been selected by venerable teachers for names of values[2] of unknown quantities, for the purpose of reckoning therewith.[3]

18. Rule for addition and subtraction: Among quantities so designated, the sum or difference of two or more which are alike must be taken: but such as are unlike,[4] are to be separately set forth.

19. Example. Say quickly, friend, what will affirmative one unknown with one absolute, and affirmative pair unknown less eight absolute, make, if addition of the two sets take place? and what will they make, if the sum be taken inverting the affirmative and negative signs?[5]

Statement: *ya* 1 *ru* 1 Answer: the sum is *ya* 3 *ru* $\dot{7}$.
ya 2 *ru* $\dot{8}$

[1] *Yávat-távat,* correlatives, quantum, tantum; quot, tot: as many, or as much, of the unknown, as this coefficient number. *Yávat* is relative of the unknown; and *távat* of its coefficient. The initial syllables of the *Sanscrit* terms enumerated in the text are employed as marks of unknown quantities; viz. *yá, cá, ní, pí, ló,* (also *ha, śwé, chi,* &c. for green, white, variegated and so forth). Absolute number is denoted by *rú*, initial of *rúpa* form, species. The letters of the alphabet are also used (ch. 6), as likewise the initial syllables of the terms for the particular things (§ 111).

[2] *Mána, miti, unmána* or *unmiti,* measure or value. See note on § 130.

[3] For the purpose of reckoning with unknown quantities. Súr. and Crĭshń.

[4] Heterogeneous: as *rúpa*, known or absolute number: *yávat-távat* (so much as) the first unknown quantity, its square, its cube, its biquadrate, and the product of it and another factor; *cálaca* (black) the second unknown quantity, its powers, and the product of it with factors: *nílaca* (blue) the third unknown, its powers, and so forth. See Crĭshń.

[5] Inverting the signs of the first set, of the second, or of both. Crĭshń.

Statement (inverting the signs in the first set): *ya* $\dot{1}$ *ru* $\dot{1}$
ya 2 *ru* $\dot{8}$

Answer: Sum *ya* 1 *ru* $\dot{9}$.

Statement (inverting the signs in the second set): *ya* 1 *ru* 1
ya $\dot{2}$ *ru* 8

Answer: Sum *ya* $\dot{1}$ *ru* 9.

Statement (inverting the signs in both sets): *ya* $\dot{1}$ *ru* $\dot{1}$
ya $\dot{2}$ *ru* 8

Answer: Sum *ya* $\dot{3}$ *ru* 7.

20. Example. Say promptly what will affirmative three square of an unknown, with three known, be, when negative pair unknown is added? and tell the remainder, when negative six unknown with eight known is subtracted from affirmative two unknown.

Statement: *ya v* 3 *ya* 0 *ru* 3[1] Answer: Sum *ya v* 3 *ya* $\dot{2}$ *ru* 3.
ya v 0 *ya* $\dot{2}$ *ru* 0

Statement: *ya* 2 *ru* 0 Answer: The remainder is *ya* 8 *ru* 8.
ya $\dot{6}$ *ru* 8

21. Rule for multiplication of unknown quantities: two and a half stanzas. When absolute number and colour (or letter) are multiplied one by the other, the product will be colour (or letter).[2] When two, three or more

[1] The powers of the unknown quantity are thus ordered: first the highest power, for example the sursolid; then the next, the biquadrate; after it the cube; then the square; next the simple unknown quantity; lastly the known species. See CRI'SHN'.

[2] Multiplication of unknown quantity denoted by colour (or letter) is threefold: namely, by known or absolute number, by homogeneous colour or like quantity, and by heterogeneous colour or unlike quantity. If the unknown quantity be multiplied by absolute number, or this by the unknown quantity, the result of the multiplication in figures is set down, and the denomination

homogeneous quantities are multiplied together, the product will be the square, cube or other [power] of the quantity. But, if unlike quantities be multiplied, the result is their *(bhávita)* 'to be' product or factum. The other operations, division and the rest,[1] are here performed like those upon number, as taught in arithmetic of known quantities.

22. The multiplicand is to be set down in as many several places as there are terms in the multiplier, and to be successively multiplied by those terms, and the products to be added together by the method above shown.[2] In this elemental arithmetic the precept for multiplying by component parts of the factor, as delivered under simple arithmetic,[3] must be understood in the multiplication of unknown quantities, of squares, and of surds.

23. Example. Tell directly, learned sir, the product of the multiplication of the unknown *(yávat-távat)* five, less the absolute number one, by the unknown *(yávat-távat)* three joined with the absolute two: and also

of the colour is retained. The continued multiplication of like quantities produces, when two are multiplied together, the square; when this is multiplied by a third such, the cube; by a fourth, the biquadrate; by a fifth, the sursolid; by a sixth, the cube of the square, or square of the cube. When heterogeneous colours, or dissimilar unknown quantities, are multiplied together, the result is a *(bhávita)* product or factum. CRĬSHN'.

Bhávita, future, or to be. It is a special designation of a possible operation, indicating the multiplication of unlike quantities. SU'R.

Like the rest of these algebraic terms, it is signified by its initial syllable *(bhá)*. Thus the product of two unknown quantities is denoted by three letters or syllables, as *yá. cá bhá*, *cá. ní bhá*, &c. Or, if one of the quantities be a higher power, more syllables or letters are requisite: for the square, cube, &c. are likewise denoted by initial syllables, *va*, *gha*, *va-va*, *va-gha* (or *gha-va*), *gha-gha*, &c. Thus *yá va. cá gha bhá* will signify the square of the first unknown quantity multiplied by the cube of the second.

A dot is, in some copies of the text and its commentaries, interposed between the factors, without any special direction, however, for this notation.

[1] Viz. square and square-root; cube and cube-root.—CRĬSHN'. Also reduction of fractions to a common denominator, rule of three, progression, mensuration of plane figure, and the whole of what is taught in simple arithmetic. SU'R.

[2] In § 18.

[3] As well as the other methods there taught.—CRĬSHN'. See *Lílávatí*, § 14—15.

the result of their multiplication inverting the affirmative and negative signs in the multiplicand, or in the multiplicator, or in both.[1]

Statement: *ya* 5 *ru* $\dot{1}$
ya 3 *ru* 2
Product: *ya v* 15 *ya* 7 *ru* $\dot{2}$.

Statement: *ya* $\dot{5}$ *ru* 1
ya 3 *ru* 2
Product: *ya v* $\dot{15}$ *ya* $\dot{7}$ *ru* 2.

Statement: *ya* 5 *ru* $\dot{1}$
ya $\dot{3}$ *ru* $\dot{2}$
Product: *ya v* $\dot{15}$ *ya* $\dot{7}$ *ru* 2.

[Statement: *ya* $\dot{5}$ *ru* 1
ya $\dot{3}$ *ru* $\dot{2}$
Product: *ya v* 15 *ya* 7 *ru* $\dot{2}$.][2]

24. Rule for division: Those colours or unknown quantities, and absolute numbers, by which the divisor being multiplied, the products in their several places subtracted from the dividend exactly balance it,[3] are here the quotients in division.

Example. Statement of the product of the foregoing multiplication, and of its multiplicator now taken as divisor: *ya v* 15 *ya* 7 *ru* $\dot{2}$. It is dividend; and the divisor is *ya* 3 *ru* 2.

Division being made, the quotient found is the original multiplicand *ya* 5 *ru* $\dot{1}$.

[1] The concluding passage is read in three different ways; the one implying, that the multiplicand, affirmative and negative, is to be inverted, or the multiplicator; the second indicating, that the terms of the multiplicand or multiplicator with their signs are to be transposed; the third signifying, that the terms of the multiplicand or multiplier must have their signs changed.—CRÍSHN'. The commentator prefers the reading and interpretation by which the signs only are reversed.

[2] This fourth example is exhibited by CRÍSHN'A-BHAT'T'A.

Multiplication is thus wrought according to the commentator, Example 1st,

ya 5 *ru* $\dot{1}$	*ya* 3	*ya v* 15 *ya* $\dot{3}$
ya 5 *ru* $\dot{1}$	*ru* 2	*ya* 10 *ru* $\dot{2}$
		ya v 15 *ya* 7 *ru* $\dot{2}$

[3] Exhaust it: leave no residue.

Statement of the second example: *ya v* $\dot{15}$ *ya* $\dot{7}$ *ru* 2 dividend, the divisor being *ya* 3 *ru* 2. Answer: The quotient found is the original multiplicand *ya* $\dot{5}$ *ru* 1.

Statement of the third example: *ya v* $\dot{15}$ *ya* $\dot{7}$ *ru* 2 dividend, with divisor *ya* $\dot{3}$ *ru* $\dot{2}$. The quotient comes out *ya* 5 *ru* $\dot{1}$, the original multiplicand.

[Statement of the fourth example: *ya v* 15 *ya* 7 *ru* $\dot{2}$ dividend, with divisor *ya* $\dot{3}$ *ru* $\dot{2}$. Answer: *ya* $\dot{5}$ *ru* 1, the original multiplicand.]

25. Example of involution.[1] Tell me, friend, the square of unknown four less known six.

Statement: *ya* 4 *ru* $\dot{6}$. Answer: The square is *ya v* 16 *ya* $\dot{48}$ *ru* 36.

26. Rule for the extraction of the square-root: Deducting from the squares which occur among the unknown quantities their square-roots, subtract from the remainder double the product of those roots two and two; and, if there be known quantities, finding the root of the known number,[2] proceed with the residue in the same manner.[3]

Example. Statement of the square before found, now proposed for extraction of the root: *ya v* 15 *ya* $\dot{48}$ *ru* 36.

Answer: The root is *ya* 4 *ru* $\dot{6}$ or *ya* $\dot{4}$ *ru* 6.

[1] The square being the product of the multiplication of two like quantities, involution is comprehended under the foregoing rule of multiplication, § 21; and therefore an example only is here given. CRÍSHN.

[2] If the absolute number do not yield a square-root, the proposed quantity was not an exact square. CRÍSHN.

[3] When the terms balance without residue, those roots together constitute the root of the proposed square. CRÍSHN.

Arithmetical Oparations with several Unknown Quantities.

27. Example. " So much as" three, " black" five, " blue" seven, all affirmative: how many do they make with negative two, three, and one of the same respectively, added to or subtracted from them?

Statement: *ya* 3 *ca* 5 *ni* 7 Answer: Sum *ya* 1 *ca* 2 *ni* 6.
ya $\dot{2}$ *ca* $\dot{3}$ *ni* $\dot{1}$ Difference *ya* 5 *ca* 8 *ni* 8.

28. Example. Negative " so much as" three, negative " black" two, affirmative " blue" one, together with unity absolute: when these are multiplied by the same terms doubled, what is the result? And when the product of their multiplication is divided by the multiplicand, what will be the quotient? Next tell the square of the multiplicand, and the root of th:s square.

Statement: Multiplicand *ya* $\dot{3}$ *ca* $\dot{2}$ *ni* 1 *ru* 1. Multiplier *ya* $\dot{6}$ *ca* $\dot{4}$ *ni* 2 *ru* 2. Answer: The product is *ya v* 18 *ca v* 8 *ni v* 2 *ya. ca bh* 24 *ya. ni bh* $\dot{1}2$ *ca. ni bh* $\dot{8}$ *ru* 2.

From this product divided by the multiplicand, the original multiplicator comes out as quotient *ya* $\dot{6}$ *ca* $\dot{4}$ *ni* 2 *ru* 2.

Statement of the foregoing multiplicand for involution: *ya* $\dot{3}$ *ca* $\dot{2}$ *ni* 1 *ru* 1. Answer: The square is *ya v* 9 *ca v* 4 *ni v* 1 *ya. ca bh* 12 *ya. ni bh* $\dot{6}$ *ca ni bh* $\dot{4}$ *ya* $\dot{6}$ *ca* $\dot{4}$ *ni* 2 *ru* 1.

From this square, the square-root being extracted, is *ya* $\dot{3}$ *ca* $\dot{2}$ *ni* 1 *ru* 1 [or *ya* 3 *ca* 2 *ni* $\dot{1}$ *ru* $\dot{1}$.][1]

[1] For both these roots being squared yield the same result. CRI'SHN'.

SECTION V.

SURDS.

29. Rule for addition, subtraction, &c. of surds:[1] Term the sum of two irrationals the great[2] surd; and twice the square root of their product, the less one. The sum and difference of these reckoned like integers are so [of the original surd roots].[3] Multiply and divide a square by a square.[4]

30. But the root of the quotient of the greater irrational number divided by the less,[5] being increased by one and diminished by one; and the sum and

[1] *Caraṅí*, a surd or irrational number. One, the root of which is required, but cannot be found without residue.—Crĭshṅ. That, of which when the square-root is to be extracted, the root does not come out exact.—Gaṅ. "A quantity, the root of which is to be taken, is named *Caraṅí*." Ná'rа́yaṅa cited by Su'r. Not generally any number which does not yield an integer root: for, were it so, every such number (as 2, 3, 5, 6, &c.) must be constantly treated as irrational. It only becomes a surd, when its root is required; that is, when the business is with its root, not with the number itself. Crĭshṅ.

A surd is denoted by the initial syllable *ca*. It will be here written *c* to distinguish it from *cá* the second unknown quantity in an algebraic expression.

[2] *Mahatí*, intending *mahatí caraṅí* a great surd, being the sum of two original irrational numbers. *Laghu*, small, is by contrast the designation of the less quantity to be connected with it. The same terms, *mahatí* and *laghví*, are used in the following stanza, § 30, with a different sense, importing the greater and less original surds. See Su'r. and Crĭshṅ.

[3] The sum and difference of the quantities so denominated are sum and difference of the two original surds. Su'r. and Crĭshṅ.

[4] This is a restriction of a preceding rule concerning multiplication of irrational numbers. § 22. —Crĭshṅ. The author in this place hints the nature of surds, under colour of giving a rule for the multiplication and division of them.—Su'r. If a rational quantity and an irrational one are to be multiplied together, the rational one is previously to be raised to the square power; the irrational quantity being in fact a square. See Su'r. and Crĭshṅ.

[5] In like manner, if the less surd divided by the greater be a fraction of which the root may be found, this, with one added and subtracted, being squared and multiplied by the greater surd, will give the sum and difference of the two surds. Crĭshṅ.

remainder, being squared and multiplied by the smaller irrational quantity, are respectively the sum and difference of the two surd roots. If there be no rational square-root [of the product or quotient], they must be merely stated apart.

31. Example. Say, friend, the sum and difference of two irrational numbers eight and two: or three and twenty-seven; or seven and three; after full consideration, if thou be acquainted with the six-fold rule of surds.

Statement: *c* 2 *c* 8. Answer: Addition being made, the sum is *c* 18. Subtraction taking place, the difference is *c* 2.[1]

Statement: *c* 3 *c* 27. Answer: Sum *c* 48. Difference *c* 12.

Statement: *c* 3 *c* 7. Answer: Since their product has no root, they are merely to be stated apart: Sum *c* 3 *c* 7. Difference *c* $\dot{3}$ *c* 7.

32. Example. Multiplicator consisting of the surds two, three, and eight; multiplicand, the surd three with the rational number five: tell quickly their product. Or let the multiplier be the two surds three and twelve less the natural number five.

Statement: Multiplier *c* 2 *c* 3 *c* 8. Multiplicand *c* 3 *c* 25.

Here, to abridge the work, previously adding together two or more surds in the multiplier, or in the multiplicand, and in the divisor or in the dividend, proceed with the multiplication and division. That being done in this case, the multiplier becomes *c* 18 *c* 3. Multiplicand as before *c* 25 *c* 3. Multiplication being made, the product is found *c* 9 *c* 450 *c* 75 *c* 54.

33. Maxim. The square of a negative rational quantity will be nega-

[1] The numbers 8 and 2 added together make 10, the *mahatí* or great surd. Their product 16 yields the root 4; which doubled furnishes 8 for the *laghu*. The sum and difference of these are 18 and 2. Or by the second method, the greater irrational 8 divided by the less 2, gives 4; the root of which is 2. This augmented and diminished by 1, affords the numbers 3 and 1; whose squares are 9 and 1. These, being multiplied by the smaller irrational number, make 18 and 2, as before.

tive, when it is employed on account of a surd; and so will the root of a negative surd be negative, when it is formed on account of a rational number.[1]

Statement of the second example: Multiplicator *c* $\dot{25}$ *c* 3 *c* 12. Multiplicand *c* 25 *c* 3. Adding together two surds in the multiplier, it becomes *c* $\dot{25}$ *c* 27. The product of the multiplication is *c* $\dot{625}$ *c* 675 *c* $\dot{75}$ *c* 81. Among these the roots of *c* $\dot{625}$ and *c* 81, namely $\dot{25}$ and 9, being added together, make the natural number $\dot{16}$: and the sum, consisting in the difference, of *c* 675 and *c* $\dot{75}$, is *c* 300. The product therefore is ***ru*** $\dot{16}$ *c* 300.

Statement of the foregoing product for dividend and the multiplier for divisor: Dividend *c* 9 *c* 450 *c* 75 *c* 54. Divisor *c* 2 *c* 3 *c* 8.

Adding together two surds, the divisor becomes *c* 18 *c* 3. Then proceeding as directed (§ 24), the quotient is the original multiplicand ***ru*** 5 *c* 3.

Statement of the second example: Dividend *c* $\dot{256}$ *c* 300. Divisor *c* $\dot{25}$ *c* 3 *c* 12.

Adding together two surds, the divisor becomes *c* $\dot{25}$ *c* 27. Here also, proceeding as before, the quotient found is the original multiplicand ***ru*** 5 *c* 3.

34—35. Or the method of division is otherwise taught: Reverse the sign, affirmative or negative, of any surd chosen in the divisor; and by such altered divisor[2] multiply the dividend and original divisor, repeating the operation [if necessary] so as but one surd remain in the divisor. The surds, which constituted the dividend, are to be divided by that single remaining surd; and if the surds obtained as a quotient be such as arise from addition,

[1] This is a seeming exception to the maxim, that a negative quantity has no square-root (§ 10). But the sign belongs to the surd root not to the entire irrational quantity. When therefore a negative rational quantity is squared to bring it to the same form with a surd, with which it is to be combined, it retains the negative sign appertaining to the root: and in like manner, when a root is extracted out of a negative rational part of a compound surd, the root has the negative sign. SU'R.

[2] Or by any number which may serve for extirpating some of the terms. Since the dividend and divisor being multiplied by the same quantity, the quotient is unchanged: and the object of the rule is to reduce the number of terms by introducing equal ones with contrary signs. See SU'R.

they must be separated by the following rule for the resolution of them,[1] in such form as the questioner may desire.'[2]

36. Rule: Take component parts at pleasure of the root of a square, by which the compound surd is exactly divisible: the squares of those parts, being multiplied by the former quotient,[3] are severally the component surds.[4]

Statement: Dividend *c* 9 *c* 450 *c* 75 *c* 54. Divisor *c* 18 *c* 3.

Here allotting the negative sign to the surd three in the divisor, it become *c* 18 *c* $\dot{3}$. Multiplying by this the dividend, and adding the surds together,[5] the dividend is *c* 5625 *c* 675. In like manner, the divisor becomes *c* 225. The dividend being divided by this, the quotient is *c* 25 *c* 3.

Example 2d. Dividend *c* 300 *c* $\dot{256}$. Divisor *c* $\dot{25}$ *c* 27.

Here assigning to the surd twenty-five the affirmative sign, multiplying the dividend, and taking the difference of affirmative and negative surds, the dividend is *c* 100 *c* 12; and the divisor *c* 4. Dividing the dividend by this, the quotient is *c* 25 *c* 3.

[1] *Viślésha-sútra*, a rule for an operation converse of that of addition: (§ 36; which compare with § 30.)

[2] They must be resolved into such portions as the nature of the question may require.

[3] By former quotient, that which is previously found under this rule is meant: the quotient of the surd by a square which measures it. See SU'R.

[4] This rule, reversing the operations directed by § 30, is the converse of that rule.—See SU'R. However, to make the contrast exact, the root of the square divisor of the surd should be resolved into parts one of which should be unity.

[5] The dividend, multiplied by the altered divisor which comprises two terms, gives the product

c 162 *c* 8100 *c* 1350 *c* 972

c $\dot{162}$ *c* $\dot{27}$ *c* $\dot{1350}$ *c* $\dot{225}$

Expunging like quantities with contrary signs, the product is *c* 8100 *c* 972 *c* $\dot{225}$ *c* $\dot{27}$; and adding together the first and third terms, and second and fourth, (that is, taking their differences by § 29—30) the product is reduced to two terms *c* 5625 *c* 675.

Again the original divisor, multiplied by the altered one, gives *c* 324 *c* 54 Expunging equal
c $\dot{9}$ *c* $\dot{54}$

quantities with contrary signs, the product is *c* 324 *c* $\dot{9}$; reducible by addition (that is, by finding the difference, § 30) to *c* 225.

The reduced dividend *c* 5625 *c* 675, divided by this divisor *c* 225, gives the quotient *c* 25 *c* 3.

In like manner, by this process in the last example, the dividend becomes *c* 8712 *c* 1452; and the divisor *c* 184. Whence the quotient *c* 18 *c* 3; resolvable by § 36 into *c* 2 *c* 8. *c* 3. SU'R.

Next in the former example, making the multiplicand a divisor, the statement is dividend *c* 9 *c* 450 *c* 75 *c* 54. Divisor *c* 25 *c* 3.

Here also, assigning to the surd three the negative sign, multiplying the dividend, and adding surds together, the dividend becomes *c* 8712 *c* 1452; and the divisor *c* 484. The dividend being divided by this, the quotient is the multiplier *c* 18 *c* 3. The original multiplicator comprised three terms. The compound surd *c* 18 (under the rule for the resolution of such: § 36) being divided therefore by the square nine, gives the quotient 2 without remainder. The square root of nine is 3. Its parts 1 and 2. Their squares 1 and 4. These, multiplied by the quotient 2, make 2 and 8. Thus the original multiplicator is again found *c* 2 *c* 8 *c* 3.

37—38. Examples of involution. Tell me, promptly, learned friend, the square of the three surds two, three, and four; that of two surds numbering two and three; and separately that of the united irrationals six, five, two, and three; as well as of eighteen, eight, and two: and the square roots of the squared numbers.

Statement 1st. *c* 2 *c* 3 *c* 5. And 2d. *c* 3 *c* 2. Also 3d. *c* 6 *c* 5 *c* 2 *c* 3. Likewise 4th. *c* 18 *c* 8 *c* 2.

Proceeding by the rule of involution[1] (*Lílávatí*, § 18—19) the squares are found, 1st. *ru* 10 *c* 24 *c* 40 *c* 60. 2d. *ru* 5 *c* 24. 3d. *ru* 16 *c* 120 *c* 72 *c* 60 *c* 48 *c* 40 *c* 24. Here also, to abridge the work, surds are to be added together when practicable, whether in squaring, or in extraction of the square root. Thus 4th. *c* 18 *c* 2 *c* 8. The sum of these is *c* 72. Its square is *ru* 72.

39—40. Rule for finding the square root: From the square of the rational numbers contained in the proposed square, subtract integer numbers[2] equal to one, two, or more of its surds; the square root of the re-

[1] With this difference however, that instead of twice the multiple of rational quantities, four times the multiple of irrational numbers is to be taken: under the rule, that a square is to be multiplied by a square (§ 29). SU'R.

[2] A rational number equal to the numbers that express the irrational terms is subtracted: and the author therefore says "subtract integer numbers (*rúpa*) equal to one or more surds," to indicate, that subtraction as of surds (§ 29) is not here intended. SU'R.

mainder is to be severally added to, and subtracted from, the rational number: the moieties of this sum and difference will be two surds in the root. The largest of them is to be used as a rational number, if there be any surds in the square remaining; and the operation repeated [until the proposed quantity be exhausted].[1]

Example. Statement of the second square, for the extraction of its root: *ru* 5 *c* 24.

Subtracting from 25, which is the square of the rational number (5) a number equal to that of the surd 24, the remainder is 1. Its square-root 1, added to, and subtracted from, the natural number 5, makes 6 and 4. The moieties of which are 3 and 2, and the surds composing the root are found, *c* 3 *c* 2.

Statement of the first square: *ru* 10 *c* 24 *c* 40 *c* 60.

From the square of the rational number (10) viz. 100, subtract numbers equal to two of the surds twenty-four and forty; the remainder is 36; and its square-root 6, subtracted from the natural number 10, and added to it, makes 4 and 16; the moieties of which are 2 and 8. The first is a surd in the root, *c* 2. Putting the second for a rational number, the same operation is again to be performed with the rest of the surds. From the square of this then treated as a rational number, 64, subtracting the number sixty, the remainder is 4; and its root 2; which, subtracted from that rational number, and added to it, severally makes 6 and 10; the moieties whereof are 3 and 5. They are surds in the root: *c* 3 *c* 5. Statement of the whole of the surds composing the root, in their order as found; *c* 2 *c* 3 *c* 5.

Statement of the third square: *ru* 16 *c* 120 *c* 72 *c* 60 *c* 48 *c* 40 *c* 24.

[1] From the involution of surds as above shown, it is evident, that the rational number is the sum of the numbers of the original surds: and the irrationals in the square are four times the product of the original terms, two and two. If they be subtracted from the square of the sum of the numbers, the remainder will be the square of the difference. Its square-root is the difference itself. From the sum and difference, the numbers are found by the rule of concurrence (*Líláratí*, § 55). The least [or sometimes the greatest] of the numbers thus found is one of the original terms; and the greater [or sometimes the less] number is the sum of the remaining irrational terms: it is used, therefore, as the rational number, in repeating the operation; and so on, until all the terms of the root are extracted. SU'R.

From the square of the rational number (16) 256, subtracting numbers equal to three surds, a hundred and twenty, seventy-two and forty-eight, making 240, and proceeding as before, two portions are found, 6 and 10. Again, from the square of the latter as a rational number, 100, subtract numbers equal to two surds twenty-four and forty, making 64, and proceed as before; two portions are found 2 and 8. Again, from the square of the latter as a rational number 64, subtract a number equal to the surd sixty; two more portions are found 3 and 5. Hence statement of the surds composing the root, in order as found, *c* 6 *c* 2 *c* 3 *c* 5.

Statement of the fourth square: *ru* 72 *c* 0.

Its square-root *c* 72. This surd-root originally consisted of three terms. Proceeding then to the resolution of it by the rule (§ 36), 72 divided by 36 gives the quotient 2. The square-root of thirty-six, 6, comprises three portions 3, 2, 1. Their squares are 9, 4 and 1; which multiplied by the former quotient (2) make 18, 8 and 2. The resolution of the surd then exhibits *c* 18 *c* 8 *c* 2.

41. Rule: If there be a negative surd-root in the square, treating that irrational quantity as an affirmative one, let the two surds in the root be found [as before]; and one of them, as selected by the intelligent calculator, must be deemed negative.[1]

42. Example. Tell me the square of the difference of the two surds three and seven; and from the square tell the root.

Statement: *c* $\dot{3}$ *c* 7 or *c* 3 *c* $\dot{7}$.

The square of either of these quantities is the same; *ru* 10 *c* $\dot{84}$.

Here treating the negative surd-root in the square as an affirmative irrational quantity, find the two surds by proceeding as before; and let either of them at pleasure be made negative. Thus the root is found *c* $\dot{3}$ *c* 7; or *c* 3 *c* $\dot{7}$.

43. Example. Let the irrational numbers two, three and five be severally affirmative, affirmative and negative; or let the positive and negative

[1] The rule is grounded on the maxim, that the square of a negative quantity is affirmative; and that there is no square-root of a negative quantity. § 10. SÚR.

signs be reversed. Tell their square; and from the square find the root; if thou know, friend, the sixfold method of surds.

Statement: *c* 2 *c* 3 *c* $\dot{5}$; or *c* $\dot{2}$ *c* $\dot{3}$ *c* 5. Their square is the same *ru* 10 *c* 24 *c* $\dot{40}$ *c* $\dot{60}$.

Here affirmative rationals equal to the negative irrationals being subtracted from the square of the rational number (10), 100, the remainder is 0. The rational number with the root added and subtracted, being halved, the surds are *c* 5 *c* 5. One is made negative *c* $\dot{5}$; and the other treated as a rational number. Statement: *ru* 5 *c* 24. Proceeding as before, the surds are found, both affirmative, *c* 3 *c* 2.

Next subtracting affirmative rationals equal to the two surds *c* 24 *c* 60, viz. 84, from the square of the rational number, and proceeding as before, the surds found are *c* 3 *c* 7. The largest of these is made negative; and, with its number taken as rational, proceeding as before, the other surds come out *c* 5 *c* 2. The greatest of these again is taken as negative, *c* $\dot{5}$.

Then, with the second example, and in the first instance, the two surds being *c* 5 *c* 5, one is taken as negative; and, its number being used as a rational one, the two portions of surds deduced from the negative one, are both negative *c* $\dot{3}$ *c* $\dot{2}$. In the second case, proceeding as directed, the surds of the root come out, *c* $\dot{2}$ *c* $\dot{3}$ *c* 5.

It might be so understood by an intelligent mathematician, though it were not specially mentioned. This matter likewise has not been explained at length by former writers. It is by me set forth, for the instruction of youth.

44—47. Rule: The number of irrational terms in the square quantity answers to the sum of the progression of the natural numbers one, &c.[1] In a square comprising three such terms, integer numbers equal to two of the terms are to be subtracted from the square of the rational number, and the

[1] The sums of the progression are for the 1st term 1; for the 2d, 3; for the 3d, 6; for the 4th, 10; for the 5th, 15.—SU'R. The rational portion of the square comprises as many terms as there were surds in the root; and the number of irrational terms in the square answers to the sum of the progression continued to one less than the number of radical terms: as the author's subsequent comment shows.

square root [of the remainder] to be then taken; in one comprising six such, integers equal to three of them; in one containing ten, integers equal to four of them; in one comprehending fifteen, integers equal to five. If in any case it be otherwise, there is error.[1] Those terms are to be subtracted from the square of the rational number, which are exactly measured[2] by four times the smaller radical surd thence to be deduced. The quotients found by that common measure are surds in the root; but, if they be not so, as not answering by the rule of remainder (§ 39)[3] that is not the root.[4]

In a square raised from irrational terms, there must necessarily be rational numbers. The square of a single surd consists of a rational number only. That of two contains one surd with a rational number; that of three comprises three; that of four comprehends six; that of five, ten; and that of six, fifteen. Hence, in the square of two, &c. terms, the number of surd terms, besides the rational numbers, answer severally to the sums of the arithmetical progressions [of natural numbers] one, &c. But, if there be not so many in the example, compound surds are to be resolved (§ 36) to complete the number of terms; and the root is then to be extracted. That is the meaning.

In a square comprising three such terms, [integer numbers equal to two, &c.] The sense of the whole passage is clear.

48. Example. Say, learned man, what is the root of a square consisting of the surds thirty-two, twenty-four, and eight, with the rational number ten?

Statement: *ru* 10 *c* 32 *c* 24 *c* 8.

Here, as the square comprises three irrational terms, first subtract integer numbers equal to two of them from the square of the rational number, and

[1] If in any supposable case an answer come out, it is not taken as the true root. It is wrong; and the question was ill proposed. SU'R.

[2] *Apavartana*, division without remainder by a common measure. § 54.

[3] By the rule for adding and subtracting the root of the remainder, &c. § 39. SU'R.

[4] As many of the irrational terms in the square, as are multiples of one of the radical irrationals, being subtracted in the first instance, they must be divisible without remainder by four times that radical term; and the quotients will be the rest of the radical terms: as is apparent from what has been said concerning the involution of a quantity consisting of surd terms. (See under § 37.) If then those quotients do not answer, as not agreeing with the terms found by the preceding (§ 39—40), the root is wrong. SU'R.

extract the root of the remainder; and afterwards work with one term. Proceeding in that manner, there is here no root. Hence it appears, that the [proposed quantity] has not an exact root consisting of surd terms. But, were it not for the restriction, a number equal to the whole of the surds might be subtracted, and a supposed root be thus found: namely, *c* 8 *c* 2. This, however, turns out wrong; for its square is *ru* 18.[1] Or summing two of the terms, thirty-two and eight, [by § 30] the expression becomes *ru* 10 *c* 72 *c* 24. Whence the root is found *ru* 2 *c* 6. But this also is wrong.[2]

49. Example. Say what will be the root of a square which contains surds equal to fifteen, eleven, and three, all multiplied by four; with the rational number ten?

Statement: *ru* 10 *c* 60 *c* 52 *c* 12.

In this square three irrational terms occur. Taking then two of them, fifty-two and twelve, and subtracting an integer equal to their amount from the square of the rational number, two surds of the root come out *c* 8 *c* 2. But four times the least of them, 8, does not measure the two terms fifty-two and twelve. These then are not to be subtracted: for the tenor of the rule (§ 46) is "Those terms are to be subtracted, which are measured by four times the smaller radical thence to be deduced." Here the rule is not rigidly restrictive to the least surd; but sometimes applies to the greater.

Putting the radical surd as a rational number, the other two irrational terms come out *c* 5 *c* 3. This too is wrong, for the square of *c* 5 *c* 3 *c* 2, is *ru* 10 *c* 24 *c* 40 *c* 60.[3]

50. Example. Say what will be the root of a square which consists of three surds eight, fifty-six, and sixty; with the rational number ten?

Statement: *ru* 10 *c* 8 *c* 56 *c* 60.

Subtracting the two first terms eight and fifty-six, and measuring those terms by four times the least surd thence deduced, 8, two terms are found 1 and 7. But these do not come out as surds of the root by the regular process of the rule of remainder (§ 39). Therefore those terms *c* 8 *c* 56, are not to be subtracted. Else the root is wrong.

[1] For the surds *c* 8 *c* 2, being added together (§ 31) make *c* 18. Its square is of course *ru* 18.

[2] For its square is *ru* 10 *c* 96.

[3] Differing from the proposed square.

51. Example. Tell the root of the square, in which are surds twelve, fifteen, five, eleven, eight, six, all multiplied by four; together with the rational number thirteen; if thou have pretensions to skill in algebra.

Statement: *ru* 13 *c* 48 *c* 60 *c* 20 *c* 44 *c* 32 *c* 24.

Here, the square comprising six surd terms, integers equal to three of them are to be first subtracted from the square of the rational term, and the root of the remainder taken; then integers equal to two; and afterwards an integer equal to one. Proceeding in this manner, no root is found. Proceeding then differently, and first subtracting from the square of the rational number, an integer equal to the first surd term; then integers equal to the second and third; and lastly equal to the rest; the root comes out *c* 1 *c* 2 *c* 5 *c* 5. This, however, is wrong; for its square is *ru* 13 *c* 8 *c* 80 *c* 160.

Defect then is imputable to those authors, who have not given a limitation to this method of finding a root.

In the case of such irrational squares, the operation must be conducted by taking the approximate roots of the surd terms, and adding them to the rational terms: whence the square root is to be deduced.[1]

Largest is not rigidly intended (§ 40). Sometimes, therefore, the least is to be used.

52. Example. Say what is the root of a square, in which are the surds forty, eighty, and two hundred, with the rational number seventeen?

Statement: *ru* 17 *c* 40 *c* 80 *c* 200.

Subtracting the two last terms from the square of the rational number, the two portions found are *c* 10 *c* 7. Again treating the smaller surd as a rational number, the result is *c* 5 *c* 2. Thus the root is *c* 10 *c* 5 *c* 2.

[1] A rule of approximation for the square-root is given in the Chapter on Algebra, in the *Sidd'hánta-sundara* of JNYÁNA-RÁJA, cited by his son SÚRYADÁSA; "The root of a near square, with the quotient of the proposed square divided by that approximate root, being halved, the moiety is a [more nearly] approximated root; and, repeating the operation as often as necessary, the nearly exact root is found." Example 5. This, divided by two which is first put for the root, gives $\frac{5}{2}$ for the quotient: which added to the assumed root 2, makes $\frac{9}{2}$; and this, divided by 2, yields $\frac{9}{4}$ for the approximate root.—SÚR. [Repeating the operation, the root, more nearly approximated, is $\frac{161}{72}$.]

CHAPTER II.

PULVERIZER.[1]

53—64. [2]Rule: In the first place, as preparatory to the investigation of the pulverizer, the dividend, divisor, and additive quantity are, if practicable, to be reduced by some number.[3] If the number, by which the dividend and divisor are both measured, do not also measure the additive quantity, the question is an ill put [or impossible] one.[4]

54—55—56. The last remainder, when the dividend and divisor are mutually divided, is their common measure.[5] Being divided by that common

[1] This is nearly word for word the same with a chapter in the *Lílávatí* on the same subject. (*Lil.* Ch. 12.) See there, explanations of the terms.

The method here taught is applicable chiefly to the solution of indeterminate problems that produce equations involving more than one unknown quantity. See ch. 6.

[2] Ten stanzas and two halves.

[3] If the dividend and divisor admit a common measure, they must be first reduced by it to their least terms; else unity will not be the residue of reciprocal division; but the common measure will; (or, going a step further, nought.)—GÁN. on *Lil.* CRĬSHN′. on *Víj.*

[4] If the dividend and divisor have a common measure, the additive also must admit it; and the three terms be correspondently reduced: for the additive, unless it be [nought or else] a multiple of the divisor, must, if negative, equal the residue of a division of the dividend taken into the multiplier by the divisor; and, if affirmative, must equal the complement of that residue to the divisor. Now, if dividend and divisor be reducible to less terms, the residue of division of the reduced terms, multiplied by the common measure, is equal to the residue of division of the unreduced terms. Therefore the additive, whether equal to the residue, or to its complement, must be divisible by the common measure. CRĬSHN′.

[5] The common measure may equal, but cannot exceed, the least of the two numbers: for it must divide it. If it be less, the greater may be considered as consisting of two terms, one the quotient taken into the divisor, the other the residue. The common measure cannot exceed that residue; for, as it measures the divisor, it must of course measure the multiple of the divisor, and

measure, they are termed reduced quantities. Divide mutually the reduced dividend and divisor, until unity be the remainder in the dividend. Place the quotients one under the other; and the additive quantity beneath them, and cipher at the bottom. By the penult multiply the number next above it, and add the lowest term. Then reject the last and repeat the operation until a pair of numbers be left. The uppermost of these being abraded by the reduced dividend, the remainder is the quotient. The other [or lowermost] being in like manner abraded by the reduced divisor, the remainder is the multiplier.[1]

could not measure the remaining portion or residue, if it were greater than it. When therefore the greater number, divided by the less, yields a residue, the greatest common measure, in such case, is equal to the remainder, provided this be a measure of the less. If again the less number, divided by the remainder, yield a residue, the common measure cannot exceed this residue; for the same reason. Therefore no number, though less than the first remainder, can be a common measure, if it exceed the second remainder: and the greatest common measure is equal to the second remainder, provided it measure the first; for then of course it measures the multiple of it, which is the other portion of the second number. So, if there be a third remainder, the greatest common measure is either equal to it, if it measure the second; or is less. Hence the rule, to divide the greater number by the less, and the less by the remainder, and each residue by the remainder following, until a residue be found, which exactly measures the preceding one; such last remainder is the common measure. (§ 54). CRĬSHN'.

[1] The substance of CRĬSHN'A's demonstration is as follows: When the dividend, taken into the multiplier, is exactly measured by the divisor, the additive must either be null or a multiple of the divisor. (See § 63). If the dividend be such, that, being multiplied by the multiplicator and divided by the divisor, it yields a residue, the additive, if negative, must be equal to that remainder; (and then the subtractive quantity balances the residue;) or, if affirmative, it must be equal to the difference between the divisor and residue; (and so the addition of that quantity completes the amount of the divisor;) or else it must be equal to the residue, or its complement, with the divisor or a multiple of the divisor added. Let the dividend be considered as composed of two portions or terms: 1st, a multiple of the divisor; 2d, the overplus or residue. The first multiplied by the multiplier (whatever this be), is of course measured by the divisor. As to the second, or overplus and remainder, the additive being negative, both disappear when the multiplier is quotient of the additive divided by the remainder, (the additive being a multiplier of the residue.) But, if the additive be not a multiple of the remainder, should unity be the residue at the first step of the reciprocal division, the multiplier will be equal to the additive, if this be negative, or to its complement to the divisor, if it be positive; and the corresponding quotient will be equal to the quotient of the dividend by the divisor multiplied by the multiplicator, if the additive be negative; or be equal to the same with addition of unity, if it be affirmative: and, generally, when reciprocal division has reached its last step exhibiting a remainder of one, the multiplier, answering to the preceding residue, become the divisor, as serving for that next before it become dividend, is equal to

57. Thus precisely is the operation when the quotients are an even number. But, if they be odd, the numbers as found must be subtracted from

the additive, if this be negative, or to its complement, if it be positive; and the corresponding quotient is equal to the quotient of the dividend by the divisor multiplied by this multiplicator; but with unity superadded, if the additive were affirmative. From this, the multiplicator and quotient answering for the original dividend and divisor are found by retracing the steps in the method of inversion. Take the following example:

Given Dividend 1211, Divisor 497, Additive 21; or, reduced to least terms, § 53 and 54, Dividend 173, Divisor 71, Additive 3.

The reciprocal division (§ 55) exhibits the following results:

Dividends.	Divisors.	Quotients.	Residues.
173	71	2	31
71	31	2	9
31	9	3	4
9	4	2	1

Consider last dividend (9) as composed of two terms; a multiple of divisor (4) and the residue; (in the instance 8 and 1). Then the multiplier is equal to (3) the additive (this being negative); and quotient is equal to the multiplier (3) taken into the quotient of the simple dividend (9) by the divisor (4): (in the instance 6). Thus, observing the directions of the rule (§ 55, 56) the last term in the series is the multiplier for the last dividend, and its product into the term next above it is the quotient of the last divisor; and the series now is

2	deduced from the series (§ 55)	2	Quot.
2		2	Quot.
3		3	Quot.
6	Quotient.	2	Quot.
3	Multiplier.	3	Add.
		0	

Hence to find the multiplier for the next superior dividend and divisor (31 and 9) consider the dividend as comprising two portions or terms; viz. 27 and 4. Any multiple of the first being divisible by the divisor (9) the multiplier is to be sought for the second portion; that is, for dividend 4 and divisor 9; being the former divisor and dividend reversed: wherefore multiplier and quotient will here be transposed; and will answer for the affirmative additive: and the series now becomes

2
2
3
6 Multiplier.
3 Quotient.

But the quotient of the first portion of the dividend (27) after multiplication by this multiplicator, will be the quotient (3) of the simple dividend taken into the multiplicator (6); which, as is apparent, is the term of the series next beneath it: to which add the quotient of the second portion, which is last term in the series, and the sum is the entire quotient (21). And the lowest term (3), being of no further use, may be now expunged: as is directed accordingly (§ 56). Thus the series now stands

2
2
21 Quotient.
6 Multiplier.

their respective abraders, the residues will be the true quotient and multiplier.[1]

The next step is to find the multiplier and quotient (the additive being still the same) for the next preceding dividend and divisor; viz. 71 and 31: and here the dividend consists of two parts 62 and 9; to the last of which only the multiplier needs to be adapted; viz. to dividend 9 and divisor 31; which again are the former divisor and dividend inverted: wherefore the multiplier and quotient are here also transposed; and the quotient of the first portion is to be added: and is the quotient (2) of the simple dividend taken into the multiplier (21) the two contiguous terms in the series. The entire quotient therefore is 48 answering to the same additive but negative: and the lowest term being no longer required may now be rejected: the series consequently exhibits

2
48 Quotient.
21 Multiplier.

Lastly, to find the multiplier and quotient for the next superior, which are the final dividend and divisor 173 and 71. Taking the dividend as composed of 142 and 31; and seeking a multiplier which will answer for the second portion 31 with the divisor 71; the multiplier and its quotient are the former transposed: and the entire quotient is completed by adding the product of the upper terms of the series, (and answers to the same additive but affirmative); after which the lowest term is of no further use: and the series is now reduced by its rejection to two terms, viz.

117 Quotient.
48 Multiplier.

Thus, according to the tenor of the rule, the work is to be repeated as many times as there are quotients of the reciprocal division; that is, until two terms remain (§ 56). In all these operations, except the first, the multiplier is last term but one in the series; and the quotient of the second portion of the dividend is the last. But, in the first operation, there is no quotient of a second portion to be added. Therefore, for the sake of uniformity in the precept, a cipher is directed to be added at the foot of the series (§ 55), that the multiplier may always be penultimate.

If the multiplier be increased by the addition of any multiple of the divisor, the corresponding quotient will be augmented by an equi-multiple of the dividend (§ 64); and, in like manner, if the multiplier be lessened by subtraction of any multiple of the divisor, the quotient is diminished by the like multiple of the dividend. Wherefore it is directed to divide the pair of numbers remaining in the series, by the dividend and divisor, and the remainders are the quotient and multiplier in their least terms. (§ 56.) CRÍSHN.

[1] The multiplier for the last dividend, being put equal to the additive, is adapted, as has been observed, (see preceding note.) to a negative additive; and thence proceeding upwards, the multiplier and quotient, which are transposed at each step, are alternately adapted to positive and negative additives; that is, at the uneven steps to a negative one; and at the even, to a positive one. If then the number of dividends, or, which is the same, that of the quotients of reciprocal division, be even, the multiplier and corresponding quotient are adapted to a positive additive; if it be odd, they are so to a negative one. In the latter case, therefore, the complement of each to the divisor and dividend respectively, is taken, to convert them into multiplier and quotient adapted to an affirmative additive. For the dividend, being multiplied by the divisor and divided by the same, has no remainder, and the quotient is equal to the dividend: therefore when it is multiplied by a

58. The multiplier is also found by the method of the pulverizer, the additive quantity and dividend being either reduced by a common measure, [or used unreduced.][1] But, if the additive and divisor be so reduced, the multiplier found, being multiplied by the common measure, is the true one.[2]

59. The multiplier and quotient, as found for an additive quantity, being subtracted from their respective abraders, answer for the same as a subtractive quantity.[3] Those deduced from an affirmative dividend, being treated in the same manner, become the results of a negative dividend.[4]

60. A half stanza. The intelligent calculator should take a like quotient [of both divisions] in the abrading of the numbers for the multiplier and quotient [sought].[5]

number less than the divisor, and separately by the complement of this multiplier to the divisor, both products being divided by the divisor, should the one have a positive remainder, the other will want just as much to complete the amount of the divisor; and the quotient of the one added to that of the other [completed] will be equal to the dividend. Wherefore, if the quotient and multiplier for a negative additive be subtracted from their respective abraders, (the dividend and divisor,) the differences will be the quotient and multiplier for a positive additive, and conversely, (§ 57 and 59). CRÍSHN'.

[1] GAN'ÉS'A on *Lílávatí.*

[2] The quotient at the same time found will be the true one.—GAN'. *on Líl.* In the former instance, the quotient as found was to be multiplied by the common measure.—*Ibid.* If the dividend and additive be abridged, while the divisor remains unchanged, it is plain, that the quotient will be an abridged one, and must be multiplied by the common measure to raise the quotient for the original numbers. In like manner, if the divisor and additive be reduced to least terms, while the dividend is retained unaltered, the multiplier thence deduced must be taken into the common measure. If separate common measures be applicable to both, viz. dividend and additive, the multiplier and quotient, as thence found in an abridged form, must be multiplied by the common measures respectively. CRÍSHN'. on *Víj.*

SU'RYADA'SA directs the multiplier alone to be found by this abbreviated method, and then to use the multiplier thence deduced for finding the quotient. See SU'R. on *Líl.*

[3] See the beginning of note ([1]) to § 56; and the note ([1]) to § 57. See also the author's remark after § 67.

A change of the sign in the dividend has the like effect on the results; and the complement of the multiplier to the divisor, and that of the quotient to the dividend, are the multiplier and quotient adapted to the dividend with an altered sign. See the sequel of this stanza. § 59.

[4] This second half of the stanza is not inserted in the *Lílávatí.* CRÍSHN'A, the commentator of the *Víja-Gan'ita,* notices with censure a variation in the reading of the text; "Those deduced from a negative dividend, being treated in the same manner, become the results of a negative divisor."

[5] The rule is applicable when the additive quantity exceeds the dividend and divisor.

61. But the multiplier and quotient may be found as before, the additive quantity being [first] abraded by the divisor; the quotient, however, must have added to it the quotient obtained in the abrading of the additive. But, in the case of a subtractive quantity, it is subtracted.

62.[1] Or the dividend and additive being abraded by the divisor, the multiplier may thence be found as before; and the quotient from it, by multiplying the dividend, adding the additive, and dividing by the divisor.[2]

63. If there be no additive quantity, or if the additive be measured by the divisor, the multiplier may be considered as cipher, and the quotient as the additive divided by the divisor.[3]

64. Half a stanza. The multiplier and quotient, being added to their

[1] This stanza, omitted in the greatest part of the collated copies of the *Lílávatí* and by most of its commentators, occurs in all copies of the *Víja-gańita,* and is noticed by the commentators of the algebraic treatise.

[2] If the divisor be contained in the additive, this is abraded by it, and the remainder is employed as a new additive (§ 61). Here the additive is composed of two portions or terms: one a multiple of the divisor; the other the remainder or new additive: from the latter the multiplier is found; such, that, multiplying the dividend by it, and adding the reduced additive, the sum, divided by the divisor, yields no remainder. The other portion of the additive, being a multiple of the divisor, of course yields none: but the quotient is increased by as many times as the divisor is contained in it, if it be positive; or reduced by as much, if it be negative.

If both dividend and additive contain the divisor, abrade both by it, and use the remainder as dividend and additive: whence find the multiplier: which will be the same as for the whole numbers: and the proof is similar, grounded on considering the dividend as composed of two portions. The quotient, however, is regularly deduced by the process at large of multiplying the dividend by the multiplier, adding the additive, and dividing by the divisor (§ 62). Or it may be deduced from the quotient that is found with the multiplier, by adding to that quotient, or subtracting from it, the sum or the difference (according as the additive was positive or negative) of the dividend taken into the multiplier and the additive, both divided by the divisor. This last mode is unnoticed by the author, being complex. CRÍSHŃ.

[3] If the additive be nought, multiply the dividend by nought, the product is nought, which being divided by the divisor, the quotient is nought, and no remainder. If the additive be a multiple of the divisor, multiply the dividend by nought, the product is nought; and the operation is confined to the division of the additive by the divisor. Being a multiple of it, there is no remainder; and the quotient of this division is the quotient sought. CRÍSHŃ.

respective [abrading] divisors multiplied by assumed numbers,[1] become manifold.[2]

65. Example. Say quickly, mathematician, what is that multiplier, by which two hundred and twenty-one being multiplied, and sixty-five added to the product, the sum divided by a hundred and ninety-five becomes cleared (giving no residue)?

Statement: Dividend 221 Additive 65.
Divisor 195

Here the dividend and divisor being divided reciprocally; the dividend, divisor and additive, reduced to their least terms by the last of the remainders 13, become Dividend 17 Additive 5.
Divisor 15

The reduced dividend and divisor being mutually divided, and the quotients put one under the other, the additive under them, and cipher at the bottom, the series which results is 1
7
5
0

Multiplying by the penult the number above it and proceeding as directed [§ 56], the two quantities obtained are 40
35

These being abraded by the reduced dividend and divisor 17 and 15, the quotient and multiplier are found 6 and 5. Or, adding to them arbitrary multiples of their abraders, the quotient and multiplier are 23, 20; or 40, 35, &c.

66. Example. If thou be expert in the investigation of such questions,

[1] To arbitrary multiples of the divisors used in abrading the pair of terms, from which they are deduced as residues of a division; in other words, multiples of the reduced dividend and divisor which had been used as divisors of the pair of terms.

[2] Additive apart, if the multiplier be equal to a multiple of the divisor, the quotient will be an equimultiple of the dividend. Wherefore, if additive be null, the multiplier is cipher (§ 63) with or without a multiple of the divisor added; and the corresponding quotient will be cipher with a like multiple of the dividend: and generally, the multiplier and quotient having been found for any given additive, dividend and divisor, equimultiples of the divisor and dividend may be respectively added to the multiplier and quotient. See CRĪSHN'.

tell me the precise multiplier, by which a hundred being multiplied, with ninety added to the product or subtracted from it, the sum or the difference may be divisible by sixty-three without a remainder.

Statement: Dividend 100
Divisor 63 Additive or subtractive 90.

Here the series is 1
1
1
2
2
1
90
0

And the quotient and multiplier found as before are 30 and 18.

Or the dividend and additive being reduced by the common measure ten, the statement is Dividend 10
Divisor 63 Additive 9.

The series is 0
6
3
9
0

And the multiplier comes out 45. The quotient is here not to be taken. As the quotients in this series are an odd number, the multiplier 45 is to be subtracted from its abrader 63; and the multiplier thus found is the same 18. The dividend being multiplied by that multiplier, and the additive quantity being added, and the sum divided by the divisor, the quotient found is 30.

Or the divisor and additive are reduced by the common measure nine: Dividend 100
Divisor 7 Additive 10. The series then is 14
3
10
0

The multiplier thence deduced is 2: which multiplied by the common measure 9, makes the same 18.

Or, the dividend and additive are reduced, and further the divisor and additive, by common measures. Dividend 10
Divisor 7 Additive 1.

Proceeding as before, the series is 1
2
1
0

Hence the multiplier is found 2;

which multiplied by the common measure of the divisor and additive (viz. 9) becomes the same 18. Whence, by multiplication and division, the quotient is found 30.

Or, adding to the quotient and multiplier arbitrary multiples of their divisors, the quotient and multiplier are 130, 81; 230, 144, &c.

67. Example. Tell me, mathematician, the multipliers severally, by which the negative number sixty being multiplied, and three being added to the product, or subtracted from it, the sum or difference may be divided by thirteen without remainder.[1]

Statement: Dividend $\dot{6}0$ Divisor 13 Additive (or subtractive) 3.

Found as before[2] for an affirmative dividend and positive additive quantity, the multiplier and quotient are 11 and 51. These, subtracted from their abraders 13, 60, give for a negative dividend and positive additive [§ 59] 2, $\dot{9}$. These again, subtracted from their abraders 13, 60, give for a negative dividend and negative additive 11, $\dot{5}1$. "Those (the multiplier and quotient) deduced from an affirmative dividend, being treated in the same manner, become results of a negative dividend." (§ 59). This has been by me specified to aid the comprehension of the dull: for it followed else from the rule, "The multiplier and quotient, as found for an additive quantity, being subtracted from their respective abraders, answer for the same as a subtractive quantity:" [ibid.] since the addition of negative and affirmative is precisely subtraction. Accordingly taking the dividend, divisor and additive as all positive, the multiplier and quotient are to be found: they are results of an additive quantity. Subtracting them from their abraders, they are to be rendered results of a negative quantity.

If either the dividend or its divisor become negative, the quotients of reciprocal division would be to be stated as negative: which is a needless trouble. Were it so done, one (either dividend or divisor) becoming negative,

[1] This stanza differs from one in the *Lílávatí* (§ 257) in the amount of the additive or subtractive quantity; and in specifying the sign of the dividend. It comprises two examples: the additive being either negative or positive.

[2] The series is 4 1 1 1 1 3 0; whence the pair of numbers 69 15: which abraded give 9 2; and, the quotients being uneven in number, they are subtracted from their abraders 60 13 and yield the quotient and multiplier 51, 11. CRÍSHN.

there would be error in the quotient [and multiplier[1]] under the last mentioned rule (§ 64).

68. Example. By what number being multiplied will eighteen, having ten added to the product, or ten subtracted from it, yield an exact quotient, being divided by the negative number eleven?[2]

Statement. Dividend 18 Additive (or subtractive) 10.
Divisor $\dot{11}$

Here the divisor being treated as affirmative, the multiplier and quotient are 8, 14. The divisor being negative, they are the same: but the quotient must be considered to have become negative, since the divisor is so; 8, $\dot{14}$. The same, being subtracted from their abraders, become the multiplier and quotient for the negative additive; 3, $\dot{4}$.

69. Example. What is the multiplier, by which five being multiplied, and twenty-three added to the product, or subtracted from it, the sum or difference may be divided by three without remainder?

Statement: Dividend 5 Additive (or subtractive) 23.
Divisor 3

Here the series is 1, 1, 23, 0 and the pair of numbers found as before is 46, 23

These are to be abraded by the dividend and divisor. The lower number being abraded by three, the quotient is seven. The upper one being so by five, the quotient would be nine. This, however, is not accepted: but, under the rule for taking a like quotient (§ 60), seven only. Thus the multiplier and quotient are found 2, 11. By the former rule (§ 59) the multiplier and quotient answering to the same as a negative quantity come out 1, $\dot{6}$. Added to arbitrary multiples of their abraders (§ 63), so as the quotient may be affirmative, the multiplier and quotient are 7, 4, &c. So in every [similar] case.

Or, applying another rule (§ 61), the statement is Divd. 5 Divr. 3 Abraded Additive 2.

[1] The error would be in the multiplier as well as the quotient. CRĬSHṄ.

[2] An example not inserted in the *Lílávatí*; being algebraic.

The multiplier and quotient hence found as before are 2, 4. These, subtracted from their respective divisors, give 1, 1; as answering to the subtractive quantity. The quotient obtained in abrading the additive being added, the result is 2, 11, answering to the additive quantity; or subtraction being made, 1, $\dot{6}$, answering to the subtractive; or (adding thereto twice the divisors, to obtain an affirmative quotient,) 7, 4.

70. Example. Tell me, promptly, mathematician, the multiplier, by which five being multiplied and added to cipher, or added to sixty-five, the division by thirteen shall in both cases be without remainder.

Statement: Dividend 5 / Divisor 13 Additive 0.

There being no additive quantity, the multiplier and quotient are 0, 0; or 13, 5.

Statement: Dividend 5 / Divisor 13 Additive 65.

By the rule (§ 63) the multiplier is cipher, and the quotient is the additive divided by the divisor, 0, 5; or 13, 10, &c.

71. Rule for a constant pulverizer:[1] Unity being taken for the additive quantity, or for the subtractive, the multiplier and quotient, which may be thence deduced, being severally multiplied by an arbitrary additive or subtractive,[2] and abraded by the respective divisors, will be the multiplier and quotient for such assumed quantity.[3]

In the first example (§ 65) the statement of the reduced dividend and divisor, with additive unity, is Dividend 17 / Divisor 15 Additive 1. Here the multiplier and

[1] A rule which is of especial use in astronomy.—CRI'SHN'. SU'R. See Algebra of BRAHMEGUPTA, § 9—12, and § 35.

[2] If the arbitrary additive be positive, the multiplier and quotient, as found for additive unity, are to be multiplied by the arbitrary affirmative additive. If it be negative, those found for subtractive unity are to be multiplied by the arbitrary subtractive, or negative additive. CRI'SHN'.

[3] The rule may be explained by that of proportion: if unity as the additive (or subtractive) quantity give this multiplier and this quotient, what will the assumed additive (or subtractive) quantity yield? CRI'SHN.

quotient are found 7, 8. These, multiplied by an assumed additive five, and abraded by the respective divisors, give for the additive 5, the multiplier and quotient 5, 6.

Next, unity being the subtractive quantity, the multiplier and quotient, thence found, are 8, 9. These, multiplied by five and abraded by their respective divisors, give 10, 11. So in every [similar] case.

Of this method of investigation great use is made in the computation of planets.[1] On that account something is here said [by way of instance.]

72. Let the remainder of seconds be made the subtractive quantity, sixty the dividend, and terrestrial days the divisor. The quotient deduced therefrom will be the seconds; and the multiplier will be the remainder of minutes. From this again the minutes and remainder of degrees are found: and so on upwards. In like manner, from the remainder of exceeding months and deficient days, may be found the solar and lunar days.[2]

[1] It is less employed in popular questions, where the dividend and divisor are variable. But, in astronomy, where additive or subtractive quantities vary, while the dividend and divisor are constant, this method is in frequent use. See CRÍSHṄ.

[2] By the rule for finding the place of a planet (*Śirómaṅi*, § 50) the whole number of elapsed days, multiplied by the revolutions in the great period *calpa*, and divided by the number of terrestrial days in a *calpa*, gives the past revolutions: the residue is the remainder of revolutions; which, multiplied by twelve and divided by terrestrial days in a *calpa*, gives the signs: the balance is remainder of signs; and multiplied by thirty, and divided by terrestrial days, gives the degrees: the overplus is remainder of degrees; and multiplied by sixty, and divided by terrestrial days, gives minutes: the surplus is remainder of minutes; and this again, multiplied by sixty, and divided by terrestrial days, gives seconds: and what remains is residue of seconds. Now, by inversion, to find the planet's place from the remainder of seconds: if the remainder of seconds be deducted from the remainder of minutes multiplied by sixty, then the difference divided by terrestrial days will yield no residue: but the remainder of minutes being unknown, its multiple by sixty is so *a fortiori:* however, remainder of minutes multiplied by sixty, and sixty multiplied by remainder of minutes, are equal; for there is no difference whether quantities be multiplicator or multiplicand to each other. Therefore sixty, multiplied by remainder of minutes, and having remainder of seconds subtracted from the product, will be exactly divisible by terrestrial days without residue; and the quotient will be seconds. Now, in the problem, sixty and the remainder of seconds [as also the terrestrial days in a *calpa*] are known: and thence to find the remainder of minutes, a multiplier is to be sought, such that sixty being multiplied by it, and the subtractive quantity (remainder of seconds) being taken from the product, the difference may be divisible by terrestrial days without residue; and this precisely is matter for investigation of *(cuttaca)* the pulverizing multiplier.

CRÍSHṄ.

The finding of the [place of the] planet and the elapsed days, from the remainder of seconds in the planet's place, is thus shown. It is as follows. Sixty is there made the dividend; terrestrial days, the divisor; and the remainder of seconds, the subtractive quantity: with which the multiplier and quotient are to be found. The quotient will be seconds; and the multiplier, the remainder of minutes. From this remainder of minutes taken [as the subtractive quantity] the quotient deduced will be minutes; and the multiplier, the remainder of degrees. The residue of degrees is next the subtractive quantity; terrestrial days, the divisor; and thirty, the dividend: the quotient will be degrees; and the multiplier, the remainder of signs. Then twelve is made the dividend; terrestrial days, the divisor; and the remainder of signs, the subtractive quantity: the quotient will be signs; and the multiplier, the remainder of revolutions. Lastly, the revolutions in a *calpa* become the dividend; terrestrial days, the divisor; and the remainder of revolutions, the subtractive quantity: the quotient will be the elapsed revolutions; and the multiplier, the number of elapsed days. Examples of this occur [in the *Śirómańi*] in the chapter of the [three] problems.[1]

In like manner the exceeding months in a *calpa* are made the dividend; solar days, the divisor; and the remainder of exceeding months, the subtractive quantity: the quotient will be the elapsed additional months;[2] and the multiplier, the elapsed solar days. So the deficient days in a *yuga*[3] are made the dividend; lunar days, the divisor; and the remainder of deficient days, the subtractive quantity: the quotient will be the elapsed fewer days;[4] and the multiplier, the elapsed lunar days.

73. Rule for a conjunct pulverizer:[5] If the divisor be the same, and the multipliers various [two or more[6]]; then, making the sum of those multi-

[1] *Praśná'd'hyáya;* meaning the *Tripraśná'd'hyáya* of the astronomical portion of the *Śirómańi*.

[2] The excess of lunar above solar months.

[3] *Yuga* is here an error of the transcriber for *calpa;* or has been introduced by the author to intimate, that the method is not restricted to time calculated by the *calpa*, but also applicable when the calculation is by the *yuga* or any other astronomical period. CRÍSHN'.

This reading, however, does not occur in copies of the *Lílávatí*, though it do in all collated ones of the *Víja-gańita:* nor is it noticed by the commentators of the *Lílávatí*.

[4] Difference between elapsed lunar and terrestrial days.

[5] See *Lílávatí*, § 265.

[6] CRÍSHN'. on *Víj*. and GAŃ. on *Líl*.

pliers the dividend, and the sum of the remainders a single remainder; and applying the foregoing method of investigation, the precise multiplier so found[1] is denominated a conjunct one.

74. Example. What quantity is it, which multiplied by five, and divided by sixty-three, gives a residue of seven; and the same multiplied by ten and divided by sixty-three, a remainder of fourteen? declare the number.

Here the sum of the multipliers is made the dividend; and the sum of the residues, a subtractive quantity; and the statement is Divd. 15 Divr. 63 Subtrac. 21.

Proceeding as before, the multiplier is found 14. It is precisely the number required.

[1] As, putting the multiplicand for dividend, the multiplier is found by the investigation which is the subject of this chapter; so, making the multiplicator dividend, the multiplier found by the investigation is multiplicand, in like manner as sixty is made dividend, in the foregoing instance (§ 72). Then, as the given quantity, being lessened by subtraction of an amount equal to the residue of the division of it by the divisor after multiplication by one of the multiplicators, becomes exactly divisible; so, by parity of reasoning, it does, when lessened by the subtraction of the respective remainders, which the whole number yields, being severally multiplied by the rest of the multiplicators and divided by the divisor. And generally, if the divisor be the same, then, as the quantity, severally multiplied by the multiplicators and lessened by the respective remainders, becomes exactly divisible by the divisor; so it does, when, being severally multiplied, the multiples are added together and the sum is lessened by the aggregate of remainders. Now the quantity multiplied by the sum of the multiplicators is the same as if severally multiplied by the multiplicators and the multiples then added together. Therefore the sum of the multiplicators is taken for a multiplicator [and employed as a dividend;] and the aggregate of the remainders is received for a remainder [and employed as subtractive or additive.] CRĬSHŃ.

CHAPTER III.

AFFECTED SQUARE.[1]

SECTION I.

75—81. Six and a half stanzas. Rules for investigating the square-root of a quantity with additive unity: Let a number be assumed, and be termed the "least" root.[2] That number, which, added to, or subtracted from, the product of its square by the given coefficient,[3] makes the sum (or difference) give a square-root, mathematicians denominate

[1] *Varga-pracrĭti* or *Crĭti-pracrĭti;* from *varga* or *crĭti*, square, and *pracrĭti*, nature or principle.

'This branch of computation is so denominated, either because the square of *yávat* or of another symbol is *(pracrĭti)* the subject of computation; or because the calculus is concerned with the number which is *(pracrĭti)* the subject affecting the square of *ya* or other symbol. The number, that is *(pracrĭti)* the subject in respect of such square, is intended by the term. It is the multiplier of the square of the unknown: and therefore, in this investigation of a root, the multiplier of the square is signified by the word *pracrĭti*.' CRĬSHN'.

See § 185; the author's own comment on that and on § 187 and § 171. In one place *pracrĭti* is applied by him to the square affected by the coefficient; in the other it is declared to intend the coefficient affecting the square. The commentator SU'RYADA'SA interprets it in the first sense (note on § 195); and CRĬSHN'A, in the latter. (Vide supra).

'The method here taught subserves the solution of certain problems producing quadratic equations that involve more than one unknown term.' CRĬSHN'.

[2] *Hraśwa, canisht'ha*, or *laghu*, (múla;) the "least" root; so denominated with reference to additive quantities, though it may exceed the other root, when the quantity is subtractive (a negative additive) and is comparatively large. See CRĬSHN'.

[3] *Pracrĭti* or *guńa;* the given coefficient *(anca)* and multiplier *(guńa)* affecting the square. See a preceding note, and Chap. 7.

a positive or a negative additive;[1] and they call that root the "greatest" one.[2]

76. Having set down the "least" and "greatest" roots and the additive, and having placed under them the same or others,[3] in the same order, many roots are to be deduced from them by composition.[4] Wherefore their composition is propounded.

77. The "greatest" and "least" roots are to be reciprocally multiplied crosswise;[5] and the sum of the products to be taken for a least root. The product of the two [original] "least" roots being multiplied by the given coefficient, and the product of the "greatest" roots being added thereto, the sum is the corresponding greatest root; and the product of the additives will be the [new] additive.

78. Or the difference of the products of the multiplication crosswise of greatest and least roots may be taken for a "least" root: and the difference between the product of the two [original] least roots multiplied together and taken into the coefficient, and the product of the greatest roots multiplied together, will be the corresponding "greatest" root: and here also the additive will be the product of the two [original] additives.

79. Let the additive divided by the square of an assumed number, be a

[1] *Cshépa*, an additive either positive or negative: a quantity superinduced, either affirmative or negative, and consequently additive or subtractive. See chap. 2, § 53 et passim. *Líl.* ch. 11, § 248.

[2] *Jyésht'ha*, the "greatest" root, contradistinguished from *Canisht'ha*, the least root: although it may in some cases be less, when the *cshépaca*, or additive, is negative.—CRÍSHN. Provided this subtractive quantity be large and the coefficient small.

[3] That is, other roots for the same coefficient affecting the square. CRÍSHN.

[4] *Bhávaná*, composition, or making right* by combination. It is twofold: 1st. *yóga-bhávaná*, or *samása-bhávaná*, composition by the sum of the products (§ 77); 2d. *antara-bhávaná*, or *visésha-bhávaná*, composition by the difference (§ 78). Recourse is had to the first, when large roots are sought; to the second, when small are required. CRÍSHN.

[5] *Vajrábhyása*, multiplication crosswise or zigzag. From *vajra*, lightning or the thunderbolt, and *abhyása*, reciprocal multiplication. It is oblique multiplication (*tiryag-guṇana*). SÚR. and CRÍSHN.

* *Bhávayati, sidd'ha-caróti* (makes right). CRÍSHN.

new additive; and the roots, divided by that assumed number, will be the corresponding roots. Or the additive being multiplied [by the square], the roots must, in like manner, be multiplied [by the number put].

80—81. Or divide the double of an assumed number by the difference between the square of that assumed number and the given coefficient; and let the quotient be taken for the "least" root, when one is the additive quantity; and from that find the "greatest" root. Here [the solutions are] infinite, as well from [variety of] assumptions, as from [diversity of] composition.[1]

[1] The principle of the first rule (§ 75,) as observed by the commentator CRÍSHN'A-BHAT'T'A, is too evident to require demonstration. That of § 79 is used by him in demonstrating the others, and is thus given: A square, multiplied or divided by a square, yields still a square. If both sides of the equation (L^2. coeff. $+ A = G^2$) be multiplied or divided by the square of any assumed number, equality continues. Now, as the squares of the "least" and "greatest" roots are here multiplied by the square of the assumed number, the factor of those roots themselves will be the simple number put.

The demonstration of § 77, which is given in words at length, joined with a cumbrous notation of the algebraical expressions, may be thus abridged: To distinguish the two sets, let L, G and A represent one set; l, g and a the other; and C the given coefficient.* Then, under § 79, putting g for the assumed number, another set is deduced from the first, L.g, G.g, A.g^2. Whence C.L^2.g^2+A.g^2=G^2.g^2. Substitute for g^2 its value C.l^2+a; and the additive A.g^2 becomes A.Cl^2+A.a; and, substituting in the first term for A its value G^2—C.L^2, it becomes C.G^2.l^2—C^2.L^2.l^2+A.a. Hence the equation C.L^2.g^2+C.G^2.l^2—C^2.L^2.l^2+A.a=G^2.g^2; whence, transposing the negative term and adding or subtracting 2C.L.G.l.g; the result is C.$(L.g+l.G)^2$+A.a=$(G.g\pm C.L.l)^2$. See § 78.

The concluding rule § 80—81 is thus proved by the same commentator: 'Twice an assumed number being put for the "least" root (§ 75) its square is four times the square of that assumed number. The point is to find a quantity such, that being added to this quadruple square taken into the given coefficient, the sum may be a square. Now the difference between the square of the sum of two quantities and four times their product is the square of their difference. Therefore four times the square of the assumed number, multiplied by the given coefficient, and added to the square of the difference [between the square of the assumed number and the coefficient,] must of course give a square-root. Thus the "least" root is twice the number assumed; and the additive quantity is the square of the difference between the square of the assumed number and the coefficient. But, by the condition of the problem, the additive quantity must be unity. Divide therefore, under § 79, by the square of the difference, at the same time dividing the root by the simple difference between the square of the assumed number and the given coefficient.' CRÍSHN'.

* CRÍSHN'A-BHAT'T'A puts the symbols *pra, á ca, á jyé, á cshé, dwi ca, dwi jyé,* and *dwi cshé,* initial syllables of *pracriti* coefficient affecting the square, *ádya* first and *dwitíya* second, *canisht'ha* least, *jyésht'ha* greatest (root) and *cshépa* additive.

82. Example. What square, multiplied by eight, and having one added to the product, will be a square? Declare it, mathematician! Or what square, multiplied by eleven, and having one added to the product, will be a square, my friend?

Statement on Example 1st: C 8 A 1.

Here putting unity for the assumed "least" root, the "greatest" root is three, and additive one. Statement of them for composition:

C 8 L 1 G 3 A 1
l 1 g 3 a 1

By the rule [§ 77] the first "least" root 1, multiplied by the second "greatest" root 3, gives the product 3. The second "least" root, by the first "greatest," gives the like product. Their sum is 6. Let this be the "least" root. The product of the two "less" roots 1, being multiplied by the given coefficient 8, and added to the product of the two "greater" roots 9, makes 17. This will be the "greater" root. The product of the additives will be the additive 1.

Statement of the former roots and additive, with these, for composition:

C 8 L 1 G 3 A 1
l 6 g 17 a 1

Here, by composition, the roots are found L 35 G 99 A 1; and so on, indefinitely, by means of composition.

Statement on Example 2d: Putting unity for the assumed "least," and subtracting two from the square of that multiplied by the given coefficient 11, the "greater" root is 3. Hence the statement for composition is

C 11 L 1 G 3 A $\dot{2}$
l 1 g 3 a $\dot{2}$

Proceeding as before, the roots for additive 4 are L 6 G 20 A 4. Then, by the rule § 79, putting two for the assumed number, the roots for unity additive are found L 3 G 10 A 1. Hence, by composition of like sets,[1] the "least" and "greatest" roots are found *L* 60 *G* 199 *A* 1. In like manner, an indefinite number of roots may be deduced.

[1] *Túlya-bhávaná;* the combining of like sets. Whatever may have been the additive quantity first found, and whether it were positive or negative, the combination of like sets raises the additive to a square; and then, under § 79, assuming a number equal to the root of that square, and dividing the additive by that square, the additive is reduced to unity, and the roots answering to it are found by division.

Or, putting unity for the "least" root, the two roots for additive five are found L 1 G 4 A 5. Whence, by composition of like scts, L 8 G 27 A 25. From this, by § 79, putting five for the assumed number, the roots for additive unity are found *L* $\frac{8}{5}$ *G* $\frac{27}{5}$ *A* 1.

Statement of these with the preceding, for composition: L 3 G 10 A 1
l $\frac{8}{5}$ g $\frac{27}{5}$ a 1

From composition by the sum, roots are deduced L $\frac{161}{5}$ G $\frac{534}{5}$ A 1.

Or, under rule § 78; from composition by the difference, they come out *L* $\frac{1}{5}$ *G* $\frac{6}{5}$ *A* 1. And so on, in numerous ways.

The roots for unity as the additive, may be found by another process, under § 80. Here, putting three for the assumed number, and proceeding as directed, the "least" root comes out 6. Viz. assumed number 3. Its square 9. Given coefficient 8. Their difference 1. Twice the assumed number 6, divided by that difference, is 6; the "least" root: L 6. Its square 36; multiplied by the given coefficient 8, is 288; which, with one added, becomes 289; the root of which is 17, the "greatest" root: G 17.

So, in the second example likewise, putting three for the assumed number, and proceeding as directed, the roots are found; L 3 G 10 A 1.

Thus, by virtue of [a variety of] assumptions, and by composition either by sum or difference, an infinity of roots may be found.[1]

[1] A variety of additives is also found: but it is not noticed, because the problem is restricted to additive unity. CRÍSHN'.

SECTION II.

83—86. Rule for the cyclic method:[1] (completion of stanza 81, three stanzas, and half another.) Making the "least" and "greatest" roots and additive,[2] a dividend, additive and divisor, let the multiplier be thence found.[3] The square of that multiplier being subtracted from the given coefficient, or this coefficient being subtracted from that square, (so as the remainder be small;[4]) the remainder, divided by the original additive, is a new additive; which is reversed if the subtraction be [of the square] from the coefficient.[5] The quotient corresponding to the multiplier [and found with it] will be the "least" root: whence the "greatest" root may be deduced.[6] With these, the operation is repeated, setting aside the former roots and additive. This method mathematicians call that of the circle. Thus are integral roots found with four, two, or one [or other number,[7] for] additive:

[1] *Chacravála*, a circle; especially the horizon. The method is so denominated because it proceeds as in a circle: finding from the roots ("greatest" and "least") a multiplier and a quotient (by Chapter 2); and thence new roots; whence again a multiplier and a quotient, and roots from them; and so on in a continued round. SU'R.

[2] Previously found by § 75. CRI'SHN'.

[3] By the method of the pulverizer (*cut't'aca*). Ch. 2.

[4] If the coefficient exceed the square of the multiplier, subtract this from the coefficient; but, if the coefficient be least, subtract it from the square: but so, as either way the residue be small. —SUR. Else another multiplier is to be sought, by Ch. 2.

[5] If the square of the multiplier were subtracted from the coefficient, the sign of the new additive is reversed: if affirmative, it becomes negative; if negative, it is changed to positive. SU'R. and CRI'SHN'.

[6] It is deduced from the "least" root and additive by the conditions of the problem: or, if required, without the extraction of a root, by this following rule. 'The original "greatest" root, multiplied by the multiplier, is added to the "least" root multiplied by the given coefficient; and the sum is divided by the additive.' CRI'SHN'.

[7] With four, two or one, additive or subtractive; or with some other number. CRI'SHN'.

and composition serves to deduce roots for additive unity, from those which answer to the additives four and two [or other number.][1]

87. Example: What is the square, which, being multiplied by sixty-seven, and one being added to the product, will yield a square-root? and what is that, which multiplied by sixty-one, with unity added to the product, will do so likewise? Declare it, friend, if the method of the affected square be thoroughly spread, like a creeper,[2] over thy mind.

Statement of Example 1st: (Putting unity for the "least" root, and negative three for the additive.) C 67 L 1 G 8 A 3.

Making the "least" root the dividend, the "greatest" root the additive, and the additive the divisor, the statement for the operation of finding the multiplier (Ch. 2) is Dividend 1 / Divisor 3 Additive 8.

Here, by the rule § 61, the series is 0 / 2 / 0; and the quotient and multiplier are found 0 / 2; which, as the number of quotients [in the series] is uneven, must be subtracted from the abraders (§ 57) leaving 1 / 1; and the quotient obtained in the abrading of the additive is to be added (§ 61) to the quotient here found; making the quotient and multiplier 3 / 1 Since the divisor is negative, the quotient is considered so too (§ 68); and the quotient and multiplier are 3 / 1 Then the square of the multiplier 1, being subtracted from the given coefficient 67, leaves 66; which, however, is not a small remainder. Putting therefore negative two for the assumed number by § 64, and multiplying by that the negative divisor 3, and adding the product to the multiplier, a new multiplier is found: viz. 7. Its square 49 being subtracted

[1] If the additive be already a square integer, the problem of finding the roots that answer to additive unity is at once solved by § 79. Else raise it to a square by the combination of like sets, and then proceed by that rule. If the roots so found be not integral, repeat the method of the circle, until the roots come out in whole numbers. CRÍSHN'.

[2] As a climbing plant spreads over a tree.

from the coefficient 67, the remainder 18, divided by the original additive $\dot{3}$, yields $\dot{6}$; the sign of which is reversed, as the subtraction was of the square of the multiplier from the coefficient;. and it thus becomes 6 positive. The quotient answering to the multiplier, viz. $\dot{5}$,* is the "least" root. Whether this be negative or affirmative, makes no difference in the further operation. It is noted then as 5 positive. Its square being multiplied by the coefficient, and six being added to the product, and the square-root being extracted, the "greater" root comes out 41.

Statement of these again for a further investigation of a pulverizer:

Dividend 5
Divisor 6 Additive 41.

Here the multiplier is found, 5. Its square, subtracted from the coefficient, leaves 42; which, divided by the original additive 6, yields 7; the sign whereof is reversed because the subtraction was from the coefficient; and the new additive comes out 7. The quotient answering to the multiplier is the "least" root, 11. Hence the "greatest" root is deduced, 90.

Statement of these again for a further pulverizer: Divd. 11
Div. $\dot{7}$ Add. 90.

By the rule § 61, the abraded additive becomes 6, and the multiplier is found 5; and, since the products in the series are uneven, it is subtracted from its abrader, and the multiplier becomes 2. Its negative divisor (the former additive) being negative ($\dot{7}$) is multiplied by negative one ($\dot{1}$) assumed by § 64 and added to that multiplier, for a new multiplier 9; from the square of which 81, subtracting the given coefficient 67, the remainder 14, divided by the original additive $\dot{7}$, gives the new additive $\dot{2}$. The quotient answering to the multiplier is the "least" root 27: whence the "greatest" root is found 221. From these, others are to be deduced by combination of like sets.

Statement: L 27 G 221 A $\dot{2}$
l 27 g 221 a $\dot{2}$

Proceeding as directed, the roots are found *L* 11934 *G* 97684 *A* 4.

* $-3+(1\times-2)$.

These roots divided by the root of the additive four, viz. 2, give roots which answer to additive unity: L 5967 G 48842 A 1.

Statement of Example 2d. C 61 L 1 G 8 A 3.

Statement for a pulverizer: Dividend 1 Divisor 3 Additive 8.

Proceeding as before, by §61, and putting two for the assumed number (§64) the multiplier is found 7. Whence roots, answering to the negative additive four, are deduced L 5 G 39 A 4. Thence, by §79, roots are found for subtractive unity, L $\frac{5}{2}$ G $\frac{39}{2}$ A $\dot{1}$. Statement of these for composition L $\frac{5}{2}$ G $\frac{39}{2}$ A $\dot{1}$
l $\frac{5}{2}$ g $\frac{39}{2}$ a $\dot{1}$.

From them are deduced roots answering to additive unity L $\frac{195}{2}$ G $\frac{1523}{2}$ A 1.

Statement of these again, with roots answering to subtractive unity, for composition L $\frac{5}{2}$ G $\frac{39}{2}$ A $\dot{1}$
l $\frac{195}{2}$ g $\frac{1523}{2}$ a 1

Hence integral roots answering to subtractive unity are obtained L 3805 G 29718 A $\dot{1}$. From these, by combining like sets, roots for additive unity come out (in whole numbers) L 226153980 G 1766319049 A 1.

SECTION III.

MISCELLANEOUS RULES.

88—89.[1] Rule:[2] If the multiplier [that is, coefficient affecting the square] be not the sum of [two] squares, when unity is subtractive, the instance proposed is imperfect.[3] The instance being correctly put, let unity twice set down be divided by the roots of the [component] squares: and the quotients be taken as two "least" roots answering to subtractive unity:[4]

[1] Conclusion of a preceding stanza § 86; one complete stanza; and beginning of another.

[2] Where unity is subtractive, to discriminate impossible cases; and to solve the problem by another method, in those which are possible. CRÍSHN'.

[3] Undeserving of regard.—SÚR. The square of no number multiplied by such a coefficient, can, after subtraction of unity, be an exact square. CRÍSHN'.

The subtractive unity is a square number. Now a negative additive may be a square number if the square of the "least" root being multiplied by the coefficient comprise two squares; for then, one square being subtracted, the other remains to yield a square-root. But, for this end, it is necessary that the coefficient should have consisted of the sum of two squares; for, as a square multiplied by a square is a square, the square of the "least" root being multiplied by the two square component portions of the coefficient, the two multiples will be squares and component portions of the product. CRÍSHN'.

In explanation of the principle of this rule, SÚRYADÁSA cites a maxim, that taking contiguous arithmeticals, or next following terms in arithmetical progression increasing by unity, twice the sum of the squares less one will be a square number.

[4] The square of a "least" root, [putting any number for the root at pleasure;] multiplied by either component square portion of the coefficient, will answer for a negative additive: for, the square of the "least" root being severally multiplied by the squares of which the sum is the coefficient, the two products added together are the square of the "least" root multiplied by the coefficient; and, if from that be subtracted the square of the same multiplied by either portion of the coefficient, the remainder will be the square of the same multiplied by the other square portion of the coefficient; and of course will yield a square-root. Now to deduce from this, roots answering to subtractive unity; put for the assumed number by § 79 the "least" root [any how assumed as above] multiplied by the root of either component square portion of the coefficient, and

and the correspondent "greatest" roots may thence be deduced. Or two roots serving for subtractive unity may be found in the manner before shown.

90. Example. Say what square, being multiplied by thirteen, with one subtracted from the product, will be a square number? Or what square, being multiplied by eight, with one taken from the product, will yield a root?

In the first of these instances, the coefficient is the sum of the squares of two and three. Therefore let unity divided by two be a "least" root for subtractive unity, $\frac{1}{2}$. From the square of that, multiplied by the coefficient, and diminished by the subtraction of unity, the corresponding "greatest" root is deduced, $\frac{3}{2}$. Or let unity divided by three be the "least" root, $\frac{1}{3}$. Hence the "greatest" root is found $\frac{2}{3}$. Or let the "least" root be 1; from the square of which, multiplied by the coefficient, and diminished by the subtraction of four, the "greatest" root comes out 3. Statement of them, in their order, L 1 G 3 A $\dot{4}$. By the rule § 79, roots answering to subtractive unity are hence found $\frac{1}{2}$ $\frac{3}{2}$. Or subtracting nine from the square of the "least" root multiplied by the coefficient, the "greatest" root comes out: and roots are thence found [by § 75—79] L $\frac{1}{3}$ G $\frac{2}{3}$ A $\dot{1}$. Or by the cyclic method (§ 83—86) integral roots may be deduced. Thus, putting those "least" and "greatest" roots and additive (§ 83) for the dividend, additive and divisor, Dividend $\frac{1}{2}$ Divisor 1 Additive $\dot{\frac{3}{2}}$; and reducing them by the common measure half, Dividend 1 Divisor 2 Additive $\dot{3}$, the multiplier and quotient are found by investigation of the pulverizer (Ch. 2), 1 and $\dot{2}$. Here putting negative unity for an assumed number, and adding its multiple of the divisor to the multiplier, another multiplier is obtained, 3. Whence, by the rule (§ 84), the additive comes out 4; and the quotient found with the

proceeding by that rule, § 79, the root answering to subtractive unity will be the "least" root [before assumed] divided by the present assumed number, which is the same "least" root multiplied by the root of a component portion of the coefficient. Reduce the numerator and denominator of this fractional value to their least terms by their common measure, the "least" root [before assumed]; the result is, for numerator, unity; for denominator, the root of the component square portion of the coefficient. CRI'SHN'.

multiplier, becomes the "least" root 3; and from these the "greatest" root is deduced, 11. Hence also, by repeating the cyclic operation (§ 83—84), integral roots for subtractive unity are found, L 5 G 18 A $\dot{1}$. Here, as in every instance, an infinity of roots may be deduced by composition with roots answering to unity.

In like manner, in the second example, where the given coefficient is eight, the "least" and "greatest" roots, found as above, are L $\frac{1}{2}$ G 1 A 1.*

91. Example: What square, being multiplied by six, and having three added to it, will be a square number? or having twelve added? or with the addition of seventy-five? or with that of three hundred?

Here, putting unity for the "least" root, the statement is C 6 L 1 G 3 A 3. Then, by rule § 79, multiplying the roots by two, [and the additive by its square four,] the roots answering for additive twelve come out L 2 G 6 A 12. So, by the same rule, multiplying by five, [and the additive by twenty-five,] they are found for additive seventy-five, L 5 G 15 A 75. Also, multiplying by ten, [and the additive by a hundred,] they are deduced for additive three hundred, L 10 G 30 A 300.

92.[2] Many being either additive or subtractive, corresponding roots may be found [variously] according to the [operator's] own judgment: and from them an infinity may be deduced, by composition with roots answering to additive unity.[3]

93. Rule:[4] The multiplier [i. e. coefficient] being divided by a square, [and the roots answering to the abridged coefficient being thence found,[5]] divide the "least" root by the root of that square.[6]

* Roots in whole numbers may be hence deduced by the cyclic method, § 83—86. CRĬSHṆ.

[2] Completion of one stanza § 89 and half of another.

[3] The preceding rule was unrestrictive. Finding by whatever means roots which answer for the proposed additive, an infinity of them is afterwards thence deducible by composition with additive unity and its correspondent roots: as the author here shows. CRĬSHṆ.

[4] Applicable when the coefficient is measured by a square. CRĬSHṆ.

[5] CRĬSHṆ.

[6] By parity, multiplying by any square the given coefficient, and finding the "least" and "greatest" roots for such raised coefficient, the "least" root so found must be multiplied by the root of that square. CRĬSHṆ.

94. Example: half a stanza. Say what square being multiplied by thirty-two, with one added to the product, will yield a square-root?

Statement: C 32. The "least" and "greatest" roots, found as before, are L $\frac{1}{2}$ G 3 A 1. Or, by the present rule § 93, the coefficient 32 divided by four, gives 8; to which the roots corresponding are found L 1 G 3 A 1; and dividing the "least" root by the root (2) of the square (4) by which the coefficient was divided, the two roots for the coefficient thirty-two, come out L $\frac{1}{2}$ G 3 A 1.

Or, dividing the coefficient by sixteen, it gives 2; to which the roots corresponding are L 2 G 3 A 1; whence, dividing the "least" root by the square-root (4) of the divisor (16), the roots answering to the entire coefficient are deduced L $\frac{1}{2}$ G 3 A 1.

Or, by the investigation of a pulverizer (Ch. 2) integral roots are obtained (§ 83—86). L 3 G 17 A 1.

95. Rule:[1] The additive,[2] divided by an assumed quantity, is twice set down, and the assumed quantity is subtracted in one instance, and added in the other: each is halved; and the first is divided by the square-root of the multiplier [that is, coefficient.] The results are the "least" and "greatest" roots in their order.[3]

[1] Applicable when the coefficient is a square number. CRĬSHN.

[2] The rule holds equally for a subtractive quantity: but with this difference, that the subtraction and addition of the number put are transposed to yield the "least" and "greatest" roots in their order. Or the rule may be applied as it stands, observing to give the negative sign to the additive: but the "least" and "greatest" roots will in this manner come out negative. It is, therefore, preferable to transpose the operations of subtraction and addition of the assumed number. CRĬSHN'.

[3] The square of the "least" root being multiplied by a coefficient which also is a square, the product will be a square number. The additive being added, if the sum too be a square, [square of the "greatest" root;] the additive must be the difference of the squares. Now the difference of two squares, divided by the difference of the two simple quantities, will be their sum. Hence, putting any assumed number for the difference, and dividing by it the additive equal to the difference of the squares, the quotient is the sum of the two quantities. Then, by the rule of concurrence (*Líl.* § 55), the finding of the two quantities is easy. The one is the "greatest" root; the other is the "least" root taken into the root of the coefficient. Therefore, by inversion, that quantity, divided by the root of the coefficient, will be the "least" root. SU'R. and CRĬSHN'.

96. Example: What square, being multiplied by nine, and having fifty-two added to the product, will be square? or what square number, being multiplied by four, and having thirty-three added, will be square?

Here, in the first example, the additive fifty-two being divided by an assumed number two, and the quotient set down twice, diminished and increased by the assumed number and then halved, gives 12 and 14. The first of these is divided by the square-root of the given coefficient; and the "least" and "greatest" roots are found, L 4 G 14.

Or dividing the additive 52 by four, they thus come out L $\frac{3}{2}$ G $\frac{17}{2}$.

In the second example, dividing the additive thirty-three by one put for the assumed number, the "least" and "greatest" roots are in like manner deduced, L 8 G 17. Or, putting three, they are L 2 G 7.[1]

97. Example:[2] Declare what square multiplied by thirteen, and lessened by subtraction of thirteen, or increased by addition of the same number, will be a square?

In the first example, coefficient 13. The "least" and "greatest" roots found [for the subtractive quantity] are L 1 G 0. Put an assumed number 3; and, by rule § 80—81, roots answering for additive unity are found L $\frac{3}{2}$ G $\frac{11}{2}$. From these, by composition, roots answering to the negative additive thirteen are deduced L $\frac{11}{2}$ G $\frac{39}{2}$. From which roots, corresponding to the negative additive, together with these other roots L $\frac{1}{2}$ G $\frac{3}{2}$ answering to subtractive unity, by the method of composition by difference, roots suited to additive thirteen are obtained L $\frac{3}{2}$ G $\frac{13}{2}$. Or by composition by sum, they come out L 18 G 65.

98. Example:[3] Say what square, multiplied by negative five, with twenty-one added to the product, will be a square number? if thou know the method for a negative coefficient.

Statement: C $\dot{5}$ A 21.

Here, putting one, the roots are 1 and 4. Or putting two, they are 2 and 1.

[1] Thus, by varying the assumptions, an infinity of results may be obtained. CRĪSHṆ.

[2] To elucidate the case when the additive equals the coefficient. CRĪSHṆ.

[3] Showing, that roots may be found, even in cases where the coefficient is negative. CRĪSHṆ.

By composition with roots adapted to negative unity, an infinity may be deduced.

99. This computation, truly applicable to algebraic investigation, has been briefly set forth. Next I will propound algebra affording gratification to mathematicians.[1]

[1] By this conclusion it is intimated, that the contents of the preceding chapters (1—3) are introductory to the analysis, which the author proposed as the subject in the opening of the treatise (§ 2); and to which he now proceeds in the next chapters (4—8). See SU'R. and CRISHN.

CHAPTER IV.

SIMPLE EQUATION.[1]

100—102. RULE: Let "so much as" *(yávat-távat)*[2] be put for the value of the unknown quantity;[3] and doing with that precisely what is proposed in the instance, let two equal sides be carefully completed, adding or subtracting, multiplying or dividing,[4] [as the case may require.]

101. Subtract the unknown quantity of one side from that of the other; and the known number of the one from that of the other side. Then by the remaining unknown divide the remainder of the known quantity: the quotient is the distinct value of the unknown quantity.[5]

[1] *Éca-varńa-samícarańa*, equation uniliteral or involving a single unknown quantity. See note 2 in next page.

[2] See § 17.

[3] *Avyacta-rási*, indistinct quantity or unknown number *(ajnyátánca)*; the unknown is represented by *yávat-távat*; or, if there be more than one, by colours or letters (§ 17); the known, by *rúpa*, form, species, (absolute number.) See SU'R.

[4] Or by multiplying and adding; or by multiplying and subtracting; by dividing and adding; or by dividing and subtracting; or by raising to a square or other [power]. CRĬSHŃ.

This first rule is common to all algebraic analysis. *Ib.*

[5] Whatever be the unknown quantity (whether unit or aggregate of the known, or a part or fraction of such unit or aggregate,) is yet not specifically known. It is therefore denominated indistinct or unknown; while that, which is specifically known, is termed distinct or known species. The operations indicated by the enunciation of the example being performed with the designation of the unknown, if by any means, conformably with the tenor of the instance, there at once be equality of the two sides, a value of the unknown in the known species is easily deducible. Thus, if on one side, there be only known number, and on the other side the unknown quantity only, then, as being equal, those numbers are a true value in the known, of that amount of the unknown. Hence, by rule of three, the quantity sought is found: viz. 'of so many unknown

102. Under this head, for two or more unknown quantities also, [the algebraist] may put, according to his own judgment, multiples or fractions of "so much as," (that is, *yávat-távat*, multiplied by two, &c. or divided;) or the same with addition or deduction [of known quantities.] Or in some cases he may assume a known value; with due attention likewise [to the problem.]

The first analysis is an equation involving a single colour (or letter).[1] The second mode of analysis is an equation involving more than one colour (or letter).[2] Where the equation comprises one, two or more colours, raised

(yávat-távat) if so many known *(rúpa)* be the value, then of the proposed number of unknown, what is the value?' But, should there be on both sides some terms of each sort, it must be so managed, that on one side there be only terms of the unknown; and on the other, of the known. Now it is a maxim, that, if equal [things] be added to, or subtracted from, or multiply, or divide, equal [things], the equality is not destroyed: as is clear. If then, from one side, the terms of the unknown contained in it be subtracted, there remain only known numbers on that side: but, for equality's sake, the like amount of unknown must be subtracted from the other side. The same is to be done in regard to the known number on one side, which must be subtracted also from the other. This being effected, there remain only terms of the unknown on one side; and of the known on the other. Then, by rule of three, 'if by this unknown quantity this known number be had, then by the stated amount of the unknown what is obtained?' the remaining known term is to be divided by the residue of the unknown and to be multiplied by the proposed unknown. The one operation (that of division) is directed by the rule (§ 101); the other (the multiplication) is comprehended in *(utt'hápana)* the "raising" of the answer;* both being reduced to proportions in which one term is unity. Therefore, by any means, (by subtraction or some other,) the two sides of the equation are to be so treated, consistently, however, with their equality, as that known number may be on one side, and unknown quantity on the other. Else the solution will not be easy.

CRÍSHN'.

[1] Or symbol of unknown quantity.

[2] *Samícaraña, samícára, samícriyá*, equation: from *sama*, equal, and *crí*, to do: a making equal. It consists of two sides *(pacsha)*; and each may comprise several terms (*c'hañda*, lit. part).

The primary distinction of analysis *(Víja)* is twofold; 1st. uniliteral or equation involving one unknown, *éca-varña-samícaraña*; where, a single unknown quantity designated by letter or colour (§ 17) being premised, two sides are equated; 2d. multiliteral or equation involving several unknown, *anéca-varña-samícaraña*, where, more than one unknown quantity represented severally by colours or letters being premised, two sides are equated. The first comprises two, and the second three sorts: viz. 1st. equation involving a single and simple unknown quantity; 2d. equation involving a single unknown raised to a square or higher power; 3d. equation involving several simple unknown quantities; 4th. equation involving several unknown raised to the square or higher power; 5th. equation involving products of two or more unknown quantities multiplied together.

* Deducing of the answer by substitution of value. See note 1, p. 188. and gloss on § 153—156.

to the square or other [power,] it is termed *(mad'hyamáharańa)* elimination of the middle term. Where it comprises a *(bhávita)* product, it is called, *(bhávita)* involving product of unknown quantities. Thus teachers of the science pronounce analysis fourfold.

The first of these is so far explained: an example being proposed by the questioner, the value of the unknown quantity should be put once, twice, or other multiple of "so much as" *(yávat-távat)*: and on that unknown quantity so designated, every operation, as implied by the tenor of the instance,[1] whether multiplication, division, rule of three, [summing of] progression, or [measure of] plane figure, is to be performed by the calculator. Having so done, he is diligently to make the two sides equal. If they be not so in the simple enunciation [of the problem]; they must be rendered equal by adding something to either side, or subtracting from it, or multiplying by some quantity or dividing.[2] Then the unknown quantity of one of the two sides is to be subtracted from the unknown of the other side; and, in like manner, the square or other [power] of the unknown. The known numbers of one side are to be likewise subtracted from the known numbers of the other.[3] If there be surds, they too must be subtracted by the method before taught.[4] Then, by the residue of the unknown quantity, dividing

These distinctions are reducible to four, by uniting the quadratics or equations of higher degree under one head of analysis; where, a power (square or other) of an unknown quantity represented by colour or letter (or more than one such) being premised, and sides being equated, the value is found by means of extraction of the root. It is called *mad'hyamá'harańa*; and is so denominated because the middle term *(mad'hyama c'hańda)* is generally removed: being derived from *mad'hyama*, middlemost, and *áharańa*, removal or elimination. (See Chap. 5.) These four distinctions are received by former writers:* the author himself, however, intimates his own preference of the primary distinction alone. CRÍSHŃ.

1 *Álápa*, enunciation of the *prích'haca*, or of the person proposing the question; or tenor of the instance *(uddésaca)*; the condition of the problem.

2 By superadding something to the least side; or subtracting it from the greater; or multiplying by it the less side; or dividing by it the greater. CRÍSHŃ.

3 The side containing the lowest unknown has the most known; and conversely. Ordering the work accordingly, subtract the unknown in the second from that in the first side; and the known in the first from that in the second.—SÚR. If there be a square or other [power] of the unknown, that also is to be subtracted from the like term of the other side. CRÍSHŃ.

4 Ch. 1. Sect. 4. Though the unknown or its power have a surd multiplier, subtraction must take place. The residue having still a surd coefficient, divide by that surd the remainder of known

* See CHATURVÉDA on BRAHMEGUPTA, (*Brahm.* 12, § 66 and 18, § 32).

the remainder of the known numbers, the quotient thus obtained becomes the value known of one unknown: and thence the proposed unknown quantity instanced is to be "raised."[1] If in the example there be two or more unknown quantities comprised,[2] putting for one of them one "so much as", let "so much as" *(yávat-távat)* multiplied by two or another assumed number, or divided by it, or lessened by some assumed number, or increased by it, be put for the rest. Or let "so much as" *(yávat-távat)* be put for one; and known values for the others. With due attention: that is, the intelligent calculator, considering how the task may be best accomplished, should so put known or unknown values of the rest. Such is the meaning.

103—104. Examples: One person has three hundred of known species and six horses. Another has ten horses of like price, but he owes a debt of one hundred of known species. They are both equally rich. What is the price of a horse?[3]

104. If half the wealth of the first, with two added, be equal to the wealth of the second; or if the first be three times as rich as the other, tell me in the several cases the value of a horse.[4]

number whether rational or irrational; that is, "square by square" (§ 29); and extract the square-root of the quotient; which will be the value of the unknown; or, if the quotient be irrational, note it as a surd value. So, in deducing an answer from that surd value, "multiply square by square" (§ 29) and extract the root, or note the surd. CRÍSHN'.

[1] The value of the unknown being thus found in an expression of the known; the answer of the question, or quantity sought, is deducible from it by the rule of proportion; and the first term of the proportion being unity, the operation is a simple multiplication. This finding of the quantity sought, or answer to the question, being the stated unknown quantity in the instance, is termed *utt'hápana*, a "raising" of it, or substitution of a value. See CRÍSHN'A; and the author's gloss on the first rule of Chapter 6.

[2] Although such examples come of course under equations involving more than one unknown, the author has introduced the subject for gratification and exercise of the understanding. —CRÍSHN'. See Ch. 5.

Reserving one among two or more unknown quantities, if values of the rest, in expressions of that or of the known species, be assumed either equal or unequal or at pleasure, then, from the value of the unknown thence found, a true answer for the instance will be deducible. CRÍSHN'.

[3] This is an example of an equation according to the simple enunciation of the instance. CRÍSHN'.

[4] Instances of adding or subtracting, multiplying or dividing, (§ 100) to produce the equation. CRÍSHN'.

Example 1st: Here the price of a horse is unknown. Its value is put one "so much as" *(yávat-távat) ya* 1; and by rule of three, 'if the price of one horse be *yávat-távat*, what is the price of six?' Statement: 1 | *ya* 1 | 6 | . The fruit, multiplied by the demand, and divided by the argument,[1] gives the price of six horses, *ya* 6. Three hundred of known species being superadded, the wealth of the first person results; *ya* 6 *ru* 300. In like manner the price of ten horses is *ya* 10. To this being superadded a hundred of known species made negative, the wealth of the second person results; *ya* 10 *ru* $\dot{1}00$. These two persons are equally rich. The two sides, therefore, are of themselves become equal. Statement of them for equal subtraction *ya* 6 *ru* 300
ya 10 *ru* $\dot{1}00$
Then, by the rule (§ 101), the unknown of the first side being subtracted from the unknown of the other, the residue is *ya* 4. And the known numbers of the second side being subtracted from the known numbers of the first, the remainder is 400. The remainder of known number 400, being divided by the residue of unknown *ya* 4, the quotient is the value in known species, of one "so much as" *(yávat-távat)* viz. 100. 'If, of one horse, this be the value, then of six what?' By this proportion the price of six horses is found, 600; to which three hundred of known species being added, the wealth of the first person is found, 900. In like manner, that of the second also comes out 900.

Example 2d: The funds of the first and second persons are, as before, these: *ya* 6 *ru* 300
ya 10 *ru* $\dot{1}00$

Here the wealth of the one is equal to half that of the other with two added; as is specified in the example. Hence, the capital of the first being halved and two added to the moiety; or that of the second less two being doubled; the two sides become equal. That being done, the statement for subtraction is *ya* 3 *ru* 152 or *ya* 6 *ru* 300
ya 10 *ru* $\dot{1}00$ *ya* 20 *ru* $\dot{2}04$
From both of which, subtraction, &c. being made, the value of one "so much as" *(yávat-távat)* is found 36. Whence, "raising" as before, the capitals of the two come out 516 and 260.

[1] *Lílávatí*, § 70.

Example 3d: The capitals are expressed by the same terms, viz.
ya 6 *ru* 300
ya 10 *ru* $\dot{1}00$

The third part of the first person's wealth is equal to the second's; or three times the last equals the first. Statement: *ya* 6 *ru* 300 or *ya* 2 *ru* 100
ya 30 *ru* 300   *ya* 10 *ru* 100

By the equation the value of "so much as" *(yávat-távat)* is found **25.** From which "raising" the answers, the capitals come out 450 and 150.

105. Example:[1] The quantity of rubies without flaw, sapphires, and pearls belonging to one person, is five, eight and seven respectively; the number of like gems appertaining to another is seven, nine and six: one has ninety, the other sixty-two, known species. They are equally rich. Tell me quickly then, intelligent friend, who art conversant with algebra, the prices of each sort of gem.

Here the unknown quantities being numerous, the [relative] values of the rubies and the rest are put *ya* 3 *ya* 2 *ya* 1. 'If of one gem this be the price, then of the proposed gems what is the price?' The number of *(yávat-távat)* the unknown, found by this proportion, being summed, and ninety being added, the property of the first person is *ya* 38 *ru* 90. In like manner the second person's capital is *ya* 45 *ru* 62. They are equally rich. Statement of the two for equal subtraction *ya* 38 *ru* 90 Equal subtraction being made,
ya 45 *ru* 62

the value of the unknown is found 4. "Raising" from it by the proportion 'If of one *yávat-távat* this be the value, then of three (or of two) what?' the prices of a ruby and the rest are deduced: viz. 12, 8, 4. 'If of one ruby this be the price, then of five what?' the amount of rubies comes out 60. In like manner sapphires 64; pearls 7. Total of these, with the addition of the absolute number 90, gives the whole capital of the one, 242: and, in like manner, that of the other, 242.

Or let the value of a ruby be put *ya* 1; and the prices of sapphires and pearls be put in known species, 5 and 3. 'If of one *yávat-távat* this be the price, then of five what?' Thus the price of five rubies is found *ya* 5; and the amount of sapphires and pearls, 40 and 21. The sum of the two, with

[1] An instance of more than one unknown quantity, and of putting assumed values (§ 102).
CRÍSHN'.

ninety added, is *ru* 151: In like manner the capital of the second person is *ya* 7 *ru* 125. Statement for equal subtraction *ya* 5 *ru* 151 Subtraction
ya 7 *ru* 125
being made, the value of *yávat-távat* comes out 13. Hence by "raising" the answers, the equal amount of capital is deduced, 216.[1]

In like manner, by virtue of [a variety of] assumptions, a multiplicity of answers may be obtained.

106. Example:[2] One says "give me a hundred, and I shall be twice as rich as you, friend!" The other replies, "if you deliver ten to me, I shall be six times as rich as you." Tell me what was the amount of their respective capitals?

Here, putting the capital of the first *ya* 2 *ru* 100, and that of the second, *ya* 1 *ru* 100; the first of these, taking a hundred from the other, is twice as rich as he is: and thus one of the conditions is fulfilled. But taking ten from the first, the capital of the last with the addition of ten is six times as great as that of the first: therefore multiplying the first by six, the statement is *ya* 12 *ru* 6̇60 Hence by the equation, the value of "so much as" *(yávat-*
ya 1 *ru* 110
távat) is found, 70. Thence, by "raising" the answer, the original capitals are deduced 40 and 170.[3]

107. Example:[4] Eight rubies, ten emeralds and a hundred pearls, which are in thy ear-ring, my beloved, were purchased by me for thee, at an equal amount; and the sum of the rates of the three sorts of gems was three less than half a hundred: tell me the rate of each, if thou be skilled, auspicious woman, in this computation.

Here put the equal amount *ya* 1. Then by the rule of three 'If this be the price of eight rubies, what is the price of one?' and, in like manner, [for the rest,] the rates in the several instances are *ya* $\frac{1}{8}$, *ya* $\frac{1}{10}$, *ya* $\frac{1}{100}$. The sum of these is equal to forty-seven.

[1] See the solution conducted with more than one symbol of unknown quantity. Ch. 6.

[2] Instance of putting multiples of the unknown with addition or subtraction of known quantities (§132).—CRÍSHN'. The question, however, requires no arbitrary assumption.

[3] See the solution otherwise managed in Ch. 6.

[4] This and the following examples are introduced for the gratification of learners. CRÍSHN'.

Statement for like subtraction $\begin{matrix} ya \frac{47}{200} & ru\ 0 \\ ya\ 0 & ru\ 47 \end{matrix}$ Reducing the two sides of the equation to a common denomination and dropping the denominator, the equation gives the value of the unknown *(yávat-távat)* 200. Hence, "raising" the answer, the rates of the gems are found, rubies at 25; emeralds at 20; pearls at 2. The equal amount of purchase of each sort is 200. The cost of the gems in the ear-ring 600.

Here, having reduced the terms to a common denomination, and proceeding to subtraction, when the first side of the equation is to be divided by the other, the numerator and denominator being transposed, the denominator is both multiplier and divisor of the second side of the equation. Being equal they destroy each other. Therefore, disappearance of the denominator[1] takes place.

108. Example:[2] Out of a swarm of bees, one fifth settled on a blossom of nauclea *(cadamba)*; and one third, on a flower of *śilínd'hrí*; three times the difference of those numbers flew to the bloom of an echites *(cuṭaja)*. One bee, which remained, hovered and flew about in the air, allured at the same moment by the pleasing fragrance of a jasmin and pandanus. Tell me, charming woman, the number of bees.

Here the number of the swarm of bees is put *ya* 1. Hence the number of bees gone to the blossom of the nauclea and the rest of the flowers mentioned, is *ya* $\frac{14}{15}$. This, with the one specified bee, is equal to the proposed unknown quantity *(yávat-távat)*. The statement therefore is $\begin{matrix} ya \frac{14}{15} & ru \frac{15}{15} \\ ya\ 1 & ru\ 0 \end{matrix}$

Reducing these to a common denomination and dropping the denominator, the value of the unknown *(yávat-távat)* is found, as before, 15. This is the number of the swarm of bees.

[1] *Ch'héda-gama*; departure, or disappearance, of the denominator. Equal subtraction being made, when, conformably with the order of proceeding (§ 101), the remainder of known number is divided by the residue of unknown quantity, the transposition of numerator and denominator takes place by the rule of division of fractions, *(Líl.* § 40.*)* Thus the remainder of known number is multiplied by the denominator of the unknown in one operation, and divided by it in the other. Wherefore, the multiplier and divisor, as being equal, are both destroyed. Thus departure of the denominator takes place. SÚR.

[2] This example occurs also in the author's treatise of Arithmetic. See *Líl.* § 54.

109. Example; here adduced for an easy solution, though exhibited by another author:[1] Subtracting from a sum lent at five in the hundred, the square of the interest, the remainder was lent at ten in the hundred. The time of both loans was alike, and the amount of the interest equal. [Say what were the principal sums?][2]

Here, if the period be put *yávat-távat,* the task is not accomplished. Therefore the time is assumed five months; and the principal sum is put *yávat-távat* 1. With this, the statement for rule of five[3] is

1 5
100 *ya* 1
5

The interest comes out *ya* $\frac{1}{4}$. Its square, *ya v* $\frac{1}{16}$, being subtracted from the principal sum after reducing to a common denominator, the second principal sum is found *ya v* $\dot{\frac{1}{16}}$ *ya* $\frac{16}{16}$. Here also, for [the interest for] five months, by the rule of five, the statement is

1 5
100 *ya v* $\dot{\frac{1}{16}}$ *ya* $\frac{16}{16}$
10

Answer: the interest is *ya v* $\dot{\frac{1}{32}}$ *ya* $\frac{16}{32}$. This is equal to the interest before found, namely *ya* $\frac{1}{4}$. Reducing the two sides of the equation to their least terms by their common measure *yávat-távat,* the statement of them for equal subtraction is

ya 0 *ru* $\frac{1}{4}$
ya $\dot{\frac{1}{32}}$ *ru* $\frac{16}{32}$

Proceeding as before,[4] the value of the unknown *yávat-távat* is found 8. It is the principal sum.

Or else [it may be solved in this following manner.[5]] The rate of interest for the second loan being divided by the rate of interest for the first, the quotient is a multiplier, by which the second principal sum being multiplied will be equal to the first. For, else, how should the interest in equal times be equal? The multiplier of the second sum is, therefore, in the present instance 2. The second sum, multiplied by one, and taken into the multiplier less one, is equal to the square of the interest. Hence the square

[1] It has been inserted by certain earlier writers in their treatises; and is introduced by the author for a display of his skill in the solution of the problem. It is a mixt one, and solved [as an indeterminate] by an equation involving one unknown.—SU'R. It is cited for the purpose of exhibiting an easy solution. CRÍSHN'.

[2] CRÍSHN'.

[3] To find the interest.—CRÍSHN'. See *Líl.* § 79.

[4] That is, reducing to a common denomination and dropping the denominator. CRISHN'.

[5] Without putting an algebraic symbol for the unknown quantity. CRÍSHN'.

of an assumed amount of interest, being divided by the multiplier less one, the quotient will be the second principal sum: and this, added to the square of the interest, will be the first sum.[1] Let the square of the interest then be put 4. Hence the first and second sums are found 8 and 4, and the interest 2. 'If the interest of a hundred be five, then of eight what?' By this proportion the interest of eight for one month comes out $\frac{2}{5}$. 'If by this, one month, by two how many are had?' The number of months is thus found, 5.

110. Example:[2] From a sum lent at the interest of one in the hundred, subtracting the square of the interest, the remainder was put out at five in

[1] The amount of interest on a hundred principal at the rate of one per cent. is the same with that on fifty, at two; on twenty-five, at four; on twenty, at five; and on ten, at ten. Therefore, by the same number, which multiplying the rate of the first loan raises it to the rate of the second, the principal of the first being divided equals the second. For how else should unequal principal sums produce an equal amount of interest in equal times? But the multiplier is the quotient obtained by dividing the rate of the second loan by that of the first. For the first rate is multiplicand; and the second rate, product. Therefore the second principal, multiplied by the quotient of the second rate by the first, will be the first rate. But the second principal is not known. A method of finding it follows. Were it arbitrarily assumed, the first principal would be found from it by multiplying it by the multiplier, and the amount of interest on the two sums would be equal in equal times: but the difference will not be equal to the square of the interest; [another condition of the problem.] It must be therefore treated differently. The square of the interest being subtracted from the first principal, the remainder is the second; and conversely the square of the interest added to the second is equal to the first. Consequently, to find the first sum, the second is to be added to the square of the interest; or it is to be multiplied by the multiplier. The multiplication may be by portions: thus, putting unity for one portion of the multiplicator, the other will be the multiplier less one: and the second principal multiplied by one, added to the second principal multiplied by the multiplicator less one, or second principal added to the square of the interest, will be equal to the first principal. That is, the square of the interest is equal to the second principal multiplied by the multiplicator less one. Hence the square of the interest, being divided by the multiplier less one, will be the second principal. Though the square of the interest be not known, it may be had by arbitrary assumption: and thereby the example may be solved completely. Thus interest being assumed, and its square being divided by the multiplicator less one, the quotient is the second principal. That, added to the square of the interest, is the first principal. And, from the principal sum and the amount of the interest, the time is found. And thus the solution of question is easy without putting *(yávat-távat)* a symbol for an unknown quantity. CRÍSHN'.

[2] This example is the author's own; varied but little from the preceding cited one. It is omitted by SU'RYADA'SA, but noticed by CRÍSHN'A, who observes, that it is designed to show the applicableness of the plain solution just exhibited by the author.

the hundred. The period of both loans was alike; and the amount of interest equal.

Here the multiplier is 5. The square 16, of an assumed value of the amount of interest (4) being divided by the multiplier less one, 4, the second principal sum is found, 4. This being added to the square of the interest, the first principal sum comes out 20. Hence, by a couple of proportions,[1] the time is obtained, 20.

Thus it is rightly solved by the understanding alone: what occasion was there for putting (*yávat-távat*) a sign of an unknown quantity? Or the intellect alone is analysis (*víja*). Accordingly it is observed in the chapter on Spherics, 'Neither is algebra consisting in symbols, nor are the several sorts of it, analysis. Sagacity alone is the chief analysis: for vast is inference.'[2]

111. Example: Four jewellers, possessing respectively eight rubies, ten sapphires, a hundred pearls and five diamonds, presented, each from his own stock, one apiece to the others in token of regard and gratification at meeting: and they thus became owners of stock of precisely equal value. Tell me, friend, what were the prices of their gems respectively?[3]

Here the rule for putting *yávat-távat*, and divers colours, to represent the unknown quantities,[4] is not exclusive. Designating them by the initials of their names, the equations may be formed by intelligent calculators, in this manner: having given to each other one gem apiece, the jewellers become equally rich: the values of their stocks, therefore, are

r 5	*s* 1	*p* 1	*d* 1
s 7	*r* 1	*p* 1	*d* 1
p 97	*r* 1	*s* 1	*d* 1
d 2	*r* 1	*s* 1	*p* 1

If equal be added to, or subtracted from, equal, the equality continues. Subtracting then one of each sort of gem from those several stocks, the remainders are equal: namely *r* 4, *s* 6, *p* 96, *d* 1. Whatever be the price of one diamond, the same is the price of four rubies, of six sapphires, and of ninety-six pearls. Hence, putting an assumed value for the equal amount of [remaining] stock, and dividing by those remainders severally, the prices

[1] By the rule of five; or else by two sets of three terms. CRÍSHṆ.

[2] *Gól.* 11. § 5.

[3] Already inserted in the *Lílávatí*, § 100. 'It is a further instance of a solution by putting several sums equal.'—SU'R. The problem is an indeterminate one.

[4] See the author's gloss on the rules at the beginning of Ch. 6.

are found. Thus, let the value be put 96, the prices of the rubies and the rest are found, 24, 16, 1, and 96.[1]

112. Example: A principal sum, being lent at the interest of five in the hundred [by the month], amounted with the interest, when a year was elapsed, to the double less sixteen. Say what was the principal?

Here the principal is put *ya* 1. Hence by the rule of five

1	12
100	*ya* 1
5	

the interest is found *ya* $\frac{3}{5}$. This, added to the principal, makes *ya* $\frac{8}{5}$. It is equal to sixteen less than the double of the principal, namely *ya* 2 *ru* $\dot{1}\dot{6}$. By this equation the principal sum is found 40; and the interest 24.

113. Example:[2] The sum of three hundred and ninety was lent in three portions, at interest of five, two and four in the hundred; and amounted in seven, ten, and five months respectively, to an equal sum on all three portions, with the interest. Say the amount of the portions.

The equal amount of each portion with its interest is put *ya* 1. 'If, for one month, five be the interest of a hundred; then, for seven months, what is the interest of the same?' Thus the interest of a hundred is found 35. This, added to a hundred, makes 135. 'If of this amount with interest, the principal be a hundred, then, of the amount with interest, that is measured by *yávat-távat*, what is the principal?' The quantity of the first portion is thus found *ya* $\frac{20}{27}$. Again, 'if, for a month, two be the interest of a hundred, then, for ten months, what is the interest of the same?' Proceeding with the rest of the work in the manner above shown, the second portion is *ya* $\frac{5}{6}$. In like manner, the third portion is *ya* $\frac{5}{6}$. Their total *ya* $\frac{65}{27}$ is equal to the whole original sum 360. Whence the value of *yávat-távat* is had 162. By this the portions sought are "raised:" namely 120, 135 and 135. The equal amount of principal with interest is 162.

114. Example: A trader, paying ten upon entrance into a town, doubled his remaining capital, consumed ten [during his stay] and paid ten on his

[1] And the amount of each man's stock, after interchange of presents, is 233. SU'R.

[2] Varied from an example in arithmetic, partly set forth in similar terms. *Lil.* §91.

departure. Thus, in three towns [visited by him] his original capital was tripled. Say what was the amount?[1]

Here the capital is put *ya* 1. Performing on this, all that is set forth in the question, the capital becomes on his return from three towns *ya* 8 *ru* 280. Making this amount equal to thrice the original capital, *ya* 3, the value of *yávat-távat* comes out 56.

115. Example: If three and a half *mánas* of rice may be had for one *dramma*, and eight of kidney-beans for the like price, take these thirteen *cácinís*, merchant! and give me quickly two parts of rice, and one of kidney-beans: for we must make a hasty meal and depart, since my companion will proceed onwards.[2]

Here the quantity of rice is put *ya* 2, and that of kidney-beans *ya* 1. 'If, for these three and a half *mánas* one *dramma* be obtained; then, for this quantity *ya* 2, what is had?' The price of the rice is thus found *ya* $\frac{4}{7}$. 'If, for eight *mánas* one *dramma*, then for this *ya* 1 what?' The price of the kidney-beans is thus found *ya* $\frac{1}{8}$. The sum of these, *ya* $\frac{39}{56}$, is equal to thirteen *cácinís*; or in *drammas* $\frac{13}{64}$. From this equation the value of *yávat-távat* comes out $\frac{7}{24}$. Whence, "raising" the answers, the prices of the rice and kidney-beans are deduced $\frac{1}{6}$ and $\frac{7}{192}$; and the quantity of rice and of kidney-beans, in fractions of a *mána*, $\frac{7}{12}$ and $\frac{7}{24}$.

116. Example: Say what are the numbers, which become equal, when to them are added respectively, a moiety, a fifth, and a ninth part of the number itself; and which have sixty for remainder, when the two other parts are subtracted.

Here the equal number is put *ya* 1. Hence, by the rule of inversion,[3] the

[1] This is according to the gloss of CRÍSHN'A, and conformable with collated copies of the text. But SU'RYA, reading *daśa-yucta-nirgamé* instead of *daśa-bhuc cha nirgamé*, omits the consumption of ten, during stay; and confines the disbursement, after doubling the principal, to ten for duties on export. The equation, according to this commentator, is *ya* 8 *ru* 210; and the value of *ya*, 42.
ya 3 *ru* 0

[2] Spoken by a pious native of *Gurjara*, going to *Dwáricá* to visit holy CRÍSHN'A; and stopping by the way for refreshment, but in a hurry to proceed, under apprehension of being separated from his fellow traveller.—SU'R. The same example has been already inserted, word for word, in the author's arithmetic. *Líl.* § 97.

[3] *Líl.* § 48. The fractions $\frac{1}{2}$, $\frac{1}{5}$, $\frac{1}{9}$, become negative; and the denominator being increased by

several numbers are *ya* $\frac{2}{3}$, *ya* $\frac{5}{6}$, *ya* $\frac{9}{10}$. In this case, all the numbers, diminished by the subtraction of the other two parts, will be brought to the same remainder, *ya* $\frac{2}{5}$.[1] This being made equal to sixty, the value of *yávat-távat* is obtained 150. Whence, by " raising" the answers, the numbers are deduced 100, 125 and 135.

117. Example:[2] Tell me quickly the base of the triangle, the sides of which are the surds thirteen and five, and the base unknown, and the area four?

In this instance, if the base be assumed *yávat-távat*, the solution is tedious.[3] Therefore the base is put in the triangle any way at pleasure,[4] since it makes no difference in the result. Accordingly the triangle is here put thus:

ya 1
c 5
c 13

Now, by the converse of the rule " half the base multiplied by the perpendicular, is in a triangle the exact area;" (*Lil.* § 164.) the perpendicular is deduced from the area divided, by half the base: viz. *c* $\frac{64}{13}$. Subtracting the square of the perpendicular from the square of the side, the square-root of the difference is the segment *c* $\frac{1}{13}$. This, subtracted from the base, leaves the other segment *c* $\frac{144}{13}$. The square of this being added to the square of the perpendicular, the square-root of the sum is the side: viz. 4.* This [the triangle being turned] is the base sought.

the numerator for a new denominator, they become $\frac{1}{3}$, $\frac{1}{6}$, $\frac{1}{10}$; which, being subtracted from *ya* 1 rendered homogeneous, leave the several original numbers *ya* $\frac{2}{3}$, *ya* $\frac{5}{6}$, *ya* $\frac{9}{10}$. SU'R.

[1] *Ya* $\frac{1}{6}$ and *ya* $\frac{1}{10}$ (making *ya* $\frac{16}{60}$), subtracted from *ya* $\frac{2}{3}$, leave *ya* $\frac{72}{180}$; *ya* $\frac{1}{3}$ and *ya* $\frac{1}{10}$ (making *ya* $\frac{13}{30}$), subtracted from *ya* $\frac{5}{6}$, leave *ya* $\frac{72}{180}$; *ya* $\frac{1}{3}$ and *ya* $\frac{1}{6}$ (making *ya* $\frac{9}{18}$), subtracted from *ya* $\frac{9}{10}$, leave *ya* $\frac{72}{180}$: which, reduced to least terms, is *ya* $\frac{2}{5}$.

[2] This example and the following are introduced to show, that the method of performing arithmetical operations, as taught in a preceding section (Ch. 1. Sect. 4), are not useless trouble. SU'R.

[3] It requires the resolution of quadratic equations. CRĬSHN.

[4] Any one of the sides is made the base.

* Half the base *c* 13 is *c* $\frac{13}{4}$. The area *ru* 4 or *c* 16, divided by that, is *c* $\frac{64}{13}$. Its square *ru* $\frac{64}{13}$, subtracted from the square *ru* 5 of the side *c* 5, or reduced to a common denomination *ru* $\frac{65}{13}$, leaves the square *ru* $\frac{1}{13}$; of which the square-root is *c* $\frac{1}{13}$. This, subtracted from *c* 13, leaves *c* $\frac{144}{13}$. Its square *ru* $\frac{144}{13}$, added to the square of the perpendicular *ru* $\frac{64}{13}$, makes *ru* $\frac{208}{13}$ or *ru* 16. Its root is 4. In like manner, putting the other side *c* 5 for the base, the perpendicular is *c* $\frac{64}{5}$. Its square *ru* $\frac{64}{5}$, subtracted from the square *ru* 13 of the remaining side (*c* 13), leaves the segment *c* $\frac{1}{5}$ (the

118. Example: The difference between the surds ten and five is one side of the triangle; the surd six is the other; and the base is the surd eighteen less rational unity, tell the perpendicular.

Here, if the segment be known, the perpendicular is discovered. Put the least segment *ya* 1. The base, less that, is the value of the other segment, *ya* $\dot{1}$ *ru* $\dot{1}$ *c* 18. Thus the statement is

c 5 $\dot{c}$10 | c 6
ya 1 | ya $\dot{1}$ ru $\dot{1}$ c 18
ru $\dot{1}$ c 18

Subtracting the square of the segment from the square of its contiguous side, two expressions of the square of the perpendicular are found *ya v* $\dot{1}$ *ru* 15 *c* $2\dot{0}0$ and *ya v* $\dot{1}$ *ya* $\dot{2}$ *ya c* $7\dot{2}$ *ru* $\dot{1}3$ *c* 72. They are equal: and, equal subtraction being made, the two sides of the equation become $\frac{ru\ 2\dot{8}\quad c\ 512}{ya\ 2\quad ya\ c\ 7\dot{2}}$ Here the syllable *ya* [the symbol of the unknown], in the divisor, being useless, is excluded;[1] and the dividend and divisor are alike $\frac{ru\ 2\dot{8}\quad c\ 512}{ru\ 2\quad c\ 7\dot{2}}$ Then, by the rule for "reversing the sign of a selected surd, and multiplying both dividend and divisor by the altered divisor," (§ 34) putting the surd seventy-two affirmative, and multiplying by *c* 4 *c* 72, the dividend becomes *c* $3\dot{1}36$ *c* 2048 *c* $5\dot{6}448$ *c* 36864. Taking the difference between the first and last, and between the third and fourth, it is reduced to *c* 18496 *c* $369\dot{9}2$ (or *ru* 136 *c* $369\dot{9}2$). The divisor in like manner becomes *c* $462\dot{4}$ (or *ru* $6\dot{8}$). Thus the statement of dividend and divisor is $\frac{ru\ 136\quad c\ 369\dot{9}2}{ru\ 6\dot{8}}$. The division being made in the manner directed, the value of *yávat-távat* is found *ru* $\dot{2}$ *c* 8. This is the least segment. The base, less this, is the other segment; namely, *ru* 1 *c* 2. From the value of the unknown *yávat-távat*, "raising" the expressions of the square of the perpen-

root of *ru* $\frac{1}{5}$); whence the other segment is *c* $\frac{16}{5}$. Its square *ru* $\frac{16}{5}$ added to the square of the perpendicular *ru* $\frac{64}{5}$ is *ru* $\frac{80}{5}$ or *ru* 16; the square-root of which is 4, (the other side, or base required,) as before. CRÍSHN'.

[1] This is the case in all instances, for the proportion to find the value of the unknown, is "if this multiple of *ya* give this known number, what does *ya* give?" and thus, being alike in both the multiplier and divisor, it is excluded from both. The author, however, has not noticed its exclusion in other instances, where the algebraic solution was in this respect obvious: but, in the present case, where the sign of a selected surd is to be reversed, and the dividend and original divisor to be multiplied by the altered divisor, its presence in that multiplication would be highly disserviceable. Its exclusion is now therefore specially noticed. CRÍSHN'.

dicular, or subtracting the square of the segment from the square of its contiguous side, the square of the perpendicular is deduced *ru* 3 *c* $\dot{8}$. Its square-root is the value of the perpendicular *ru* $\dot{1}$ *c* 2.[1]

119. Example:[2] Tell four unequal[3] numbers, thou of unrivalled understanding![4] the sum of which, or that of their cubes, is equal to the sum of their squares.

Here the numbers are put *ya* 1, *ya* 2, *ya* 3, *ya* 4. Their sum is *ya* 10. It is equal to the sum of their squares *ya v* 30. Dividing the two sides by the common measure *yávat-távat*, the statement is $\begin{matrix} ya\ 30 & ru\ 0 \\ ya\ 0 & ru\ 10 \end{matrix}$ From the value of *yávat-távat* hence found as before, by equal subtraction [and division, § 101] viz. $\frac{1}{3}$, the numbers are deduced by substitution of that value, $\frac{1}{3}$, $\frac{2}{3}$, $\frac{3}{3}$, $\frac{4}{3}$.*

[1] The problem may be solved by the arithmetic of surds without algebra. (*Líl.* § 163.) The sum of the sides is *c* $\dot{5}$ *c* 10 *c* 6. Their difference *c* 5 *c* $\dot{10}$ *c* 6. Multiplied together, the product comprises nine terms, *c* $\dot{25}$ *c* 50 *c* 30 *c* 50 *c* $\dot{100}$ *c* $\dot{60}$ *c* $\dot{30}$ *c* 60 *c* 36; wherein *c* 30 *c* $\dot{60}$ *c* $\dot{30}$ *c* 60 balance each other; *c* 50. and *c* 50 added together make *c* 200; and *c* $\dot{25}$ *c* $\dot{100}$ and *c* 36, being rational, make *ru* $\dot{5}$ *ru* $\dot{10}$ *ru* 6, or summed *ru* $\dot{9}$. The product then is *ru* $\dot{9}$ *c* 200; to be divided by the base *ru* $\dot{1}$ *c* 18. Thus the statement is $\frac{c\ \dot{81}\quad c\ 200.}{c\ \dot{1}\quad c\ 18}$ Proceeding with this by the rule, § 34, and putting *c* $\dot{1}$ positive, the dividend becomes by multiplication *c* $\dot{81}$ *c* 200 *c* 1458 *c* 3600, reducible by the difference of the roots of the rationals 81 and 3600, and by finding the difference of the irrationals 200 and 1458, to *ru* 51, *c* 578. The divisor by similar multiplication is *c* $\dot{1}$ *c* 18 *c* $\dot{18}$ *c* 324; wherein the middle terms balance each other, and the remaining two are rational, giving the difference *ru* 17 or in the surd form, *c* 289. Hence the quotient of *ru* 51 *c* 578 by *ru* 17 (or *c* 289) is *ru* 3 *c* $\dot{2}$; which added to, and subtracted from, the base *ru* $\dot{1}$ *c* 18, gives the sum and difference *ru* 2 *c* 8 and *ru* $\dot{4}$ *c* 32; the moieties whereof *ru* 1 *c* 2 and *ru* $\dot{2}$ *c* 8, are the two segments: and from these the perpendicular is found as before, *ru* $\dot{1}$ *c* 2. CRÍSHN'.

[2] These and several following examples are instances of the resolution of equations involving the square, cube, or other [power] of the unknown, by any practicable depression of both sides by some common divisor, without elimination of the middle term. CRÍSHN'.

[3] Unequal or dissimilar; unalike.—SU'R. This is a necessary condition. Else unity repeated would serve for an answer to the question.—CRÍSHN'.

[4] SU'RYADA'SA reads and interprets *asama-prajnya*, of unrivalled understanding! CRÍSHN'A-BHAT'T'A notices that reading, as well as the other, *samach'hédán* having like denominators: rejecting, however, this; as it is not necessary, that it be made a condition of the problem, though it rise out of the solution.

* Sum $\frac{10}{3}$. Sum of the squares $\frac{30}{9}$. SU'R.

In the second example, the numbers are also put *ya* 1, *ya* 2, *ya* 3, *ya* 4: the sum of their cubes is *ya gh* 100, equal to the sum of the squares *ya v* 30. Depressing the two sides by the common divisor, square of *yávat-távat;* and proceeding as before, the numbers are deduced by substituting the value of *yávat-távat* ($\frac{3}{10}$); namely, $\frac{3}{10}$, $\frac{6}{10}$, $\frac{9}{10}$, $\frac{12}{10}$.*

120. Example: Tell the [sides of a] triangle, of which the area may be measured by the same number with the hypotenuse; and [of] that, of which the area is equal to the product of side, upright and hypotenuse multiplied together.

In this case, the statement of an assumed figure is

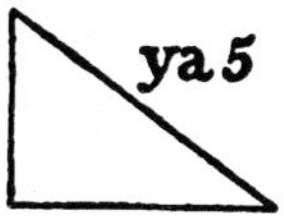

ya 3
ya 5
ya 4

Here half the product of the side multiplied by the upright is the area, *ya v* 6. It is equal to the hypotenuse *ya* 5. Depressing both sides by the common measure *yávat-távat*, and proceeding as before, the side, upright and hypotenuse, deduced from the value found of *yávat-távat* (viz. $\frac{5}{6}$) are $\frac{10}{3}$, $\frac{5}{2}$ and $\frac{25}{6}$. In like manner, by virtue of [various] assumptions, other values also may be found.[2]

In the second example, the same figure is assumed. Its area is *ya v* 6. This is equal to the product of side, upright and hypotenuse, *ya gh* 60. Depressing both sides by the common divisor, square of *yávat-tavat*, the side, upright and hypotenuse, found as before, from the equation, are $\frac{2}{5}$, $\frac{3}{10}$ and $\frac{1}{2}$. By virtue of assumptions, other values likewise may be obtained.[2]

121. Example: If thou be expert in this computation, declare quickly two numbers, of which the sum and the difference shall severally be squares; and the product of their multiplication, a cube.

Here the two numbers are put *ya v* 4 *ya v* 5; so assumed, that being added or subtracted, the sum or difference may be a square.[3] The product of their multiplication is *ya v v* 20. It is a cube. By making it equal to the

* Sum of the cubes $\frac{2700}{1000}$; of the squares, $\frac{270}{100}$. SÚR.

[2] That is, both problems are indeterminate. So likewise were those proposed in the preceding stanza, § 119.

[3] *Ya v* 9; or *ya v* 1.

cube of ten times the assumed *yávat-távat*, and depressing the two sides of the equation by the common divisor, cube of *yávat-távat*, and proceeding as before, the two numbers are found 10000 and 12500.[1]

122. Example: If thou know two numbers, of which the sum of the cubes is a square, and the sum of their squares is a cube, I acknowledge thee an eminent algebraist.

In this instance the two numbers are put *ya v* 1, *ya v* 2. The sum of their cubes is *ya v gh* 9. This of itself is a square as required. Its root is *ya gh* 3.

Is not that quantity the cube of a square, not the square of a cube? No doubt the root of the square of a cube is cube. But how is the root of the cube of a square, a cube? The answer is, the cube of the square is precisely the same with the square of the cube.[2] Hence if squares be raised twice, or four, or six, or eight times, their roots will be so once, twice, thrice, or four times, respectively. It must be so understood in all cases.

Now the sum of the squares of those quantities is *ya v v* 5. It must be a cube. Making it equal to the cube of five times *yávat-távat*, and depressing the two sides of the equation by the common divisor, cube of *yávat-távat*, and proceeding as before, the two numbers are found 625 and 1250.[3]

[1] From the depressed equation *ya* 20 *ru* 0
ya 0 *ru* 1000
the value of *ya* is found 50. Its square is 2500, of which the multiples *ya v* 4 and *ya v* 5 are 10000 and 12500. In like manner, putting other quantities, as *ya v* 16 and *ya v* 20; and making their product *ya v v* 320 equal to cube of *ya* 20, (*ya gh* 8000;) the equation depressed by the common measure *ya gh*, is *ya* 320 *ru* 0
ya 0 *ru* 8000
Whence the value of *ya* is 25; the square of which is 625; and its multiples 10000 and 12500 are the two numbers. By varying the suppositions, a multiplicity of answers is obtained. SÚR.

[2] The cube of the square is the sixth product of the quantity. It is the third product of the second product of equal quantities multiplied. As the cube of the second product, so is the second product of like multiplication of a third product. Therefore, it is also the square of the third product. CRÍSHN'.

[3] The value of *ya* being 25. Or, putting the two numbers *ya gh* 5, and *ya gh* 10, the sum of their squares is a cube *ya gh v* 125. Its cube root is *ya v* 5. The sum of the cubes of the same quantities is *ya gh gh* 1125. It is a square. Make it equal to the square of *ya v v* 75, viz. *ya v v* 5625. Reduce the two sides of the equation by the common measure *ya v v v*. The equation is *ya* 1125 *ru* 0
ya 0 *ru* 5625
Whence the value of *ya* is found 5; and the two numbers 625 and 1250. In like manner a multiplicity of answers may be obtained. CRÍSHN'.

Thus it is to be considered, how practicably the unknown quantity [or its power] may be made a common measure.

123. Example: Tell me, friend, the perpendicular in a triangle, in which the base is fourteen, one side fifteen, and the other thirteen.[1]

If the segment be known, the perpendicular is so. Put therefore *yávat-távat* for the least segment, *ya* 1: the other is the base less *yávat-távat*, *ya* $\dot{1}$ *ru* 14. Statement:

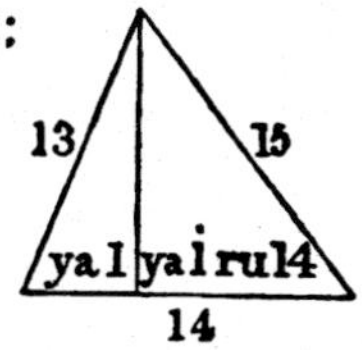

13 15
ya 1 ya $\dot{1}$ ru 14
14

The squares of the sides, less the squares of the contiguous segments, are the square of the perpendicular. They are equal consequently. Statement of them for equal subtraction: *ya v* $\dot{1}$ *ya* 0 *ru* 169. From these the equal
ya v $\dot{1}$ *ya* 28 *ru* 29
square vanishes; and then, proceeding as before, the value of *yávat-távat* is found, 5. From which the two segments are deduced 5 and 9; and the square of the perpendicular being "raised" [by substitution of that value] in both expressions, it is deduced alike both ways: viz. 12.

Here the substitution for a square, is by a square; and for a cube, by a cube: as is to be understood by the intelligent calculator.

124. Example:[2] If a bambu, measuring thirty-two cubits, and standing upon level ground, be broken in one place by the force of the wind; and the tip of it meet the ground at sixteen cubits; say, mathematician, at how many cubits from the root is it broken?

In this case, the lower portion of the bambu is the upright. Its value is put *yávat-távat*. Thirty-two, less that, is the upper portion, and is the hypotenuse. The interval between the root and tip is the side. See

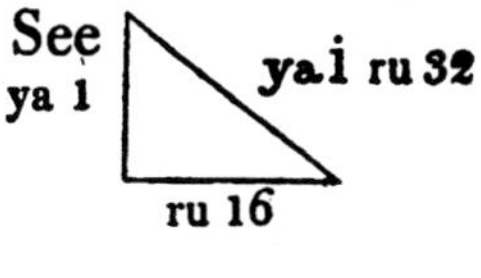

ya 1 ya $\dot{1}$ ru 32
ru 16

[1] There was not much occasion for this example.—CRÍSHN'. For the finding of the perpendicular had been already exemplified by § 118. That, however, was performed by the arithmetic of surds: and this is done by a plain algebraic calculation.

[2] The base, as well as the sum of the hypotenuse and upright, being given, to discriminate them. SUR. and CRÍSHN'. See *Líl.* § 148; where this example has been already inserted.

The sum of the squares of the side and upright, *ya v* 1 *ru* 256, is equal to the square of the hypotenuse, *ya v* 1 *ya* 64 *ru* 1024. The equal squares vanishing, and [the usual process being pursued] as before, the value of *yávat-távat* is found 12. Whence the upright and hypotenuse are deduced by substitution of that value, 12, 20. In like manner, if the sum of the hypotenuse and side be given, as in the example "A snake's hole is at the foot of a pillar,"[1] [they may be discriminated] also.

125. Example:[2] In a certain lake swarming with ruddy geese and cranes, the tip of a bud of lotus was seen a span above the surface of the water. Forced by the wind, it gradually advanced, and was submerged at the distance of two cubits. Compute quickly, mathematician, the depth of water.

In this case, the length of the stalk of the lotus is the depth of water. Its value is put *ya* 1. It is the upright. That, added to the bud of the lotus, is the hypotenuse, *ya* $\frac{2}{2}$ *ru* $\frac{1}{2}$. The side is two cubits. See

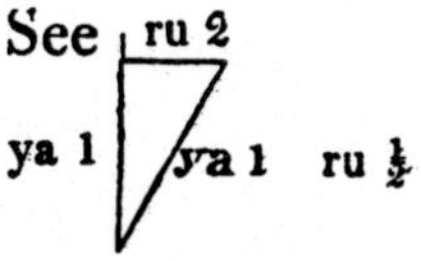

ru 2

ya 1 ya 1 ru $\frac{1}{2}$

Here also, the sum of the squares of the side and upright, *ya v* 1 *ru* 4, being made equal to the square of the hypotenuse *ya v* 1 *ya* 1 *ru* $\frac{1}{4}$, the depth of water, which is the value of the upright, is found $\frac{15}{4}$; and the hypotenuse $\frac{17}{4}$.

126. Example:[3] From a tree a hundred cubits high, an ape descended and went to a pond two hundred cubits distant: while another ape, vaulting to some height off the tree, proceeded with velocity diagonally to the same spot. If the space travelled by them be equal, tell me quickly, learned man, the height of the leap, if thou have diligently studied calculation.

The equal distance travelled is 300. The measure of the leap is put *ya* 1. The height of the tree added to this is the upright: the equal distance travelled, less *yávat-távat*, is the hypotenuse. The interval between the tree

[1] *Líl.* § 150. See further a note on § 140 of this treatise.

[2] The difference between the hypotenuse and upright, as well as the side (base) being given, to find the hypotenuse and upright.—SU'R. and CRI'SHN'. See *Líl.* § 153, where the same example is inserted. See likewise *Líl.* § 152.

[3] This also is found in the *Líláváti*, § 155. It is borrowed with some variation from BRAHME-GUPTA or his commentator. BRAHM. 12. § 39.

and the well is the side. See

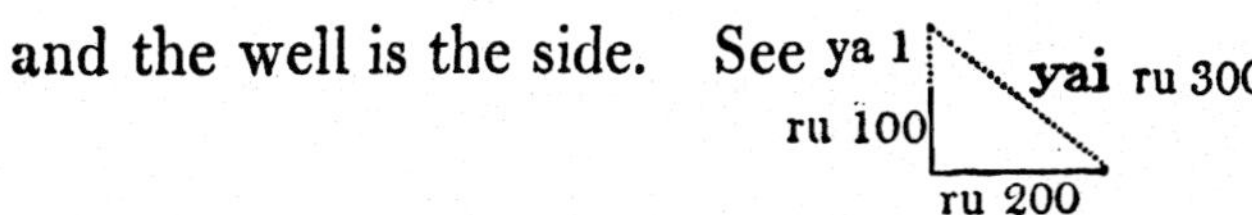

Making the sum of the squares of upright and side equal to the square of the hypotenuse, the measure of the leap is found 50.[1]

127. Example:[2] Tell the perpendicular drawn from the intersection of strings stretched mutually from the roots to the summits of two bambus fifteen and ten cubits high standing upon ground of unknown extent.[3]

Here, to effect the solution, the value of the ground intercepted between the bambus is arbitrarily assumed: say 20. Put the value of the perpendicular let fall from the intersection of the strings *ya* 1. See

'If, to the upright fifteen, the side be twenty, then to one measured by *yávat-távat* what will be the side?' Thus the segment contiguous to the less bambu is found *ya* $\frac{4}{3}$. 'If, to the upright ten, the side be twenty, then, to one measured by *yávat-távat*, what will be the side?' Thus the segment contiguous to the greater bambu is found *ya* 2. Making the sum of these, *ya* $\frac{10}{3}$ equal to twenty, *ru* 20, the perpendicular comes out 6; the value of *yávat-távat*. Whence, by substitution of this value, the segments are deduced 8 and 12.

Or the segments are relative [that is, proportionate] to the bambus; and their sum is the base.[4] As the sum of the bambus (25) is to the sum of the segments (20); so is each bambu (15 and 10) to the segments respectively (12 and 8). They are thus found; and, from them, by a proportion,[5] the per-

[1] And hence the value of the hypotenuse, 250.

[2] Like the preceding, this too is repeated from the *Lílávatí*, § 160.

[3] The ground intercepted between the bambus is expressly said to be of unknown extent, to intimate that the distance is not necessary to the finding of the perpendicular. CRÍSHN.

[4] As the bambu becomes greater or less, so does the segment. It may be found therefore by the rule of three: 'as the sum of the bambus is to whole base, so is one bambu to the particular segment.' CRÍSHN.

[5] As a side, equal to the base, is to an upright, which is the less bambu, so is a side, equal to the greater segment, to the perpendicular. The less bambu is the upright, the base is the side, and the string passing from the tip of the less bambu to the root of the other bambu is the hypotenuse; by virtue of the figure: and [in the small triangle] the greater segment is side, and the perpendi-

pendicular is deduced (6). What occasion then is there for [putting a symbol of the unknown] *yávat-távat?*

Or the product of the two bambus multiplied together, being divided by their sum, will be the perpendicular, whatever be the distance between the bambus.[1] What use then is there in assuming a base? This will be clearly understood by the intelligent, after stretching strings upon the ground [to exhibit the figure].

cular is upright. In like manner, proportion is to be made with the less segment and greater bambu. CRÍSHN'.

[1] As the sum of the bambus is to the base, so is the greater bambu to the segment contiguous to it? Then, 'as the base is to the less bambu, so is the greater segment to the perpendicular.' Here the base, being both dividend and divisor, vanishes; and the product of the two bambus, divided by their sum, is the perpendicular. CRÍSHN'.

CHAPTER V.

QUADRATIC, &c. EQUATIONS.

NEXT equation involving the square or other [power] of the unknown is propounded. [Its re-solution consists in] the elimination of the middle term,[1] as teachers of the science[2] denominate it. Here removal of one term, the middle one, in the square quantity, takes place: wherefore it is called elimination of the middle term. On this subject the following rule is delivered.

128—130. Rule: When a square and other [term] of the unknown is involved in the remainder; then, after multiplying both sides of the equation by an assumed quantity, something is to be added to them,[3] so as the side[4] may give a square-root. Let the root of the absolute number again be

[1] *Mad'hyamáharańa;* from *mad'hyama* middlemost, and *áharańa* a taking away or *(apanayana)* removal,—CRÍSHŃ. or *(nása)* a destroying,—SU'R. that is, elimination. The resolution of these equations is so named, because it is in general effected by making the middle term (twice the product of the roots, § 26) disappear from between two square terms.—SU'R. and CRÍSHŃ.; and note on "equation,", Ch. 4.

[2] *Áchàryas;* ancient mathematicians *(ádya-ganaca):* as ÁRYA-BHAŤŤA and the son of JISHŃU [BRAHMEGUPTA] and CHATURVÉDA [PRÍT'HU'DAÇA-SWÁMÍ]. SU'R.

[3] This is not exclusive: in some cases the two sides are to be reduced by some common divisor; and in some instances an assumed quantity is to be subtracted from both sides. CRÍSHŃ.

[4] So as each side of the equation may yield a square-root. Both being rendered such, the root of the known is then to be made equal to the root of the unknown side.—SU'R. However, if the absolute number be irrational, its root may be put in the form of a surd. See CRÍSHŃ. cited in note 1 next page.

made equal to the root of the unknown: the value of the unknown is found from that equation.[1] If the solution be not thus accomplished, in the case of cubes or biquadrates, this [value] must be elicited by the [calculator's] own ingenuity.

130. If the root of the absolute side of the equation be less than the number, having the negative sign, comprised in the root of the side involving the unknown, then putting it negative or positive, a two-fold value[2] is to be found of the unknown quantity: this [holds] in some cases.[3]

[1] When on one side is the unknown quantity; and on the other the absolute number; then, since they are equal, the absolute number is the value of the unknown: as already shown. But, for the purpose of a division of the remaining absolute number, it is requisite that the unknown should stand separate. The equation, therefore, must be so treated; as that such may be the case. If to equal sides then equal addition be made, or from them equal subtraction; or if equal multiplication or equal division of them take place; or extraction of their roots, or squaring of them, or raising to the cube or other [power]; there is no loss of equality: as is clear. Now, if a square or other [power] of the unknown be on one side, and number only on the other, the unknown cannot stand separate without extraction of the root. The roots, therefore, of both sides are to be extracted consistently with their equality: and, that being done, the roots also will be equal. Therefore it is said, "after multiplying both sides by an assumed quantity, something is to be added to them." CRÍSHN'.

An equation comprising a single colour (or symbol of unknown quantity) being prepared as before directed, when, after equal subtraction, division is next to be made (§ 101), if on the side of the unknown the square and other term of the unknown remain, then both sides are to be multiplied by some quantity, and something is to be added; so as both sides may furnish squares. The roots of both being then extracted, let equal subtraction be again made; and, that being done, the value of the unknown is obtained by division. The principle of the process is, that, if the residue (or prepared equation) comprise the square and other term of the unknown, the solution can be then only effected, when the square is the only term of the unknown; and that can only be when the root of the square is extracted. Accordingly it is said "so as the side may give a square-root." Thus, the root of the side of the equation comprising the square of the unknown being extracted, the remainder (§ 26) of the compound square, that is, the middle term, which stands between the square of the unknown and square of number, is exterminated. SU'R.

[2] *Mána, miti, unmána,* or *unmiti;* measure, value; root of the equation. See § 17.

[3] Making the known number on the side of the equation involving the unknown term, positive, a value of the unknown is to be found: then, making the same negative, another value of the unknown is to be deduced: and thus, both ways, the condition of the question is answered. Bearing this in mind, the author has said in another place, 'There are two hypotenuses, &c.' SU'R.

Demonstration: When equal subtraction from equal sides takes place, if there be a

131. Śríd'hara's rule on this point: 'Multiply both sides of the equation by a number equal to four times the [coefficient] of the square, and

number, negative, on the side of the unknown; then, by the rule for changing the sign of the subtrahend (§ 5), the sum of the absolute numbers on the two sides will be the value: but, if the number be positive on the unknown side, then, by same rule, the difference of the quantities is the value. When it is thus discovered how many times the sum or difference of the absolute quantities is greater than the difference of the terms of the unknown, the value of the unknown is obtained, being that which the sum or difference so many times measures. Thus both answer the conditions of the question; because such multiple of the divisor balances the dividend. [*Lil.* § 17.] But, if the value of *yávat-távat* be too little with reference to the given number specified in the example, it is unsuitable. With a view to this, the author has said at the close of the rule, It holds "in some cases." (§ 130). Súr.

If the root of the known quantity exceed the absolute number comprised in the root on the side of the unknown, why should not there be a two-fold value in this instance also, by the same reasoning as in the other case? Hear the reply. In the instance, where the number being negative, the unknown is positive, that number must be subtracted from the absolute side, that the remainder of the unknown may be positive: the number becomes therefore affirmative; and there is no incongruity. But in the second case, where, the number being positive, the unknown is negative, the unknown must be subtracted from the other side of the equation, to become affirmative; and the number on the absolute side, being subtracted from that comprised in the root extracted from the unknown side, becomes negative: and, if it be the greater of the two, the value is negative. The second value consequently is every way incongruous. Hence the rule (§ 130). When the tenor of the condition is "unknown less number," if that is to be squared, the term of the simple unknown comes out negative, because the number [which multiplies it] is so. In such case when the square-root is extracted, number only is negative, not the unknown: for it is certain that the number is negative in the condition as proposed. If the unknown were put negative, the side of the equation would be negative; for it cannot be affirmative while the greater quantity is subtractive. Or, admit that it is affirmative in some cases; still it would differ from the side of equation that is consistent with the condition of the problem. Such being the case, how can it be equal to the second side which is one consistent with that condition? Therefore a value, coming from such an equation, must be incongruous; because it is negative: for people do not approve a negative absolute number. Hence, in such an example, although the root of the known quantity be less than the negative number in the root of the unknown side (§ 130), there cannot be a two-fold value: for a value, grounded on the assumption of the number being affirmative [contrary to the tenor of the condition] must be incongruous. In like manner, if the tenor be "the number less the unknown squared," the unknown alone, not the number, will be the negative in the root: by parity of reasoning. Therefore, in this instance also, there cannot be a two-fold value: for a value, grounded on the assumption of the number being negative, will be incongruous. It is thus in many various ways. Sometimes, by subtraction of the addition or other means it is the reverse. Sometimes, by reason of the unknown being naturally negative, though the root might possibly be two-fold, the second is incongruous. Accordingly the author has said indefinitely, "This holds in some cases." See an example of the incongruity of the second value, § 140. See likewise an

add to them a number equal to the square of the original [coefficient] of the unknown quantity.[1] (Then extract the root.[2])[3]

instance in the chapter on the three problems (*Śirómańi*, Ch. 3, § 100), where the question is proposed as one requiring the resolution of a quadratic equation; and the answer (*ib.* § 101) shows, that, in taking the roots of the two sides of the equation, the unknown has been taken negative and the absolute number positive: for, if the number were taken negative, the answer would come out differently. Thus by the reasoning here set forth, the congruity or incongruity of a two-fold value is to be every where understood: and the author's remark, it holds in some cases (§ 130), is justified.
CRĪSHŃ.

[1] That is, multiply both sides by the quadruple of the number belonging to the square of the unknown; and add to both sides the square of the number which belonged to the unknown previous to that multiplication. This being done, the side involving the unknown must furnish a square-root: and the second side of the equation, being equal to it, should do so likewise. Else of course the instance is an imperfect one.—CRĪSHŃ. If the known side, nevertheless, do not furnish a root, it must be taken in the form of a surd.—*Ibid.* ŚRÍD'HARA's rule, having reference both to the unknown and to its square, is applicable only in the case where the same side of equation comprises the square of the unknown, and the unknown too. In any other case means of obtaining a root must be devised by the intelligent calculator exercising his own ingenuity.*—*Ibid.*

[2] This insertion is according to the reading which occurs and is expounded in the commentary of SŨRYADÁSA, *avyacta-varga-rupair yuctau pacshau tatŏ múlam:* "both sides are to have added, numbers equal to the square [of the coefficient] of the unknown. Then the root [is to be extracted]." CRĪSHŃA-BHAṬṬA's reading and RÁMA-CRĪSHŃA's, with which their exposition too is conformable, and which has been followed in the preceding part of the version of the text, differs widely: *púrvá'vyactasya crĭtéh sama-rupáńi cshipét tayór éva.* Collated copies of BHÁSCARA's text agree with this: but the variation is marked in the margin of one exemplar.

[3] Demonstration: After preparing the equation by equal subtraction, if the square of the unknown and the simple unknown be on one side, and absolute number only on the other; the first side of the equation cannot by any means furnish a root without some addition. For, if the unknown alone be squared, the square of the unknown will be the single result: but, if the unknown with number added to it, be squared, the result will be [three terms] square of the unknown, unknown, and number: but, in the proposed case, two terms only being present, namely square of the unknown, and unknown, it is not the square of any quantity. Therefore, number must be added. By subtraction of the unknown, the square of the unknown remaining a single term, would furnish a root: but then the other side consisting of [two terms] unknown and number, would not afford one. If the term, containing the square of the unknown, give a root, the addition only of a number is needed. But, if it do not afford a root, that term must be multiplied or divided by some quantity. Addition to it will not answer; for the other side, with equal addition, will yield no root. Nor should a term of the unknown be added to both sides: for that would be troublesome. Besides, if the coefficient of the unknown be two, what multiple of the square shall

* The concluding remark must be taken as relating to equations of a higher degree: for the other case of quadratics is the simple one; ŚRÍD'HARA's rule sufficing for affected quadratic equations.

132. Example: The square-root of half the number of a swarm of bees is gone to a shrub of jasmin; and so are eight-ninths of the whole swarm: a female is buzzing to one remaining male, that is humming within a lotus,

be added? If 2, 7, 14, 23, 34, 47, or 62, times the square of the unknown be added, the one side will furnish a root, and not the other. If 1, 4, or the like multiple be the addition, the first side will give no root. Nor will the subtraction of the square serve: for the square, being negative on the second side, will not there afford a root. Nor should once, or four times, the square be added, when the square had 3 or 5 [for coefficient]; but multiplication by 2, 6, &c. take place, when it has 2, 6 and so forth: for the solution will not be uniform, but very troublesome; since both sides will comprise [three terms] square of the unknown, unknown, and number. Therefore the square of the unknown is to be only multiplied by some quantity. When that square, however, affords a root, number only needs to be added. The question is, what this additive number should be. If the root of the square unknown term have 1 [for coefficient], the addition of the square of half [the coefficient of] the unknown term must make that side of equation yield a root. For the product of one and of half will be half the unknown; and twice that will be equal to the unknown; and the extraction (§ 26) will be equal without remainder. So, if [the coefficient] of the root be two, the additive should, by the same analogy, be equal to the square of a quarter of the [coefficient of the] unknown term. If it be three, the additive should be the square of the sixth part. Consequently the number to be added must be equal to the square of the quotient of the *(anca)* coefficient of the unknown term by twice the *(anca)* coefficient of the root which the square unknown term affords. But, if the coefficient of the square unknown term do not yield a square-root, it must do so when the term is multiplied by that same coefficient. Therefore both sides of the equation are to be multiplied by the coefficient of the square unknown term. Here, by the preceding analogy, to find the additive number, the coefficient of the unknown term is to be divided by twice the coefficient of the root of the square unknown term. Now the coefficient of the root is the unmultiplied (or original) coefficient of the square unknown term. Therefore the coefficient of the unknown term is to be divided by twice the original coefficient of the square; and is to be multiplied by the original coefficient of the square, as common multiplier of the two sides of equation. Abridging the multiplier and divisor by the common measure, the original coefficient of the square unknown term, the result is two for the divisor of the original coefficient of the unknown term. In like manner, where the coefficient of the square unknown term yields a root without previous multiplication, there also both sides being multiplied by that coefficient, and a number equal to the square of half the previous coefficient of the simple unknown term being added, a root is had: for the reasoning holds indifferently. Both sides, then, by this process being made to afford roots, if they be further multiplied by four to avoid a fraction, there is no detriment to their square nature. The square unknown term then, being multiplied by four, the coefficient of its root is doubled. The coefficient of the simple unknown term is to be divided by that. The divisor then is four times the original coefficient of the square. Now the common multiplier of the equation is just so much. Multiplier and divisor then being equal, both vanish; and the additive number is the square of the original coefficient of the simple unknown term. CRISHN.

in which he is confined, having been allured to it by its fragrance at night. Say, lovely woman, the number of bees.[1]

Put the number of the swarm of bees *ya v* 2. The square-root of half this is *ya* 1. Eight-ninths of the whole swarm are *ya v* $\frac{16}{9}$. The sum of the square-root and fraction, added to the pair of bees specified, is equal to the amount of the swarm, namely *ya v* 2. Reducing the two sides of the equation to a common denomination, and dropping the denominator, the equation is

$$\begin{matrix} ya\ v\ 18 & ya\ 0 & ru\ 0 \\ ya\ v\ 16 & ya\ 9 & ru\ 18 \end{matrix}$$

and, subtraction being made, the two sides are

$$\begin{matrix} ya\ v\ 2 & ya\ 9 & ru\ 0 \\ ya\ v\ 0 & ya\ 0 & ru\ 18 \end{matrix}$$

Multiplying both these by eight, and adding the number eighty-one, and extracting both roots, the statement of them for an equation is

$$\begin{matrix} ya\ 4 & ru\ \dot{9} \\ ya\ 0 & ru\ 15 \end{matrix}$$

Whence the value *yávat-távat* comes out 6. By substituting the square of this, the number of the swarm of bees is found 72.

133. Example: The son of PRĬT'HÁ, exasperated in combat, shot a quiver of arrows to slay CARN'A. With half his arrows he parried those of his antagonist; with four times the square-root of the quiver-full, he killed his horse; with six arrows he slew ŚALYA; with three he demolished the umbrella, standard and bow; and with one he cut off the head of the foe. How many were the arrows, which ARJUNA let fly?[2]

In this case put the number of the whole of the arrows *ya v* 1. Its half is *ya v* $\frac{1}{2}$. Four times the square-root are *ya* 4. The specified arrows are *ru* 10. Making the sum of these (*ya v* $\frac{1}{2}$ *ya* 4 *ru* 10) equal to this quantity *ya v* 1; reducing both sides of the equation to a common denomination, dropping the denominator, making subtraction as usual, adding sixteen to both sides of the equation

$$\left[\begin{matrix} ya\ v\ 1 & ya\ \dot{8} & ru\ 16 \\ ya\ v\ 0 & ya\ 0 & ru\ 36 \end{matrix}\right]$$

extracting the square-roots, and making these again equal

$$\left[\begin{matrix} ya\ 1 & ru\ \dot{4} \\ ya\ 0 & ru\ 6 \end{matrix}\right]$$

the value of *yávat-távat* is found 10. From which, by substitution, the number of arrows comes out 100.

134. Example: Of the period [series] less one, the half is the first term;

[1] This example is repeated from the *Lílávatí*, § 68.

[2] This also occurs in the *Lílávatí*, 67.

a moiety of the first term is the increase [common difference]; and the sum [of the progression] is the product of increase, first term and period multiplied together, and augmented by the addition of its seventh part. Tell the increase, first term, and period.[1]

Here the period is put *ya* 4 *ru* 1; the first term *ya* 2; the common difference *ya* 1. The product of their multiplication augmented by its seventh part is *ya gh* $\frac{64}{7}$ *ya v* $\frac{16}{7}$. This amount of the progression is equal to its sum found by the rule (*Líl.* § 119) viz. *ya gh* 8 *ya v* 10 *ya* 2. Depressing both sides by their common divisor *yávat-távat*, reducing them to a common denomination, and dropping the denominator, and then making subtraction, the two sides of the equation become $\begin{array}{l} ya\ v\ 8\quad ya\ \dot{5}4\quad ra\ 0 \\ ya\ v\ 0\quad ya\ 0\quad ru\ 14 \end{array}$ From these multiplied by eight, and having the square of twenty-seven added to them, the square-roots being extracted, are $\begin{array}{l} ya\ 8\quad ru\ \dot{2}7 \\ ya\ 0\quad ru\ 29 \end{array}$ Equating these, the value of *yávat-távat* is found from the equation, 7; and the substitution of that value gives the first term, common difference, and number of terms, 14, 7 and 29.

135. Example: What number being divided by cipher, and having the original quantity added to the quotient and nine subtracted from this sum,[2] and the consequent remainder being squared and having its square-root added to that square, and the whole being then multiplied by cipher, will amount to ninety?[3]

Here the number is put *ya* 1. This divided by cipher, is *ya* $\frac{1}{0}$. (The addition and subtraction being made,[4] it is still *ya* $\frac{1}{0}$. This squared is *ya v* $\frac{1}{0}$.

[1] An example of the sum of an arithmetical progression. See *Líl.* Ch. 4.

[2] The version is according to SU'RYADA'SA's reading of the text: but CRI'SHN'A-BHAT'T'A appears to have read, as does RA'MA-CRI'SHN'A, 'having (*cóti*) ten millions added or subtracted;' *rásih* cótyá *yuctó 't'havónitah*, instead of *rásir ádya-yuctó navónitah*. Collated copies differ: but the variation is noticed in the margin of one.

[3] This and the following are examples of the arithmetic of cipher. See § 12—14, and *Líl.* § 44—46.

[4] CRI'SHN'A-BHAT'T'A seems here also to have read, 'with ten millions added or subtracted:' and the quantity, being a fraction with cipher for its denominator, remains unaltered by the addition of a finite quantity (§ 16). But SU'RYADA'SA, though he cites § 16 and *Líl.* § 45, deduces from the conditions of the question the equation, $\begin{array}{l} ya\ v\ 4\quad ya\ \dot{3}4\quad ru\ 72 \\ ya\ v\ 0\quad ya\ 0\quad ru\ 90 \end{array}$ by adding to *ya v* 4 *ya* $\dot{3}6$

The root added to it makes *ya v* $\frac{1}{0}$ *ya* $\frac{1}{0}$. Multiplied by cipher, the multiplier and divisor being alike vanish, leaving *ya v* 1 *ya* 1.) Hence multiplying [the equation] by four and adding one, and proceeding as before, the number is found 9.

136. Example: Say what is the number, which having its half added to it, and being multiplied by cipher, and the product squared, and added to twice the root of that square, and this sum being divided by cipher, becomes fifteen?

The number is put *ya* 1. This, having its half added to it, becomes *ya* $\frac{3}{2}$. Being multiplied by cipher, it is not to be made nought but to be considered as multiple of cipher, further operations impending. Wherefore *ya* $\frac{3}{2}\cdot 0$, being squared, and having twice the root added, becomes *ya v* $\frac{9}{4}\cdot 0$ *ya* $\frac{12}{4}\cdot 0$. This is divided by cipher: and here, as before, the multiplier being 0, and the divisor 0, both multiplicator and divisor, as being equal, vanish; and the quantity is unaltered:[1] viz. *ya v* $\frac{9}{4}$ *ya* $\frac{12}{4}$. Equating with fifteen, reducing to a common denomination, and then dropping the denominator, the two sides of the equation by preparation become

$$\begin{array}{l} ya\ v\ 9 \quad ya\ 12 \quad ru\ 0 \\ ya\ v\ 0 \quad ya\ 0 \quad ru\ 60 \end{array}$$

Adding four,[2] and extracting the square-roots, the value of *yávat-távat* by equal subtraction comes out 2.

137. Example: a stanza and a half. What is the number, learned man, which being multiplied by twelve and added to the cube of the number, is equal to six times the square added to thirty-five?[3]

The number is put *ya* 1. This multiplied by twelve, and added to the cube of the number, is *ya gh* 1 *ya* 12. It is equal to this other quantity *ya v* 6 *ru* 35. Subtraction being made, the first side of the equation be-

ru 81, its root *ya* 2 *ru* $\dot{9}$; and thence [doubling the equation, and] proceeding by the rule § 131, he derives the equation of the roots

$$\begin{array}{l} ya\ 8 \quad ru\ \dot{34} \\ ya\ 0 \quad ru\ 38 \end{array}$$

From which, by the usual process, he finds 9 for the value of *ya*. One of the copies of the text, which have been collated for the present translation, omits the whole of this intermediate work here enclosed within a parenthesis.

[1] *Lil.* § 44—45.

[2] The additive four sufficing to make them afford square-roots. CRI'SHN'.

[3] It has been said "If the solution be not thus accomplished, [as] in the case of cubes or biquadrates, the value must be elicited by the [calculator's] own ingenuity." § 129. The present and the following are instances of the application of that aphorism. SU'R.

comes *ya gh* 1 *ya v* $\dot{6}$ *ya* 12; and the other side, *ru* 35. Adding the negative number eight to both (or subtracting eight from both sides[1]) and extracting the cube-roots,[2] they are *ya* 1 *ru* $\dot{2}$ From the equation of these
ya 0 *ru* 3
again, the amount of the number is found 5.

138. Example: If thou be conversant with operations of algebra, tell the number, of which the biquadrate, less double the sum of the square and of two hundred times the simple number, is a myriad less one.

Let the number be put *ya* 1. This, multiplied by 200, is *ya* 200. Added to the square of the number, it is *ya v* 1 *ya* 200; which, multiplied by two becomes *ya v* 2 *ya* 400. The biquadrate of the number, less that, is *ya v v* 1 *ya v* $\dot{2}$ *ya* $\dot{400}$. This is equal to a myriad less unity. The two sides of the equation are *ya v v* 1 *ya v* $\dot{2}$ *ya* $\dot{400}$ *ru* 0 Here adding to the first
ya v v 0 *ya v* 0 *ya* 0 *ru* 9999
side *yávat-távat* four hundred, with unit absolute, it yields a root (*ya v* 1 *ru* $\dot{1}$): but the other side (*ya* 400 *ru* 10000) does not; and the solution therefore is not accomplished. Hence ingenuity is in this case called for. Adding then, to both sides of the equation, square of *yávat-távat* four, *yávat-távat* four hundred, and a single unit absolute, roots of both may be extracted. Thus the first side, with the additive, becomes *ya v v* 1 *ya v* $\dot{2}$ *ru* 1. The other side, with it, exhibits *ya v* 4 *ya* 400 *ru* 10000. Their roots are *ya v* 1 *ru* $\dot{1}$ and *ya* 2 *ru* 100. From these, equal subtraction being made, the two sides of the equation are deduced *ya v* 1 *ya* $\dot{2}$ *ru* 0
ya v 0 *ya* 0 *ru* 99
Again adding unit to each side, the roots are obtained *ya* 1 *ru* $\dot{1}$ From
ya 0 *ru* 10
which equation the value of *yávat-távat* comes out 11. In like instances the value must be elicited by the sagacity of the intelligent analyst.

139. Example: The eighth part of a troop of monkeys, squared, was skipping in a grove and delighted with their sport. Twelve remaining

[1] A variation of the text is here put in a parenthesis. The effect is the same; and one reading serves to interpret the other.

[2] By the analogy of the rule for the extraction of the square-root (§ 26) taking the roots of the cube of the unknown and of the absolute number, and subtracting from the remainder thrice the products [of the square and simple quantities] two and two. SU'R.

were seen on the hill, amused with chattering to each other. How many were they in all?[1]

In this case the troop of monkeys is put *ya* 1. The square of its eighth part, added to twelve, being equal to the whole troop, the two sides of equation are

$$\begin{array}{llll} ya\,v\,\frac{1}{64} & ya\ 0 & ru\ 12 \\ ya\,v\ 0 & ya\ 1 & ru\ 0 \end{array}$$

Reducing these to a common denomination, dropping the denominator, and making equal subtraction, they become

$$\begin{array}{llll} ya\,v\ 1 & ya\ \dot{64} & ru\ 0 \\ ya\,v\ 0 & ya\ 0 & ru\ \dot{768} \end{array}$$

From these, with the square of thirty-two added to them, the roots are extracted

$$\begin{array}{ll} ya\ 1 & ru\ \dot{32} \\ ya\ 0 & ru\ 16 \end{array}$$

The number of the root on the absolute side is here less than the known number, with the negative sign, in the root on the side of the unknown. Making it then negative and positive, a two-fold value of *yávat-távat* is thence obtained, 48 and 16.

140.[2] Example: The fifth part of the troop less three, squared, had gone to a cave; and one monkey was in sight, having climbed on a branch. Say how many they were?[3]

Here the troop is put *ya* 1. Its fifth part is $ya\ \frac{1}{5}$. Less three, it is $ya\ \frac{1}{5}$ $ru\ \frac{\dot{15}}{5}$. This squared is $ya\,v\ \frac{1}{25}$ $ya\ \frac{\dot{30}}{25}$ $ru\ \frac{225}{25}$. With the one seen $(\frac{25}{25})$, it is $ya\,v\ \frac{1}{25}$ $ya\ \frac{\dot{30}}{25}$ $ru\ \frac{250}{25}$. This is equal to the troop *ya* 1. Reducing these

[1] This instance is relative to the rule (§ 130) which admits a two-fold value of the unknown, when the square-root on the absolute side is less than the known number, comprised in the square-root on the other side of the equation. SU'R. and CRÍSH'N'.

[2] SU'RYADA'SA here interposes an example, the same which is inserted in the *Lílávatí*, § 150. It is not, however, found in collated copies of the *Víja-gańita*, nor noticed in this place by CRÍSHN'A-BHAT'T'A, nor by RA'MA-CRÍSHN'A. The solution, as wrought by the first named commentator, follows: 'Put *ya* 1 for the side of the triangle or distance between the snake's hole and the point of meeting. If this side be subtracted from the sum of the side and hypotenuse, namely 27, the remainder is the hypotenuse: it is $ya\ \dot{1}$ $ru\ 27$. Its square, $ya\,v\ 1$ $ya\ \dot{54}$ $ru\ 729$, is equal to the sum of the squares of the side put *ya* 1 and upright given 9: namely $ya\,v\ 1$ $ru\ 81$. Equal subtraction being made, the value of *ya* is found, 12. Thus the distance between the hole and point of meeting comes out 12 cubits; and this, subtracted from the distance from the hole of the spot where the snake was seen, namely 27, leaves the equal progress of the two, 15.—SU'R. The example, as is apparent, is here out of place, and should have been noticed by the scholiast, where the author has himself referred to it, in his gloss on § 124.

[3] Two instances are here given to show, that the twofold value is admissible in some cases only. CRÍSHN'.

sides of equation to a common denomination, dropping the denominator, and making equal subtraction, the equation becomes

$$\begin{array}{llll} ya\ v\ 1 & ya\ 5\dot{5} & ru & 0 \\ ya\ v\ 0 & ya\ 0 & ru & 250 \end{array}$$

Multiplying by four, and adding a number equal to the square of fifty-five (3025), the roots extracted are

$$\begin{array}{ll} ya\ 2 & ru\ 55 \\ ya\ 0 & ru\ 45 \end{array}$$

Here also a two-fold value is found as before, 50 and 5. But the second is in this case not to be taken: for it is incongruous. People do not approve a negative absolute number.[1]

141. Example: The shadow of a gnomon twelve fingers high being lessened by a third part of the hypotenuse, became fourteen fingers long. Tell quickly, mathematician, that shadow.

In this case the shadow is put *ya* 1. This, less a third of the hypotenuse, becomes fourteen fingers. Hence, conversely, subtracting fourteen fingers from it, the remainder is a third of the hypotenuse, $ya\ 1\ \ ru\ 1\dot{4}$. This then, multiplied by three, is the hypotenuse; $ya\ 3\ \ ru\ \dot{4}2$. Its square is $ya\ v\ 9\ \ ya\ 2\dot{5}2\ \ ru\ 1764$; which is equal to this other value of the square of the hypotenuse,[2] $ya\ v\ 1\ \ ru\ 144$. Equal subtraction being made, the two sides of

[1] The second value being five, its fifth part, one, cannot have three subtracted from it. There is incongruity; to indicate which the author adds expressly, 'the second is in this case not to be taken.' SU'R.

Put *ya* 5 for the troop of monkeys. Its fifth part is *ya* 1. Less three, it is $ya\ 1\ \ ru\ \dot{3}$. This squared is $ya\ v\ 1\ \ ya\ \dot{6}\ \ ru\ 9$. With one added, it is $ya\ v\ 1\ \ ya\ \dot{6}\ \ ru\ 10$; and is equal to *ya* 5. Equal subtraction being made, the equation is

$$\begin{array}{llll} ya\ v\ 1 & ya\ \dot{1}1 & ru & 0 \\ ya\ v\ 0 & ya\ 0 & ru & 10 \end{array}$$

Multiplying by four, and superadding the square of 11, it becomes

$$\begin{array}{llll} ya\ v\ 4 & ya\ \dot{4}4 & ru & 121 \\ ya\ v\ 0 & ya\ 0 & ru & 81 \end{array}$$

Here, since the known number was proposed as negative [i. e. subtractive] the root should be taken, under the reasoning before stated [gloss on § 130] $ya\ 2\ \ ru\ \dot{1}1$; not $ya\ \dot{2}\ \ ru\ 11$. The root of the second side of the equation is 9.* By further equation of the roots, the value of *ya* comes out 10. Whence the number of the troop is deduced 50. But, if the known number (11) be made positive, the value of *ya* will be 1; and the whole quantity 5. From its fifth part (one), three cannot be subtracted. If indeed the enunciation of the question were, "The fifth part of the troop taken from three" [instead of "less three"], the second value, and not the first, would be taken. For the fifth part of the first value cannot be deducted from three. CRĬSHŃ.

[2] For the shadow being the side of a rectangular triangle, and the gnomon twelve fingers in length being the upright: (*Sirómańi:* Book 1, Ch. 3.) the rule, that the square-root of the sum of their squares is the hypotenuse, is universally known. SUR.

* Or else $\dot{9}$. But with this root the author would take $ya\ \dot{2}\ \ ru\ 11$ instead of $ya\ 2\ \ ru\ \dot{1}1$.

the equation become $\begin{matrix} ya\,v\;8 & ya\;2\dot{5}2 & ru\;0 \\ ya\,v\;0 & ya\;0 & ru\;1\dot{6}20 \end{matrix}$ Multiplying by two, and superadding the square of the number sixty-three, the roots are $\begin{matrix} ya\;4 & ru\;6\dot{3} \\ ya\;0 & ru\;27 \end{matrix}$ Making these again equal, and proceeding as before, the value of *yávat-távat* is obtained two-fold $\frac{45}{2}$ and 9. The second value of the shadow is less than fourteen: therefore, by reason of its incongruity, it should not be taken. Hence it was said "this holds in some cases." (§ 130.)

It is in derogation of the maxim delivered in PADMANA'BHA's algebra,[1] on this subject:

142. "When the root of the absolute side is less than the known number being negative on the other side, making it positive and negative, the value comes out two-fold."[2]

143—144. Example: What are the four quantities, friend! which with two severally added to them, yield square-roots; and of which the products, taking them two and two, contiguous,[3] become also square numbers when eighteen is added: and which are such, that the square-root of the sum of all the roots added to eleven, being extracted, is thirteen? tell them to me, algebraist!

In this case, the number, which, added to two quantities, renders them square numbers, is the additive of the [original] quantities. That additive, multiplied by the square of the difference of the roots, is the additive of the product. That is, the product of those two quantities, with the addition of this [additive] must yield a square-root. The products of the roots of the quantities, taken two and two contiguous, being lessened by a deduction of the additive of the quantities, are the roots of the products of those quantities.[4] This principle must be understood in all [similar] cases. In the present instance, the additive of the products is nine times the additive of the simple quantities. The square-root of nine is three. Therefore, the roots

[1] *Padmanábha-Víja.*

[2] The quotation, as copied by the commentator SU'RYADA'SA, contains the very reservation for which BHA'SCARA contends "the value will be two-fold sometimes:" *dwid'há mánan cwachit bhavét,* instead of *dwividhótpadyaté mitih,* the reading which occurs in collated copies.

[3] The first and second are to be multiplied together; and the third and fourth.—SU'R. First and second; second and third; third and fourth. CRĪSHN'.

[4] For the demonstration of both these positions, see note to § 145.

of the quantities are [arithmeticals] differing by the common difference three: *ya* 1 *yá* 1 *ru* 3 *ya* 1 *ru* 6 *ya* 1 *ru* 9. The products of these, two and two, less the additive of the simple quantities, being computed, are the square-roots of the products of the quantities with eighteen added. Hence the roots of the products as above described are *ya v* 1 *ya* 3 *ru* 2; *ya v* 1 *ya* 9 *ru* 16; *ya v* 1 *ya* 15 *ru* 52.[1] The sum of all these and of the original roots is *ya v* 3 *yá* 31 *ru* 84. Making this, with eleven added to it, equal to the square of thirteen; multiplying the sides of the equation (after equal subtraction) by twelve; and superadding the square of thirty-one, the roots are *ya* 6 *ru* 31 From the equation of these again,
ya 0 *ru* 43
the value of *yávat-távat* is found 2. Whence the roots of the quantities are deduced, 2, 5, 8 and 11. Of course the original quantities (being the squares of those roots, less the additive of the simple quantities,) are 2, 23, 62 and 119.

On this subject there is a maxim of an original author:[2]

145. 'So many times as the additive of the products contains the additive

[1] SU'RYADA'SA, employing only the first of the foregoing positions, as it is contained also in the maxim cited from an earlier writer (§ 145), deduces the algebraic expressions for the roots of the products from those of the simple quantities: 'The additive of the products (18) is nine times the additive of the simple quantities (2). Its square-root is 3. The simple roots then are *ya* 1 *ru* 0
ya 1 *ru* 3
ya 1 *ru* 6
ya 1 *ru* 9
And the quantities deduced by squaring these and subtracting the additive two, are *ya v* 1 *ya* 0 *ru* 2
ya v 1 *ya* 6 *ru* 7
ya v 1 *ya* 12 *ru* 34
ya v 1 *ya* 18 *ru* 79
The products of the contiguous two and two, with eighteen added, are
ya v v 1 *ya gh* 6 *ya v* 5 *ya* 12 *ru* 4
ya v v 1 *ya gh* 18 *ya v* 41 *ya* 284 *ru* 256
ya v v 1 *ya gh* 30 *ya v* 113 *ya* 1560 *ru* 2704
The square-roots of these three are *ya v* 1 *ya* 3 *ru* 2
ya v 1 *ya* 9 *ru* 16
ya v 1 *ya* 15 *ru* 52
Their sum (*ya v* 3 *ya* 27 *ru* 66) with the sum of the simple roots (*ya* 4 *ru* 18) or *ya v* 3 *ya* 31 *ru* 84, with 11 added, is equal to the square of 13, or 169. The equation then is *ya v* 3 *ya* 31 *ru* 95 and, after subtraction, be-
ya v 0 *ya* 0 *ru* 169
comes *ya v* 3 *ya* 31 *ru* 0 Proceeding as usual (§ 131), the value of *yávat-távat* comes out 2.'
ya v 0 *ya* 0 *ru* 74
SU'R.

[2] Neither BHA'SCARA, nor his commentators, intimate the name of this ancient and (*ádya*) original author, whose words are here quoted.

of the simple quantities, by the square-root of that [submultiple] as a common difference, the unknown quantities are to be put [in arithmetical progression] and to be squared, and then diminished by subtraction of the additive.'[1]

This supposition of apposite quantities required much dexterity in computations.

146. Example: Say what is the hypotenuse in a plane figure, in which the side and upright are equal to fifteen and twenty? and show the demonstration of the received mode of computation.[2]

Here the hypotenuse is put *ya* 1. Turning the triangle, the hypotenuse is made the base; and its side and upright are the sides: and the side and upright in each of the triangles situated on either side of the perpendicular

[1] The demonstration of this, and of BHA'SCARA's corresponding position with the further one subjoined by him, is given by the commentator CRI'SHN'A: 'If the square-root of a quantity increased by an additive be known, then by inversion the quantity is the square of that root, less the additive; and is also known. From the first root the first quantity is found *p v* 1 *a* $\dot{1}$; and, in like manner, from the second root, the second quantity *d v* 1 *a* $\dot{1}$. That, which, added to their product, makes it a square, is the additive of the product. Multiplied together their product is *p v. d v bh* 1 *p v. a bh* $\dot{1}$ *d v. a bh* $\dot{1}$ *a v* 1. In the second term, the square of the first root multiplied by the additive is negative; and, in the third term, the like multiple of the square of the second root is also negative. To abridge, put sum of the squares of the roots multiplied by the additive with the negative sign. The first term is the product of squares of the roots; or, which is the same, the square of the product of the roots. The statement then is *r* prod. *v* 1 *r v* sum *a* $\dot{1}$ *a v* 1. In the second of these three terms, the sum of the squares of the roots is resolvable into two parts; the square of the difference of the roots and twice the product of the roots. (*Lil.* § 135.) The second term is thus resolved into two; namely the square of the difference and twice the product multiplied by the additive: and the statement of all the terms in their order consequently is *r* prod. *v* 1 *r* diff. *v. a* $\dot{1}$ *r* prod. *a* $\dot{2}$ *a v* 1. Now the number, which, added to this product of the quantities, makes it afford a square-root, is the additive of the product. But here, if the square of the difference of the roots multiplied by the additive (*r* diff. *v. a* 1) be superadded, the remaining terms *r* prod. *v* 1 *r* prod. *a* $\dot{2}$ *a v* 1, will yield a square-root. It is therefore demonstrated, that the additive of the simple quantities, multiplied by the square of the difference of the roots, is additive of the product; and the product of the same roots, less the additive, will be the root of the product of the quantity. The same reasoning is applicable to the second and third; and to the third and fourth. Thus the root of one quantity being put *yávat-távat*, the roots of all may be rightly deduced from it' [by their common difference computed from the additives]. CRI'SHN.

[2] The question of the hypotenuse is here put, solely to inquire the principle of the solution of this problem. CRI'SHN.

let fall in the proposed triangle, are analogous to the former.[1] Hence the proportion: 'if, when *yávat-távat* is hypotenuse, this be the side (15), then, the hypotenuse being fifteen (equal to the original side), what is?' Thus the side [of the smaller triangle] is found, and is the segment contiguous to the original side: $\frac{ru\ 225}{ya\ 1}$ Again, if *yávat-távat* being hypotenuse, this be the upright (20), then, the hypotenuse being twenty (equal to the original upright), what is?' Thus the upright [of the other little triangle] is found, and is the segment contiguous to the original upright:[2] $\frac{ru\ 400}{ya\ 1}$ The sum of the segments is equal to hypotenuse: and from so framing the equation, the value of hypotenuse is deduced,[3] the square-root of the sum of the squares of the side and upright, viz. 25. Hence, substituting the value, the segments are found 9 and 16; and thence the perpendicular 12. See

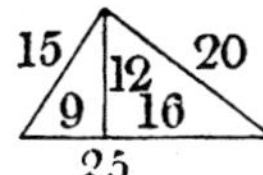

Or the solution is thus otherwise propounded. The hypotenuse is *ya* 1. Half the product of the side and upright is the area of the triangle: 150.

[1] They are proportional (*anurúpa*). CRĬSHṆ.

[2] The greater side being here named the upright, (while either side might have been so denominated; *Líl.* § 133.) in the original triangle, the greater side of the one small triangle must be taken as the upright found by the proportion 'as *yávat-távat* is to the original upright, so is an hypotenuse of that length to a quantity sought:' it is the segment, which is the greater side of this triangle; not the perpendicular, which is its less side. So the smaller side of the other little triangle must be taken as the side found by the proportion, 'as *yávat-távát* is to the original side, so is a hypotenuse of that length to a quantity sought:' and it is the segment, this being the least side of the triangle; not the perpendicular which is here the greater side. CRĬSHṆ.

[3] CRĬSHṆA gives the solution by literal symbols alone. By the first proportion *ya* 1 | *bhu* 1 | *bhu* 1 | the segment contiguous to the side is $\frac{bhu\ v\ 1}{ya\ 1}$ By the other proportion, *ya* 1 | *co* 1 | *co* 1 | the segment contiguous to the upright is $\frac{co\ v\ 1}{ya\ 1}$ The sum of the segments $\frac{bhu\ v\ 1 \quad co\ v\ 1}{ya\ 1}$ is equal to the base *ya* 1. Reducing to a common denomination and dropping the denominator, the two sides of the equation become $\begin{matrix} ya\ v\ 1 \\ bhu\ v\ 1 \quad co\ v\ 1 \end{matrix}$ Hence the square of the hypotenuse is equal to the sum of the squares of upright and side.—CRĬSHṆ. Here *bhu* is initial of *bhuja*, the side (lit. arm); and *co* of *cóti*, the upright.

With four such triangles, another figure having four sides, each equal to the hypotenuse,[1] is constructed for the purpose of finding the hypotenuse. See

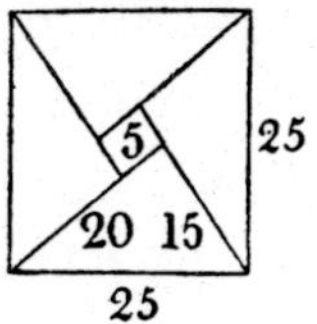

Thus another interior quadrilateral is framed; and the difference between the upright and side is the length of its side. Its area is 25. Twice the product of the upright and side is the area of the four triangles, 600. The sum of these is the area of the entire large figure; 625. Equating this with the square of *yávat-távat*, the measure of the hypotenuse is found, 25.[2] If the absolute number, however, be not an exact square, the hypotenuse comes out a surd root.

147. **Rule**: Twice the product of the upright and side,[3] being added to the square of their difference, is equal to the sum of their squares, just as with two unknown quantities.[4]

Hence, for facility, it is rightly said 'The square-root of the sum of the squares of upright and side, is the hypotenuse?'[5]

[1] The triangles are to be so placed, as that the hypotenuse may be without; and the upright of one be in contact with the side of another: else, by merely joining four rectangular triangles [with the equal sides contiguous,] a quadrilateral having unequal diagonals [that is, a rhomb] is constituted; in which one diagonal is twice the upright; and the other double the side of the triangle; instead of a square comprising five figures (four triangles and a small interior square). But, if the upright and side be equal, a square only is framed, which ever way the side is placed, since there is no difference of the upright and side: and in this case there is no interior square. CRĬSHN'.

[2] In this instance also, CRĬSHN'A exhibits the solution by literal symbols: 'Area of the triangle *bhu. co* $\frac{1}{2}$. Multiplied by four, it is the area of four such triangles, *bhu. co* 2. Difference *bhu* $\dot{1}$ *co* 1. Its square *bhu v* 1 *bhu. co* $\dot{2}$ *co v* 1. This, which is the area of the interior square, being added to the area of the four triangles, *bhu. co* 2, makes *bhu v* 1 *co v* 1; the area of the entire square.' CRĬSHN'.

[3] This is not confined to upright and side; but applicable to all quantities. (*Líl.* § 135.) CRĬSHN'.

[4] Let the two quantities be *ya* 1 *ca* 1. The square of their difference will be *ya v* 1 *ya. ca bh* $\dot{2}$ *ca v* 1. To this twice the product *ya. ca bh* 2 being added, the result is the sum of the squares *ya v* 1 *ca v* 1. CRĬSHN'.

[5] See the same rule expressed in other words; *Líl.* § 134.

Placing the same portions of figure in another form, see 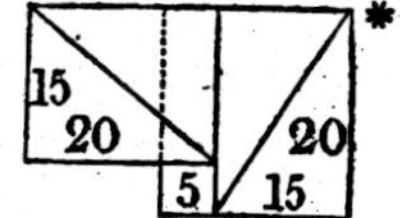

*

148. Example: Tell me, friend, the side, upright and hypotenuse in a [triangular] plane figure, in which the square-root of three less than the side, being lessened by one, is the difference between the upright and hypotenuse.

In this case the difference between the upright and hypotenuse is arbitrarily assumed: say 2. Hence, by inversion, (taking the square of that added to one and adding three to the square;) the side is obtained, 12. Its square, or the difference of the squares of hypotenuse and upright, is 144. The difference of the squares of two quantities is equal to the product of their sum and difference.[2] For a square[3] is the area of an equilateral quadrangle [and equi-diagonal[4]]. This for example, is the square of seven, 49:

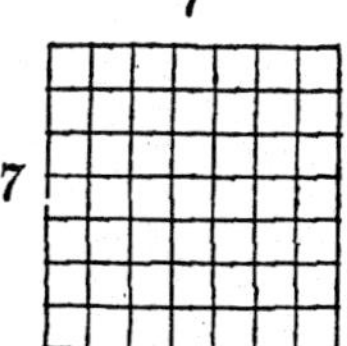

Subtracting from it the square of five, viz. 25, the remainder is 24. See

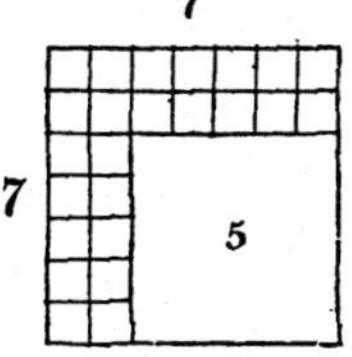

Here the difference is two; and the sum is twelve: and the product of the sum and difference consists of 24 equal compartments

12
2

Thus it is demonstrated, that the difference of the squares is equal to the product of the sum and difference. Hence, in the instance, the difference of the squares, 144, being divided by the assumed difference of the hypo-

* Producing the line, the figure is divided into two squares: one the square of the upright; the other the square of the side: and their sum is the area of the first or large square; and its square-root is the side of the quadrilateral. CRĬSHN'.

[2] *Līl.* § 135.

[3] *Varga*, or 2d power.

[4] CRĬSHN'A.

tenuse and upright, 2, is the sum, 72.[1] This sum, twice set down, and having the difference severally subtracted, and being halved according to the rule of concurrence,[2] gives the upright and hypotenuse 35 and 37. In like manner, putting one, the side, upright and hypotenuse are 7, 24, and 27. Or, supposing four, they are 28, 96, and 100.[3] So in every [similar] case.

149. Rule: The difference between the sum of the squares of two quantities whatsoever, and the square of their sum, is equal to twice their product; as in the case of two unknown quantities.[4]

For instance, let the quantities be 3 and 5. Their squares are 9 and 25. The square of their sum, 64. From this taking away the sum of the squares, the remainder is 30. See

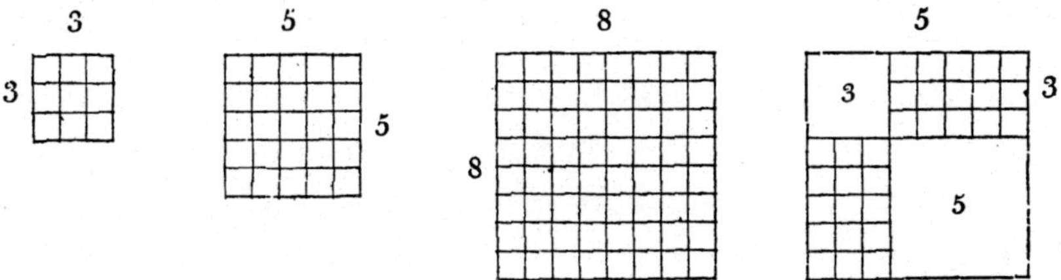

Here square compartments, equal to twice the product are apparent; and [the proposition] is proved.

150. Rule: The difference between four times the product and the square of the sum, is equal to the square of the difference of the quantities; as in the instance of unknown ones.[5]

Let the quantities be 3 and 5. From the square of their sum, taking away four products[6] at the four corners, there remain in the middle, square

[1] See *Lil.* § 57.

[2] *Lil.* § 55.

[3] The problem is an indeterminate one.

[4] Let the quantities be *ya* 1 *ca* 1. The sum of their squares is *ya v* 1 *ca v* 1; and the square of their sum *ya v* 1 *ya. ca bh* 2 *ca v* 1. The difference between which is *ya. ca bh* 2; or twice the product of the two quantities. CRÏSHN'.

[5] Let the two quantities be *ya* 1 *ca* 1. Four times the product is *ya. ca bh* 4. The square of the sum is *ya v* 1 *ya. ca bh* 2 *ca v* 1. From this square taking four times the product, the remainder is *ya v* 1 *ya. ca bh* 2̇ *ca v* 1. And this is the square of the difference of the two quantities. CRÏSHN'.

[6] Rectangles.

compartments equal to the square of the difference of the quantities; and [the proposition] is proved. See

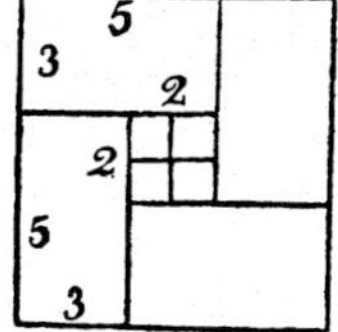

151. Example: Tell the side, upright and hypotenuse, of which the sum is forty, and the product of the upright and side is a hundred and twenty.

Here twice the product of the side and upright is 240. It is the difference between the square of their sum and the sum of their squares.[1] The sum of the squares of the side and upright is the same with the square of the hypotenuse.[2] Therefore it is the difference between the square of the sum of the side and upright, and the square of hypotenuse; and is equal to the product of sum and difference. Therefore this difference, 240, divided by the sum 40, gives the difference of hypotenuse and the sum of the side and upright, viz. 6. The sum, having the difference severally subtracted and added, and being then halved, gives by the rule of concurrence,[3] the sum of the upright and side 23, and the hypotenuse 17.[4] From the square of the sum of the upright and side, namely 529, subtract four times the product (§ 150), viz. 480, the square root 7 of the remainder (49) is the difference of the side and upright. From the sum and difference, the side and upright are found by subtraction and addition and then taking the moieties: and they come out 8 and 15.

152. Example:[5] Tell me severally the side, upright and hypotenuse, the sum of which is fifty-six; and their product seven times six hundred.

In this instance put hypotenuse *ya* 1. Its square is *ya v* 1. It is the sum of the squares of the side and upright. The sum of the three sides (hypotenuse, upright and base) less hypotenuse is the sum of the upright and side;

[1] § 149.

[2] § 146.

[3] *Líl.* § 55.

[4] For the sum of two sides must exceed the hypotenuse. (*Líl.* § 161.) CRÍSHN'.

[5] This example, though overlooked by SÚRYADÁSA, is noticed both by CRÍSHN'A and RÁMA-CRÍSHN'A; and is found in all the collated copies of the text.

ya i̇ *ru* 56. So the product of the three, divided by hypotenuse, gives the product of upright and side $\frac{ru\ 4200}{ya\ 1}$ By a preceding rule (§ 149) the difference between the sum of the squares (*ya v* 1) and the square of their sum (*ya v* 1 *ya* 112 *ru* 3136̇), namely *ya* 112̇ *ru* 3136, is equal to twice the product or $\frac{ru\ 8400}{ya\ 1}$ First reducing to least terms by the common divisor a hundred and twelve; then bringing both sides of the equation to a common denomination, dropping the denominator, making equal subtraction,[1] and superadding the square of fourteen to both sides of the equation,[2] the value of *yávat-távat* is found, 25. A second value in this case comes out by way of alternative, namely 3: but it is not to be taken because it is incongruous.[3]

In the instance, the product of the three sides, 4200, being divided by the hypotenuse 25, gives the product of the upright and side, 168. Thus the sum of the upright and side being 31 (56 less 25), the difference of the upright and side is found by the preceding rule (§ 150) namely 17. And thence, by the rule of concurrence,[4] the side and upright are deduced 7 and 24.

So in all [like] cases.

The intelligent, by a compendious method, do in some instances resolve a problem by reasoning alone. But the grand operation is by putting [a symbol] of the unknown.

[1] The text to which RÁMACRÍSHN'A's commentary is appended, here exhibits

Equation *ya v* 112 *ya* 3136 *ru* 0
ya v 0 *ya* 0 *ru* 8400

Abridged by 112, *ya v* 1 *ya* 28 *ru* 0
ya v 0 *ya* 0 *ru* 75

[2] Some copies of the text substitute the equivalent operation of multiplying by four, and adding the square of twenty-eight.

[3] See remark under § 141.

[4] *Líl.* § 55.

CHAPTER VI.

EQUATION INVOLVING MORE THAN ONE UNKNOWN QUANTITY.

NEXT, Analysis by a Multiliteral Equation is propounded.

153—156. Rule: Subtract the first colour (or letter[1]) from the other side of the equation; and the rest of the colours (or letters) as well as the known quantities, from the first side:[2] the other side being then divided by the [coefficient of the] first, a value of the first colour will be obtained.[3] If there be several values of one colour, making in such case equations of them and dropping the denominator,[4] the values of the rest of the colours are to be found from them.[5] At the last value, the multiplier and quotient, found by the method of the pulverizer,[6] are the values of both colours, dividend

[1] See the author's following comment; and the note upon it.

[2] That is, the two sides of the equation are to be so treated, as that a single colour may remain on one side: which is effected by equal subtraction of all the rest of the terms on that side from both; and of the term similar to it on the other. It is not necessary to restrict the choice of the particular colour: but, as there is no motive for passing by the first, that is selected to be retained. CRĬSHṄ.

[3] This division is the equivalent of the proportion, in which one of the unknown is the third term, and a multiple of it is the first; to find the value. See note on § 157.

[4] After reduction to a common denomination.

[5] This suffices for problems admitting but one solution. What follows, relates to indeterminate problems.

[6] An answer to an indeterminate problem being required in whole numbers. Else arbitrary values may be put for all the remaining unknown terms in the last and single value of an unknown. In such case the answer is easy: but is probably fractional. CRĬSHṄ.

and divisor:[1] if there be other colours in the dividend, put for them any arbitrarily assumed values; and so find those two. By substituting with these inversely, the values of the rest of the colours are then obtained. But, if a value be fractional, the investigation of the pulverizer is to be repeated; and, with that substituting for the last colour, deduce the values conversely from the first.[2]

This is analysis by equation comprising several colours.[3]

In this, the unknown quantities are numerous, two, three or more. For which *yávat-távat* and the several colours are to be put to represent the values. They have been settled by the ancient teachers of the science:[4] viz. "so much as" (*yávat-távat*), black (*cálaca*), blue (*nílaca*), yellow (*pítaca*), red (*lóhitaca*), green (*haritaca*), white (*śwétaca*), variegated (*chitraca*), tawny (*capilaca*), tan-coloured (*pingala*), grey (*d'húmraca*), pink (*pátalaca*), white (*śavalaca*), black (*śyámalaca*), another black (*méchaca*), and so forth. Or

[1] The colour or letter, appertaining to the divisor, is the quantity of which the algebraic expression was the value; its coefficient being the divisor or denominator.—See note to § 157. The colour belonging to the dividend or numerator, is one comprised in that algebraic expression of the value of the former. See note on the author's comment below. One unknown is a function of the other.

[2] The commentator CRÍSHN'A notices two variations, or altogether three readings of this passage. He prefers one as most consistent with the author's own explanation of his text; and interprets it thus: If, in course of substitution, the value of another colour be had fractional, investigation of a pulverizer is to be again performed; and, with that multiplier, substituting for the last colour, deduce the values inversely from the first. That is, with the particular multiplier termed pulverizer (Ch. 2), substituting for the colour contained in the two or more last values, again deduce the values inversely from the preceding: meaning from the value which contained a fractional one. Beginning thence, let inverse substitution take place.

The second reading is, 'the other colour (or letter) is to be found by repeating the investigation of the pulverizer; and with that substituting for the original [or, according to another construction, for the last, colour]; deduce the values conversely from the first;' or, as the third reading varies it, 'deduce conversely the last and the first.' It is defective in either construction: for the pronoun "that" refers to the "other colour (or letter);" the value of which is found by investigation of the pulverizer: but "the other" so referred to is the divisor in that investigation: and the "last" colour, for which substitution is to be made, is the dividend: and the sentence, therefore, according to these readings, directs a substitution of value for the wrong colour. CRÍSHN'.

[3] *Anéca-varn'a-samícaran'a víja.* See Ch. 4. note on gloss following § 100—102.

[4] *Púrváchároyáih*, by former teachers. What particular authors are intended is unexplained. BRAHMEGUPTA employs names of colours to designate the unknown, without any remark; whence it appears that the use was already familiar. See BRAHM. 18. § 52 *et seq.*

letters[1] are to be employed; that is the literal characters *c*, &c. as names of the unknown, to prevent the confounding of them.

Here also, the calculator, performing as before directed (Ch. 4) every operation implied by the conditions of the example, brings out two equal sides, or more sides, of equation. Then comes the application of the rule: 'From one of the two sides of the equation, subtract[2] the first (letter or) colour of the other. Then subtract from that other side the rest of the (letters or) colours, as well as the known quantities. Hence the one side being divided[3] by the residue of the first (letter or) colour, a value of the (letter or) colour which furnishes the divisor is obtained. If there be many such sides, by so treating those that constitute equations, by pairs, other values are found. Then, among these, if the values of one (letter or) colour be manifold, make them equal by pairs, drop the denominator,[4] and proceeding by the rule [§ 153], find values of the other (letters or) colours: and so on, as practicable.

Thereafter,[5] the number (coefficient) of the dividend (letter or) colour in the last value is to be taken as the dividend quantity; and that of the divi-

[1] *Varńa*, colour or letter: for the word bears both imports. Former writers used it in the one sense, and directed all the unknown quantities after the first to be represented by colours. But the author takes it also in the second acceptation; and directs letters to be employed, instancing the consonants in their alphabetical order. He appears to intend initial syllables. (See his solution of the problem in § 111.) His predecessors, however, likewise made use of initial syllables for algebraic symbols; for instance the marks of square, cube and other powers; and the sign of a surd root: as well as the initials of colours as tokens of unknown quantities.

[2] See § 101.

[3] Ibid.

[4] After reduction to a common denominator.

[5] In the last and single value of the unknown denoted by a colour, if one or more unknown terms denoted by colours be comprised, values might be arbitrarily put for all these terms in the dividend; and these values being summed, and divided by the denominator, would give the value of the first colour. It might be either a fraction or integer, and the values of the rest would be those arbitrarily assumed. Such a solution is facile. But, if the answer be required in whole numbers, then reserving one colour put arbitrary values for the rest; and thus a single colour with certain absolute numbers will remain in the dividend. Now such a value of that colour is to be assumed, as that the coefficient of the colour, being multiplied by the assumed quantity, and added to the absolute number, and divided by the denominator, may yield no residue: for so the value of the first colour will be an integer. This is the very problem solved in the investigation of a pulverizer (Ch. 2). If then a value of the colour in the dividend be put equal to the multiplier so found, the colour appertaining to the divisor will be the quotient, and an integer. Hence the text, "At the last value, &c." (§ 154—155.) CRĬSHṄ.

sor (letter or) colour, as the divisor quantity; and the absolute number, as the additive quantity: with which, proceeding by the rule of investigation of a pulverizer (Ch. 2), the multiplier, which is so found, is the value of the dividend (letter or) colour; and the quotient, which is obtained, is that of the divisor (letter or) colour. The reduced[1] divisor and dividend [used as abraders in the investigation] of these two values, being multiplied by some assumed[2] (letter or) colour, are to be put as additives [of the multiplier and quotient,[3] or values so found,] and thence, substituting their values with these additives, for those colours (or letters) in the value of the former colour (or letter), and dividing by the denominator, the quotient, which is obtained, is the value of the former colour (or letter). In like manner, inversely substituting [the values thus successively found], the values of the other colours (or letters) are thence deduced. But, if there be two or more colours (or letters) in the last value, then putting arbitrarily assumed values for them, and substituting by those values and adding the results to the absolute number, the investigation of a pulverizer is to be performed.

In the course of inverse substitution, if the values of a colour (or letter), in a value of a preceding one, be fractional, then the multiplier, which is found by a further investigation of the pulverizer, with the addition [of the divisor[4]] is the value of the dividend colour (or letter).[5] Then substituting with it for that colour (or letter) in the last values of colours (or letters), and proceeding by inverse substitution, in the preceding ones, the values of the other colours (or letters) are found.

In this [analysis], when the value of a (letter or) colour is found, (whether that value be a known quantity, or an unknown one, or known and un-

[1] See § 54.

[2] The assumed colour represents the arbitrary factor, introduced (§ 64) to make arbitrary multiples of the (abrading) divisors additives of the multiplier and quotient; that is, of the values here found. By substituting cipher for this assumed colour, as is frequently done in the following examples (§ 160, 161, 163, &c.), the simple values are used without the additives. It does not represent an unknown quantity which is sought; but a factor of the divisors, which is to receive an arbitrary value: and it serves to show the relation of the quantities in the manifold answers of an indeterminate problem, the solution of which is required in whole numbers. See CRÍSHN'.

[3] See § 64.

[4] Ibid.

[5] Recourse is had to this method, to clear the fraction: for the same reason holds indifferently. CRÍSHN'.

known together;) and is multiplied by the coefficient of the unknown, the removal or extermination of that (literal character or) colour is called *(Utt'hápana)* "raising" or "substitution."[1]

Example: "The quantity of rubies without flaw, sapphires and pearls," &c. (§ 105.)

In this case, putting *yávat-távat*, &c. for the rates of the rubies and the rest, and making the number of each sort of gem with its rate a multiple, and superadding the absolute number, the statement for equal subtraction is

ya 5 *ca* 8 *ní* 7 *ru* 90
ya 7 *ca* 9 *ní* 6 *ru* 62

Proceeding as directed, to "subtract the first colour, &c," (§ 153) this single value of *yávat-távat* is obtained $\frac{ca\ \dot{1}\ \ ni\ 1\ \ ru\ 28}{ya\ 2}$*

Being single, this same value is "last" (§ 154). Therefore the investigation of the pulverizer must take place. In this dividend there is a couple of colours: wherefore (§ 155) the value of *ní* is arbitrarily put unity; with which substituting for *ní*, and superadding it to the absolute number, there results $\frac{ca\ \dot{1}\ \ ru\ 29}{ya\ 2}$ Hence, by the rule of investigation of the pulverizer, the multiplier and quotient, together with the additives [deduced from their divisors] are found

pí 2 *ru* 1
pí 1 *ru* 14

Then, substituting for *pí* by putting cipher for it, the rates of the rubies and the rest come out 14, 1, 1. Putting one for *pí*, they are 13, 3, 1; assuming two, 12, 15, 1; or, supposing three, 11, 7, 1. Thus, by virtue of suppositions, an infinity of answers is obtained.

Example: "One says 'give me a hundred, and I shall be twice as rich as you,' &c. (§ 106).

[1] The value of any colour, that is found, whether expressed by absolute number, or symbol of unknown quantity, or both, and occurring in another, is deduced by the rule of three: for instance, in example § 157, the value of *ca* is found $\frac{ni\ 20\ \ pi\ 16}{ca\ 9}$; and *pí* is *lo* 4 *ru* 0; and *ní* is *lo* 31 *ru* 0. Then the proportion, *ní* 1 | *lo* 31 *ru* 0 | *ní* 20 |, gives *lo* 620 *ru* 0; and this, *pí* 1 | *lo* 4 *ru* 0 | *pí* 16 |, gives *lo* 64 *ru* 0. Their sum, *lo* 684 *ru* 0, divided by 9, gives the value of *ca*; viz. *lo* 76 *ru* 0. This is termed *utt'hápana*, raising or substitution. CRÍSHN'.

* For the reason of retaining the symbol of the unknown *ya*, in the fraction expressing its value, see note on § 157.

Let the respective capitals be *ya* 1 *ca* 1. Taking a hundred from the capital of the last, and adding it to that of the first, they become *ya* 1 *ru* 100 and *ca* 1 *ru* $1\dot{0}0$. The wealth of the first is double that of the second: therefore equating it with twice the second's capital, a value of *yávat-távat* is obtained $\frac{ca\ 2\quad ru\ \dot{3}00}{ya\ 1}$ Again, ten being taken from the first and added to the capital of the second, there results *ya* 1 *ru* 10 and *ca* 1 *ru* 10. But the second is become six times as rich as the first: wherefore making the second equal to the sextuple of the first, a value of *yávat-távat* is obtained $\frac{ca\ 1\quad ru\ 70}{ya\ 6}$. With these reduced to a common denomination and dropping the denominator, an equation is formed; from which, as being one containing a single colour (or character of unknown quantity), the value of *ca* comes out by the foregoing analysis (Ch. 4); viz. 170.[1] With which substituting for *ca*, in the two values of *yávat-távat*, and adding it to the ab solute number, and dividing by the appertinent denominator, the value of *yávat-távat* is found, 40.

157. Example: The horses belonging to these four persons respectively are five, three, six and eight; the camels appertaining to them are two, seven, four and one; their mules are eight, two, one and three; and the oxen owned by them are seven, one, two and one. All are equally rich. Tell me severally, friend, the rates of the prices of horses and the rest.

Here put *yávat-távat*, &c. for the prices of the horses and the rest. The number of horses and cattle, being multiplied by those rates, the capitals of the four persons become

$$\begin{array}{llll} ya\ 5 & ca\ 2 & ní\ 8 & pí\ 7 \\ ya\ 3 & ca\ 7 & ní\ 2 & pí\ 1 \\ ya\ 6 & ca\ 4 & ní\ 1 & pí\ 2 \\ ya\ 8 & ca\ 1 & ní\ 3 & pí\ 1 \end{array}$$

These are equal. From the equation of the first and second, the value of *yávat-távat* is obtained $\frac{ca\ 5\quad ní\ \dot{6}\quad pí\ \dot{6}}{ya\ 2}$ From that of the second and third, it is $\frac{ca\ 3\quad ní\ 1\quad pí\ \dot{1}}{ya\ 3}$ In like manner, from that of the third and fourth, it is $\frac{ca\ 3\quad ní\ \dot{2}\quad pí\ 1}{ya\ 2}$.[*]

[1] The commentator, CRÍSHN'A, quotes from his preceptor, VISHN'U CHANDRA, a rule for examples of this nature; abridged, as he observes, from the algebraic solution.

[*] In these fractional values of *ya* deduced from the preceding equations, by equal subtraction

Reducing these[1] to a common denomination,[2] and dropping the denominator, the value of *cálaca* is found from the equation of the first and second, (*ní* 20 *pí* 16) / (*ca* 9); and from that of the second and third, (*ní* 8 *pí* 5̇) / (*ca* 3). From an equation of these two reduced to a common denomination, the value of *nílaca* is had (*pí* 31) / (*ní* 4). This being "last" value (§ 154) the investigation of the pulverizer gives, (as there is no additive;[3]) multiplier 0, or, with, the addition [of its divisor[4]] *lo* 4 *ru* 0. It is the value of *pítaca*. Also, (for the same reason) the quotient 0,[3] with the addition [of its divisor[4]] *lo* 31 *ru* 0. It is the value of *nílaca*. Substituting for *ní* and *pí* by their respective values in that of *ca*, adding them together, and dividing by the appertinent denominator, the value of *cálaca* is obtained *lo* 76 *ru* 0. Substituting for *ca* and the rest by their own values in that of *yávat-távat*, adding them together, and dividing by the appropriate denominator, the value of *yávat-távat* comes out *lo* 85 *ru* 0. Then, *lóhitaca* being replaced by unity arbitrarily assumed, the values of *yávat-távat* and the rest are found, 85, 76, 31, 4. Or putting two for it, they are 170, 152, 62, 8. Or, supposing three, they are 255, 228, 93, 12. Thus, by virtue of suppositions, an infinity of answers may be obtained.

158—159. Example by ancient authors:[5] Five doves are to be had for three *drammas*; seven cranes,[6] for five; nine geese, for seven; and three

and then by proportion, *ya* 2 | *ca* 5 *ní* 6̇ *pí* 6̇ | *ya* 1 and *ya* 3 | *ca* 3 *ní* 1 *pí* 1̇ | *ya* 1 also *ya* 2 | *ca* 3 *ní* 2̇ *pí* 1 | *ya* 1, the syllable *ya* is inserted in the denominator to indicate that the value is of *ya*; not to include it as a factor of the denominator: for the first and third terms containing it were reduced by it as a common divisor; and, if that were not done, the numerator would be a multiple of it. CRISHN'.

[1] Other values of *ya* might be found by combining the first and third; first and fourth; and second and fourth. But that is not done, as there is no occasion. CRISHN'.

[2] The foregoing operations to find the value of the first colour were in fulfilment of the rule § 153. The work now proceeds to the finding of the values of the rest by § 154.

[3] See § 63.

[4] See § 64.

[5] *Ádya*, original or early writers. The commentators do not here specify them; nor hint whence the quotation comes. SÚRYADÁSA only says "certain writers;" and observes, that it is a well known instance. The rest are silent.

[6] *Sárasa*, the Siberian crane: Ardea Siberica.

peacocks, for nine: bring a hundred of these birds for a hundred *drammas*, for the prince's gratification.

In this case, putting *yávat-távat*, &c. for the prices of the doves, &c. find the number of birds of each kind by proportion; and make the equation with a hundred. Or else multiply the rates three, five, &c. and the number of birds five, seven, &c. by *yávat-távat*, and severally make an equation with a hundred.[1] Thus, *ya* 3, *ca* 5, *ní* 7, *pí* 9, being the prices, make their sum equal to a hundred; and the value of *yávat-távat* is found $\frac{ca\ \dot{5}\ \ ní\ \dot{7}\ \ pí\ \dot{9}\ \ ru\ 100}{ya\ 3}$. Again, making the birds *ya* 5, *ca* 7, *ní* 9, *pí* 3, equal to a hundred, the value of *yávat-távat* is obtained $\frac{ca\ \dot{7}\ \ ní\ \dot{9}\ \ pí\ \dot{3}\ \ ru\ 100}{ya\ 5}$.

From the equation of these, reduced to a common denomination, and dropping the denominator, the value of *cálaca* is had $\frac{ní\ \dot{2}\ \ pí\ \dot{9}\ \ ru\ 50}{ca\ 1}$. The dividend here contains two colours; therefore the value of *pítaca* is arbitrarily assumed four. With this substituting for *pítaca*, there results $\frac{ní\ \dot{2}\ \ ru\ 14}{ca\ 1}$.

Hence by investigation of the pulverizer, the quotient and multiplier, with their additives [multiples of their divisors], are
lo 2 *ru* 14 value of *ca*
lo 1 *ru* 0 — of *ní*
lo 0 *ru* 4 — of *pí*

Substituting for *cálaca* and the rest, by these their values, in the value of

[1] In the argument of proportion the sum of the rates as well as of the birds is twenty-four: and the requisition is a hundred. From some multiple of the argument or money, the number of birds is to be found. If the birds be found from the argument multiplied by an equal factor, the sum of both will not be a hundred: for the sum of the *drammas*, which are the arguments, multiplied by four, is ninety-six; and so is that of the birds multiplied by the same; and the sum, multiplied by five, is a hundred and twenty. If indeed they be multiplied by a proportional factor, namely twenty-five sixths, the sum will no doubt be a hundred; but the birds will not be entire. Therefore unequal factors must be used: a different one for the price of the doves; and another for that of the cranes; one for the rate of the geese; and another for that of the peacocks. Those factors are unknown; and therefore *yávat-távat*, &c. are put for them. The rates multiplied are the prices paid. Then, as three *drammas* are to five doves, so is the price *ya* 3 to the number of doves bought, *ya* 5. In like manner the numbers of the other birds are found by proportion.

Or put *yávat-távat*, &c. for the prices paid, which are unknown: and thence compute the number of the birds of each kind by proportion: viz. *ya* $\frac{5}{3}$ *ca* $\frac{7}{5}$ *ní* $\frac{9}{7}$ *pí* $\frac{3}{9}$. The solution will be the same, with this difference, however, that the sum of the birds must be taken by reduction of the fractions to a common denomination. CRÍSHN.

yávat-távat, and dividing by its denominator, the value of *yávat-távat* is brought out *lo* 1 *ru* $\dot{2}$. Substituting for *lôhitaca* with three arbitrarily assumed for it, the values of *yávat-távat* and the rest come out 1, 8, 3, 4. With these "raising" the birds and their prices, the answer is Prices 3 40 21 36 / Birds 5 56 27 12. Or, by putting four, the values are 2, 6, 4, 4; and the answer is Prices 6 30 28 36 / Birds 10 42 36 12. Or, by supposition of five, the values are 3, 4, 5, 4; and the answer Prices 9 20 35 36 / Birds 15 28 45 12.

Thus, by means of suppositions, a multitude of answers may be obtained.

160. Example:[1] What number is it, which, being divided by six, has five for a remainder; or divided by five, has a residue of four; or divided by four, has a remainder of three; or divided by three, leaves two?

Let the number be *ya* 1. This, divided by six, has five for a remainder: division by six being therefore made, the quotient is *ca.* The divisor, multiplied by *ca,* with its remainder five added to it, is equal to *ya.* From this equation the value of *ya* is obtained $\frac{ca\,6\ \ ru\,5}{ya\,1}$. In like manner, *ní,* &c. are quotients answering to the divisors five and so forth; and values of *ya* are thence obtained: $\frac{ní\,5\ \ ru\,4}{ya\,1}$ $\frac{pí\,4\ \ ru\,3}{ya\,1}$ $\frac{lo\,3\ \ ru\,2}{ya\,1}$. From the equation of the first and second of these values, a value of *ca* is deduced, $\frac{ní\,5\ \ ru\,1}{ca\,6}$; from the equation of the second and third, a value of *ní,* viz. $\frac{pí\,4\ \ ru\,\dot{1}}{ní\,5}$; and from that of the third and fourth, a value of *pí,* namely $\frac{lo\,3\ \ ru\,\dot{1}}{pí\,4}$.

Hence,[2] by investigation of the pulverizer, values of *lo* and *pí* are brought out; which, with the additives [derived from the divisors[3]], are *ha* 4 *ru* 3 value of *lo* / *ha* 3 *ru* 2 — of *pí*. Substituting for *pí* by that value, in the value o

[1] To illustrate the rule for exterminating a fraction, 156. CRÍSHN.

[2] It is "last" value: wherefore investigation of the multiplier takes place. § 154. CRÍSHN.

[3] See § 64.

ní, this becomes $\frac{ha\ 12\ \ ru\ 7}{ní\ 5}$:* and here dividing it by its denominator, the value of *ní* comes out a fraction.[2] Removing then the fraction by investigation of the pulverizer, the multiplier, with its additive [borrowed from the divisor,] as found by that method, is *swe* 5 *ru* 4. It is the value of *ha*. Substituting with this for *ha* in the values of *lo* and *pí*,[3] they become *swe* 15 *ru* 14 value of *pí* / *swe* 20 *ru* 19 — of *lo*. Now substituting for *pí*, with this value, in the value of *ní*,[4] and dividing by its denominator, the value of *ní* is brought out, without a fraction, *swe* 12 *ru* 11. Substituting for *ní* with this value, in the value of *ca*, and dividing by the denominator, the value of *ca* is obtained *swe* 10 *ru* 9. Substituting for *ca* and the rest, by these values in the several values of *ya*, it comes out *swe* 60 *ru* 59.[5]

Or [putting *ya* 1 for the quantity] divided by six and having five for a remainder (§ 160), the quantity is *ca* 6 *ru* 5, as before. This, divided by five, has a residue of four (§ 160): put *ní* for the quotient; and, by the equation with the divisor multiplied by that quotient and added to the residue (*ní* 5 *ru* 4), there results $\frac{ní\ 5\ \ ru\ 1}{ca\ 6}$ the value of *ca* in a fractional expression. By investigation of the pulverizer, that value, in an expression not fractional, becomes *pí* 5 *ru* 4. Substituting for *ca* with this, in the original value, *ca* 6 *ru* 5, it is *pí* 30 *ru* 29. This again, divided by four, has a remainder of three (§ 160): an equation then being made as before, there results $\frac{lo\ 4\ \ ru\ \dot{26}}{pí\ 30}$. Here also, by investigation of the pulverizer, the value of *pí* is converted into *ha* 2 *ru* 1. Whence, substituting with it, in the expression *pí* 30 *ru* 29, the quantity is found *ha* 60 *ru* 59. This again,

* By the rule of three terms: *pí* 1 | *ha* 3 *ru* 2 | *pí* 4 | *ha* 12 *ru* 8. This value of *pí* 4, with *ru* $\dot{1}$, makes *ha* 12 *ru* 7. CRÍSHN'.

[2] Division by the denominator does not succeed exactly. SU'R.

[3] Being the two "last" values. CRÍSHN'.

[4] This is inverse substitution, commencing from the "first" or preceding (§ 156), which is here *ní*. CRÍSHN'.

[5] It comes out the same in all the expressions of the value of *ya*: and putting nought for *swe* (and thus exterminating the unknown term) the conditions of the question are all answered with the remaining number.—SU'R. And the quotients or values of *ca*, &c. are 9, 11, 14, 19. By the supposition of one for *swe*, the number is 119; and the quotients are 19, 23, 29, and 39. CRÍSHN'.

divided by three, leaves two (§ 160): and the quantity here comes out the same. By substituting nought, one, two, &c. a multiplicity of answers may be obtained.

161. Example: What numbers, being multiplied respectively by five, seven, and nine, and divided by twenty, have remainders increasing in progression by the common difference one, and quotients equal to the remainders.

In this case put the residues *ya* 1, *ya* 1 *ru* 1, *ya* 1 *ru* 2. They are the quotients also. Let the first number be *ca* 1. From this multiplied by five, subtracting the divisor taken into the quotient, the remainder is *ca* 5 *ya* $\dot{20}$. Making this equal to *ya* 1, a value of *yávat-távat* is obtained $\frac{ca\ 5}{ya\ 21}$. Let the second number be put *ní* 1. From this multiplied by seven, subtracting the divisor taken into *ya* added to one, the result is *ní* 7 *ya* $\dot{20}$ *ru* $\dot{20}$; and making this equal to *ya* 1 *ru* 1, a value of *yávat-távat* is had $\frac{ní\ 7\ \ ru\ \dot{21}}{ya\ 21}$. Let the third number be *pí* 1. From this multiplied by nine, subtracting the divisor taken into *ya* added to two, the residue is *pí* 9 *ya* $\dot{20}$ *ru* $\dot{40}$; and making this equal to *ya* 1 *ru* 2, a value of *yávat-távat* is found $\frac{pí\ 9\ \ ru\ \dot{42}}{ya\ 21}$. From the equation of the first and second of these, the value of *cálaca* is $\frac{ní\ 7\ \ ru\ \dot{21}}{ca\ 5}$; and from that of the second and third, the value of *nílaca* is $\frac{pí\ 9\ \ ru\ \dot{21}}{ní\ 7}$. This being "last" value, the investigation of the pulverizer takes place: and quotient and multiplier, with additives [derived from their divisors], are by that method found, *lo* 9 *ru* 6 value of *ní*; *lo* 7 *ru* 7 — of *pí*. Here the additive is designated *lóhitaca*;[1] and the expressions in their order, are values of *nílaca* and *pítaca*. Substituting for *ní* by this value, in that of *ca*, and dividing by its denominator, the value of *ca* comes out fractional $\frac{lo\ 63\ \ ru\ 21}{ca\ 5}$.

To make it integer by investigation of the pulverizer, reduce the divi-

[1] The commentator SÚRYADÁSA pursues the operation, without introduction of this symbol of an unknown: remarking, that it would serve to embarrass and mislead the student.

dend and additive to their least terms by the common measure twenty-one [§ 58], and the values *cálaca* and *lóhitaca* are found *ha* 63 *ru* 42 value of *ca*
ha 5 *ru* 3 — of *lo*
Substituting for *lóhitaca* by its value, in the values of *nílaca* and *pítaca*, these are brought out *ha* 45 *ru* 33 value of *ní* Again, with these values,
ha 35 *ru* 28 — of *pí*

ha 63 *ru* 42 for *ca* substituting for *cálaca* and the rest in the values of *yávat-*
ha 45 *ru* 33 for *ní*
ha 35 *ru* 28 for *pí*

távat, and dividing by the appertinent denominators, the value of *ya* is obtained *ha* 15 *ru* 10. Here, as the quotient is equal to the residue, and the residue cannot exceed the divisor, substitute nought only[1] for *haritaca*, and the quotients are found 10, 11, 12. Deducing *cálaca* and the rest from their values, the quantities are brought out in distinct numbers, 42, 33, 28.

162. Example: What number, being divided by two, has one for remainder; and, divided by three, has two; and, divided by five, has three: and the quotients also, like itself?

Let the number be put *ya* 1. This, divided by two, leaves one; and the quotient also, divided by two, has a remainder of one. Let the quotient be *ca* 2 *ru* 1. The divisor multiplied by this, with addition of the residue, being equal to *ya* 1, the value of *yávat-távat* is obtained, *ca* 4 *ru* 3. It answers one of the conditions. Again, the number, being divided by three, has a residue of two: and so has the quotient. Put *ní* 3 *rú* 2. This, multiplied by the divisor, and added to the residue, is *ní* 9 *ru* 8; which is equal to *ca* 4 *ru* 3; whence the value of *ca* is fractional. Cleared of the fraction by means of the pulverizer, it becomes *pí* 9 *ru* 8; with which, substituting for *ca*, the number is found *pí* 36 *ru* 35. This answers two of the conditions. Again, the same number, divided by five, has a remainder of three; and so has the quotient. Put *lo* 5 *ru* 3. This, multiplied by the divisor, and added to the residue, is *lo* 25 *ru* 18. Making it equal to *pí* 36 *ru* 35, the value of *pí* is fractional. Clearing it of the fraction by the pulverizer, the result is *ha* 35 *ru* 3. Substituting with this for *pí*, the number is found *ha* 900 *ru* 143. Substituting for *ha* with nought, the number comes out

[1] Supposing unity, the quotients would come out 25, 26 and 27.—RÁM. And would exceed the divisor 20.

143.[1] Division being made conformably with the conditions of the problem, the quotients are 71, 47 and 28; and by these the conditions are fulfilled.

163. Example: Say what are the numbers, except six and eight,[2] which, being divided by five and six respectively, have one and two for remainder; and the difference of which, divided by three, has a residue of two; and their sum, divided by nine, leaves a remainder of five; and their product, divided by seven, leaves six? if thou can overcome conceited proficients in the investigation of the pulverizer, as a lion fastens on the frontal globes of an elephant.

In this case, the two numbers, which being divided by five and six, leave one and two respectively, are put *ya* 5 *ru* 1 and *ya* 6 *ru* 2. The difference of these, divided by three, gives a residue of two. Put *ca* for the quotient; and let the divisor multiplied by that added to the residue (*ca* 3 *ru* 1) be equated with the difference *ya* 1 *ru* 1. The value of *ya* is obtained *ca* 3 *ru* 1. The two numbers deduced from substitution of this value are *ca* 15 *ru* 6 and *ca* 18 *ru* 8. Again, the sum of these, divided by nine, leaves five. Put *ní* for the quotient; and let the divisor, multiplied by that and added to the remainder (*ní* 9 *ru* 9) be equated with the sum *ca* 33 *ru* 14. The value of *ca* is had $\frac{ní\ 9\ \ ru\ \dot{9}}{ca\ 33}$; and is a fraction. Rendering it integer by the pulverizer, it becomes *pí* 3 *ru* 0. From which the two quantities, deduced by substitution, are *pí* 45 *ru* 6 and *pí* 54 *ru* 8. Again, proceeding to the product of these, as it rises to a quadratic, the operation is a grand one.[3] Wherefore, substituting with unity for *pí*, the first quantity is made an absolute number, 51. Again, the product of these, abraded[4] by seven, yields *pí* 3 *ru* 2. Put *lo* for the quotient of this dividend by seven to leave six. The divisor multiplied by that quotient and added to the residue [*lo* 7 *ru* 6] is equal to the abraded product (*pí* 3 *ru* 2). Thence, by inves-

[1] This sentence, which is wanting in two of the collated copies, is found in the margin of one, and in the text of that which is accompanied by the gloss of Ráma-Crishna; where alone the subsequent sentence occurs. Both are repeated in his commentary.

[2] These, furnishing too obvious an answer to the question, (for they fulfil all its conditions,) are excepted. Crishn.

[3] It is vain; for the equation rises to cubic and biquadratic. Súr.

[4] See § 56.

tigation of a pulverizer as before, the value of *pí* is found *ha* 7 *ru* 6.[1] The number deduced by substitution of this comes out *ha* 378 *ru* 332. The additive of the former number (*pí* 45) multiplied by this (*ha* 7) is its present additive (*ha* 315): and thus the first number or quantity with its additive is brought out *ha* 315 *ru* 51.

Or else putting an absolute number for the first, the second is to be sought.[2]

164. Example: What number is it which multiplied severally by nine and seven, and divided by thirty, yields remainders, the sum of which, added to the sum of the quotients, is twenty-six?

As the divisor is the same, and the sum of the remainders and quotients is given, the sum of the multiplicators is for shortness made the multiplier; and the number is put *ya* 1. This, multiplied by the sum of the multiplicators, is *ya* 16. Put *ca* for the sum of the quotients of the division by thirty. Subtracting the divisor taken into that (*ca* 30) from the number multiplied by the [sum of the] multiplicators (*ya* 16); and equating the difference added to the quotient, with twenty-six,[3] the value of *ya* found by the pulverizer is *ní* 29 *ru* 27. As the sum of the remainders and quotients is restricted, the additive is not to be applied. Substituting therefore with nought for *ní*, the value of *ya* is 27: and this is the number sought.

165. Example: What number being severally multiplied by three, seven, and nine, and divided by thirty, the sum of the remainders too being divided by thirty, the residue is eleven?

In this case also, the sum of the multiplicators is made the multiplier, as before (§ 164): viz. 19. The number is put *ya* 1. The quotient *ca* 1. Subtracting the divisor multiplied by this from the number taken into the

[1] Equation *pí* 3 *lo* 0 *ru* 2 / *pí* 3 *lo* 7 *ru* 6. Whence, by subtraction, $\frac{lo\ 7\ \ ru\ 4}{pí\ 3}$: and, clearing the fraction by means of the pulverizer, the quotient and multiplier are 6 and 2. Whence the values
ha 7 *ru* 6 of *pí*
ha 3 *ru* 2 of *lo*
SU'R. and RA'M.

[2] Putting 6, it is *ha* 126 *ru* 8. Or putting 36, it is *ha* 126 *ru* 104. RA'M.

[3] *ya* 16 *ca* 2̇9 *ru* 0
ya 0 *ca* 0 *ru* 26
SU'R. and RA'M.

multiplicator, the remainder is *ya* 19 *ca* 30. The sum of the remainders, abraded by thirty, leaves a residue of eleven. The second condition therefore being comprehended in the first, this is equal to eleven; and from such equation, proceeding as before,[1] the number comes out *ní* 30 *ru* 29.

166. Example: What number being multiplied by twenty-three, and severally divided by sixty and eighty, the sum of the remainders is a hundred? Say quickly, algebraist.

167. Maxim: If more than one colour represent, in a dividend, quotients of a numerator, an arbitrary value is not to be assumed, lest the solution fail.[a]

Therefore it must be treated otherwise. In this instance the solution is to be managed by distributing the sum of the residues, so as these may be less than their divisors and nothing be imperfect. Accordingly the remainders are assumed 40 and 60. The number is put *ya* 1. This, multiplied by twenty-three and divided by sixty, gives a quotient: for which put *ca*. The divisor taken into that and added to the remainder being equated with this term *ya* 23, a value of *ya* is obtained, $\frac{ca\ 60\ \ ru\ 40}{ya\ 23}$. In like manner, another value is had $\frac{ní\ 80\ \ ru\ 60}{ya\ 23}$. From the equation of these, the values of *ca* and *ní* are found by the pulverizer, *pí* 4 *ru* 3 value of *ca*; *pí* 3 *ru* 2 —— of *ní*. Substituting

[1] Equation *ya* 19 *ca* 30 *ru* 0 / *ya* 0 *ca* 0 *ru* 11. Whence value of *ya*, $\frac{ca\ 30\ \ ru\ 11}{ya\ 19}$. By the pulverizer, the multiplier and quotient are 18 and 13. Making these the values, and changing the letter, the number is found *ní* 30 *ru* 29. SUR.

[a] Putting *ya* 1 for the number, and *ca* 1 and *ní* 1 for the quotients, the value of *ya* is $\frac{ca\ 60\ \ ní\ 80\ \ ru\ 100}{ya\ 46}$, or reduced to least terms $\frac{ca\ 30\ \ ní\ 40\ \ ru\ 50}{ya\ 23}$. This is to be cleared of the fraction: and, by the rule (§ 155), as there is more than one colour, either *ca* or *ní* may be put arbitrarily any number. But they are quotients of the same dividend or numerator by the divisors 60 and 80. If an absolute number be put for *ca* the quotient of 60, then *ní*, the quotient of 80, is absolute too; being a quarter less. So likewise, if any number be put for *ní*, the quotient of 80, then *ca*, the quotient of 60, is absolute also, being a third more. Such being the case, the solution would not conform to the sum of the remainders given at a hundred. Nor would the answer agree with the question; if the assumption be arbitrarily made. CRÍSHN. and RÁM.

with these, the value of *ca* is brought out, a fraction, $\frac{pí\ 240\ \ ru\ 220}{ya\ 23}$. Clearing it of the fraction by the pulverizer, it becomes *lo* 240 *ru* 20.[1] Or let the remainders be put 30 and 70. From these the number is deduced *lo* 240 *ru* 90.[2] In like manner, a multiplicity of answers may be found.

168. Example: Say quickly what is the number, which, added to the quotient by thirteen of its multiple by five, becomes thirty?

Put *ya* 1 for the number. This, multiplied by five and divided by thirteen, gives a quotient: for which put *ca*. The quotient and original number, added together, *ya* 1 *ca* 1, are equal to thirty. But this equation does not answer. For there is no ground of operation, since neither multiplier, nor divisor, is apprehended. Accordingly, it is said -

169. In a case in which operation is without ground or in which it is restricted, do not apply the operation: for how should it take effect?[3]

The solution therefore is to be managed otherwise in this case. If then the number be put equal to the divisor in the instance, viz. 13, the proportion "as this sum of number and quotient, 18, is to the quotient 5, so is 30 to what?" brings out the quotient $\frac{25}{3}$; and subtracting this from thirty, the remainder is the number sought, which thus is found $\frac{65}{3}$.

170. Example instanced by ancient authors: a stanza and a half. Three traders, having six, eight, and a hundred, for their capitals respectively, bought leaves of betle[4] at an uniform rate; and resold [a part] so; and disposed of the remainder at one for five *pańas;* and thus became equally rich. What was [the rate of] their purchase? and what was [that of] their sale?

[1] Substituting nought for *ló*, the conditions are answered. SU'R.

[2] Put *ya* 1 for the number; and *ca* 4 and *ca* 3 for the quotients. Subtract the quotient taken into its divisor, from the dividend, the remainders are found $\begin{matrix} ya\ 23 & c\dot{a}\ 240 \\ ya\ 23 & ca\ \dot{2}40 \end{matrix}$ They are alike; and, as their sum is a hundred, each is equal to fifty. From this equation, the values of *ya* and *ca* are brought out, by means of the pulverizer, $\begin{matrix} ní\ 240 & ru\ 190 & \text{value of } ya. \\ ní\ 22 & ru\ 18 & \text{value of } ca. \end{matrix}$ CRI'SHN'.

[3] Very obscure: but not rendered more intelligible by the commentators.

[4] RA'MACRI'SHN'A reads and interprets *dala* leaves of (*Nágavallí*) piper betle. Another reading is *phala*, fruit.

Put *ya* 1 for the [rate of] purchase; and let the [rate of] sale be assumed a hundred and ten. The purchase, multiplied by six and divided by the sale, gives a quotient; for which put *ca* 1. Subtracting the divisor multiplied by this, from the quantity multiplied by six, the remainder is *ya* 6 *ca* 110. This, multiplied by five, and added to the quotient, gives the number of *paṅas* belonging to the first trader. In like manner the money of the second and of the third is to be found. Here the quotient is deduced by the proportion 'as six is to *ca*, so is eight (or a hundred) to what?' The quotient of eight comes out $ca\ \frac{4}{3}$; and that of a hundred, $ca\ \frac{50}{3}$. Subtracting the divisor taken into the quotient, from the dividend, the remainder, multiplied by five and added to the quotient, gives the *paṅas* appertaining to the second, $ya\ \frac{120}{3}\quad ca\ \frac{\dot{2}196}{3}$. In like manner the third's money is found, $ya\ \frac{1500}{3}\quad ca\ \frac{2\dot{7}450}{3}$. These are all equal. Reducing them to a common denomination, and dropping the denominator, and taking the equation of the first and second, and that of the second and third, the value of *ya* comes out, alike [both ways], $\frac{ca\ 549}{ya\ 30}$: And, by the pulverizer, it is found *ní* 549 *ru* 0. Substituting with unity for *ní*, the rate of purchase is brought out, 549.[1]

This, which is instanced by ancient writers as an example of a solution resting on unconfined ground, has been by some means reduced to equation;

[1] Equation of the 1st and 2d $\begin{matrix} ya\ 30 & ca\ \dot{5}49 \\ ya\ \frac{120}{3} & ca\ \frac{2196}{3} \end{matrix}$, reduced to a common denomination $\begin{matrix} ya\ 90 & ca\ 164\dot{7} \\ ya\ 120 & ca\ 2196 \end{matrix}$ Whence value of *ya* $\frac{ca\ 549}{ya\ 30}$.

Equation of 2d and 3d, $\begin{matrix} ya\ 120 & ca\ 2\dot{1}96 \\ ya\ 1500 & ca\ 27450 \end{matrix}$ Whence value of *ya* $\frac{ca\ 23254}{ya\ 1380}$ and, abridging by 46, $\frac{ca\ 549}{ya\ 30}$.

Equation of 1st and 3d, $\begin{matrix} ya\ 90 & ca\ 1647 \\ ya\ 1500 & ca\ 27450 \end{matrix}$ Whence value of *ya* $\frac{ca\ 25803}{ya\ 1410}$ and, abridging by 47, $\frac{ca\ 549}{ya\ 30}$ Proceeding by the pulverizer, the quotient and multiplier, briefly found under the rule (§ 63), are $\begin{matrix} ní\ 549 \\ ní\ 0\ \ ru\ 0 \end{matrix}$ And the value of *ya* the colour of the divisor, comes out *ní* 549 *ru* 0; and that of *ca*, the colour of the dividend, *ní* 30 *ru* 0. At 549 betle leaves for a *paṅa*, six bring 3294; eight, 4392; and a hundred, 54900: which sold at the rate of 110, fetch 29, 39 and 499; leaving remainders 104, 102 and 10; and these at the rate of one for five, bring 520, 510 and 50. Added together, in their order, they make the amount of the sale 549. RÁM.

and such a supposition introduced, as has brought out a result in an unrestricted case as in a restricted one. In the like suppositions, when the operation, owing to restriction, disappoints; the answer must by the intelligent be elicited by the exercise of ingenuity. Accordingly it is said,

'The conditions, a clear intellect, assumption of unknown quantities, equation, and the rule of three, are means of operation in all analysis.'

CHAPTER VII.

VARIETIES OF QUADRATICS.[1]

NEXT, varieties of the solution involving extermination of the middle term are propounded.

171—174. Rule beginning with the latter half of the concluding stanza [in the preceding rule, § 156]: three and a half stanzas. Equal subtraction[2] having been made, when the square and other terms of the unknown remain, let the square-root of the one side be extracted in the manner before directed;[3] and the root of the other, by the method of the affected square,[4] and then, by the equation of the two roots, the solution is to be completed.

173. If the case be not adapted to the rule of the affected square, make the second side of equation equal to the square of another colour, and find the value of the colour, and so the value of the first, through the affected square. By ingenious algebraists many different ways are to be devised: so as to render the case fit for the application of that method.

174. For their own elemental sagacity (assisted by various literal symbols) which has been set forth by ingenious ancient authors,[5] for the in-

[1] *Mad'hyamáharana-bhéda:* varieties of quadratic, &c. equations. See Ch. 5.

[2] *Sama-śód'hana, tulya-sudd'hi,* equal subtraction; or transposition, with other preparations of the equation. See § 101. Ch. 4.

[3] See Ch. 5. § 128 and 131.

[4] Ch. 3.

[5] BRAHMEGUPTA and CHATURVÉDA, and the rest, —SÚR. meaning CHATURVÉDA PRÍT'HÚDACA SWÁMÍ the scholiast of BRAHMEGUPTA.

struction of men of duller intellect, irradiating the darkness of mathematics, has obtained the name of elemental arithmetic.[1]

After equal subtraction has been made, if a square of the unknown with other terms remain, then the square-root of the one side of the equation is to be extracted in the manner before taught (§ 128). If the square of an unknown with unity stand on the other side, two roots are to be found for this side of the equation by the method of the affected square (Ch. 3). Here the number, which stands with the square of the colour, is *(pracríti)* the coefficient affecting it.[2] And the absolute number is to be made the additive. In this manner, the "least" root is the value of the colour standing with the coefficient, and the "greatest" is that of the root of the whole square. Making, therefore, an equation to the root of the first side, the value of the preceding colour is to be thence brought out.

But, if there be on one side of the equation, the square of the unknown with the [simple] unknown, or only the [simple] unknown with absolute number, or without it; such is not a case adapted for the method of the affected square: and how then is the root to be found? The text proceeds to answer 'If the case be not adapted, &c.' (§ 173). Making it equal to the square of another colour, the root of one side of the equation is to be extracted as before; and two roots, by the rule of the affected square, to be investigated, of the other side: and here also, the "least" is the value of the colour belonging to the coefficient, and the "greatest" is square-root of the side of the equation. Then duly making an equation of the roots, the values of the colours are to be thence found.

If nevertheless, though the second side be so treated, the case be still not adapted to the rule, the intelligent, devising by their own sagacity, means of bringing it to the form to which the rule is applicable, must discover values of the unknown.

If they are to be discovered by the mere exercise of sagacity, what occasion is there for algebra? To this doubt, the text replies (§ 174). Because sagacity alone is the paramount elemental analysis: but colours (or symbols)

[1] *Víja-mati,* causal sagacity: for nothing can be discovered, unless by ingenuity and penetration. *Víja-gańita,* causal calculus: from *víja,* primary cause, and *gańita,* computation. SU'R.

[2] The number *(anca)* or coefficient is the *pracríti,* or subject affecting the colour or symbol that is squared. See Ch. 3. under § 75.

are its associates; and therefore ancient teachers, enlightening mathematicians as the sun irradiates the lotus, have largely displayed their own sagacity, associating with it various symbols: and that has now obtained the name of *(Víja-gańita)* elemental arithmetic. This indeed has been succinctly expresssed by a fundamental aphorism in the *Siddhánta;*[1] but has been here set forth at somewhat greater length for the instruction of youth.

175—176. Rule: When the square-root of one side of the equation has been extracted, if the second side of it contain the square of an unknown quantity together with unity (or absolute number); in such case "greatest" and "least" roots are then to be investigated by the method of the affected square. Making the "greatest" of these two equal to the square-root of the first side of the equation, the value of the first colour is thence to be found, in the manner which has been taught. The "least" will be the value of the colour that stands with the coefficient. Thus is the rule of affected square to be here applied by the intelligent.

The meaning has been already explained.

177. Example: What number, being doubled and added to six times its square, becomes capable of yielding a square-root? tell it quickly, algebraist!

Put *ya* for the number. Doubled, and added to six times its square, it becomes *ya v* 6 *ya* 2. It is a square. Put it equal to the square of *ca*; and the statement of equation is *ya v* 6 *ya* 2 *ca v* 0 Equal subtraction being
ya v 0 *ya* 0 *ca v* 1
made and the two sides being multiplied by six, and superadding unity, the square-root of the first side found as before is *ya* 6 *ru* 1. The roots of the second side, investigated by the rule of the affected square. are L 2 G 5 or *L* 20 *G* 49.[2] Here the "greatest" of two roots is the square-root of the second side of the equation. From the equation of that value (5 or 49) with the root of the first side *ya* 6 *ru* 1, the value of *ya* is found $\frac{4}{3}$ or 8.

[1] See quotation from Chapter on Spherics under § 110.

[2] Assume the least root 2. Its square 4, multiplied by the coefficient 6, is 24. Added to 1, it affords the root 5. Statement: C 6 L 2 G 5 A 1 Whence, by composition (§ 77), l 20 g 49.
L 2 G 5 A 1
SÚR. and RÁM.

The "least" of the pair of roots (either 2 or 20) is the value of *ca*, the symbol standing with the coefficient. The number sought then is the integer 8, or the fraction $\frac{2}{3}$; and, in like manner, by the variety of "least" and "greatest" roots, a multiplicity of answers may be obtained.

178. Example from ancient authors: The square of the sum of two numbers, added to the cube of their sum, is equal to twice the sum of their cubes. Tell the numbers, mathematician!

The quantities are to be so put by the intelligent algebraist, as that the solution may not run into length. They are accordingly put *ya* 1 *ca* 1̇ and *ya* 1 *ca* 1.[1] Their sum is *ya* 2. Its square *ya v* 4. Its cube *ya gh* 8. The square of the sum added to the cube is *ya gh* 8 *ya v* 4. The cubes of the two quantities respectively are *ya gh* 1 *ya v. ca bh* 3̇ *ca v. ya bh* 3 *ca gh* 1̇ cube of the first; and *ya gh* 1 *ya v. ca bh* 3 *ca v. ya bh* 3 *ca gh* 1 cube of the second; and the sum of these is *ya gh* 2 *ca v. ya bh* 6; and doubled, *ya gh* 4 *ca v. ya bh* 12. Statement for equal subtraction:
ya gh 8 *ya v* 4 *ca v. ya bh* 0
ya gh 4 *ya v* 0 *ca v. ya bh* 12
After equal subtraction made, depressing both sides by the common divisor *ya*, and superadding unity, the root of the first side of equation is *ya* 2 *ru* 1. Roots of the other side (*ca v* 12 *ru* 1) are investigated by the rule of the affected square,[2] and are L 2 G 7 or *L* 28 *G* 97. "Least" root is a value of *ca*. Making an equation of a "greatest" root with *ya* 2 *ru* 1, the value of *ya* is obtained: viz. 3 or 48. Substitution being made with the respective values, the two quantities come out 1 and 5, or 20 and 76, and so forth.

179—180. Rule: a stanza and a half. Depressing the second side of the equation by the square, if practicable, let both roots be investigated: and then multiply "greatest" by "least." Or, if it were depressed by the biquadrate, multiply "greatest" by the square of "least." The rest of the process is as before.

[1] They are so put, as that one condition of the problem be fulfilled. SU'R. and RA'M.

[2] Put 2 for "least" root. Its square 4, multiplied by the coefficient 12, is 48: which added to 1, yields a square-root 7. Statement: C 12 L 2 G 7 A 1
L 2 G 7 A 1
Whence, by composition (§ 77), 128 g 97. SU'R. and RA'M.

The rule is clear in its import.

181. Example: Tell me quickly, mathematician, the number, of which the square's square, multiplied by five and lessened by a hundred times the square, is capable of yielding a square-root.

Here the number is put *ya* 1. Its biquadrate, multiplied by five, and lessened by a hundred times the square of the number, is *ya v v* 5 *ya v* $\dot{1}00$. It is a square. Put it equal to the square of *ca*, and the root of the square of *ca* is *ca* 1. Depressing the second side of the equation, namely *ya v v* 5 $\dot{y}$*a v* 100, by the common divisor, square of *ya*, the roots, investigated by the rule of the affected square,[1] come out L 10 G 20 or *L* 170 *G* 380. Depression by the square having taken place, multiply "greatest" root by "least" (§ 179); and thus "greatest" is brought out 200 or 64600. This is the value of *ca*. "Least" root is the value of the colour joined with the coefficient: and that is the number sought: viz. 10 or 170.

182. Example: Most learned algebraist! tell various pairs of integer numbers, the difference of which is a square, and the sum of their squares a cube.

Put the two numbers *ya* 1 and *ca* 1. Their difference is *ya* 1 *ca* $\dot{1}$. Making it equal to the square of *ní*, the value of *ya* is had, *ca* 1 *ní v* $\dot{1}$. Substituting with this for *ya*, the two quantities become *ca* 1 *ní v* $\dot{1}$ and *ca* 1. The sum of their squares is *ca v* 2 *ní v. ca bh* $\dot{2}$ *ní v v* 1. It is a cube. Make it then equal to the cube of the square of *ní*;[2] and, subtraction taking place, there results, in the first side of equation, *ní v gh* 1 *ní v v* $\dot{1}$; and, in the second, *ca v* 2 *ní v. ca bh* $\dot{2}$. Multiplying both sides by two and superadding the biquadrate of *ní*, the square-root of the second side of

[1] Assume the "least" root 10. Its square 100, multiplied by the coefficient 5, is 500. This, added to the number 100 with the negative sign, makes 400. Its root 20 is "greatest" root. Statement C 5 L 10 G (20, or, depression by the square having previously taken place,) 200 A $\dot{1}00$. SÚR. and RÁM.

So from the above (C 5 L 10 G 20 A $\dot{1}00$), by § 77, there results c 5 l 1 g 2 a $\dot{1}$: whence, by composition of like (§ 77), *L* 4 *G* 9 *A* 1; and, by composition of unalike (ib.)
C 5 L 10 G 20 A $\dot{1}00$ roots are deduced *l* 170 *g* 380 *a* $\dot{1}00$.
L 4 *G* 9 *A* 1

[2] This is a limitation more than is contained in the problem.

the equation is *ca* 2 *ní v* $\dot{1}$. Depressing the first side by the biquadrate of *ní* as common divisor, to *ní v* 2 *ru* $\dot{1}$, the roots investigated by the rule of affected square,[1] are L 5 G 7; or *L* 29 *G* 41. Then multiplying "greatest" by square of "least," conformably to the rule (§ 180), it comes out *G* 175, or 34481. "Least" root is the value of *ní*. Substituting with that, the former root becomes *ca* 2 *ru* 25; or *ca* 2 *ru* 841. Making an equation of this with "greatest" root, the value of *ca* is obtained 100 or 17661. Substituting these values respectively, the pair of numbers is brought out 75 and 100; or 16821 and 17661; and so forth.

183. Rule: comprised in a stanza and a half. If there be the square of a colour together with the simple unknown quantity and absolute number,[2] making it equal to the square of another colour, find the root; and, on the other side, investigate two roots, by the method of affected square, as has been taught. Consider the "least" as equal to the first root; and "greatest" as equal to the second.

The root of the first side of the equation having been taken, if there be on the other side the square of the unknown with the simple unknown, and with or without absolute number, make an equation of that remaining side with the square of another colour and take the root. Then let the roots of this other be investigated by the rule of affected square. Of the two roots so investigated, making "least" equal to the root of the first side of equation, and "greatest" equal to the root of the second, let the values of the colours be sought.

[1] Put 5 for "least" root. Its square 25, multiplied by the coefficient 2, makes 50. Subtracting one (for the negative additive) the remainder 49 yields a square-root 7; and the two roots are 5 and 7.—SU'R. RA'M. By § 88—89, the roots are 1 and 1. By composition with the above C 2 L 1 G 1 A 1 They are l 12 g 17 a 1; and by further composition C 2 L 1 G 1 A 1
L 5 *G* 7 *A* $\dot{1}$ l 12 g 17 a 1
they are *l* 29 *g* 41 *a* $\dot{1}$.

[2] A variation in the reading of this passage is noticed by SU'RYADA'SA: viz. *avyacta-rúpah* instead of *sávyacta-rúpah*. The meaning, as this is interpreted by him and by RA'MACRISHN'A, is, if there be both a term of the unknown and absolute number besides the square of the unknown. The other reading may be explained as confined to one term (the unknown) besides the square. See SU'R. and RA'M. The author himself in his comment dispenses with the third term, or absolute number, which is indeed not necessary to bring the form within the operation of the rule.

184. Example: Say in what period (or number of terms) is the sum of a progression continued to a certain period tripled; its first term being three and the common difference two?

In this case the statement of the two progressions is I3 D2 P*ya*1* / I3 D2 P*ca*1 The sums of these progressions are *ya v* 1 *ya* 2† / *ca v* 1 *ca* 2 Making three times the first equal to the second, the two sides of equation are *ya v* 3 *ya* 6 / *ca v* 1 *ca* 2 Tripling both, and superadding nine, the root of the first is found *ya* 3 *ru* 3. Making the second side, namely *ca v* 3 *ca* 6 *ru* 9, equal to the square of *ní*, the two sides of equation become *ca v* 3 *ca* 6 / *ní v* 1 *ru* 9̇ Tripling these and superadding nine, the root of the first of them is found *ca* 3 *ru* 3. Roots of the second (*ní v* 3 *ru* 18) investigated by the rule of affected square,[3] are L 9 G 15 or L 33 G 57.[4] Making equations of "least" with the first root, namely *ya* 2 *ru* 3; and of "greatest" with the second, *ca* 3 *ru* 3; the values of *ya* and *ca* are brought out 2 and 4; or 10 and 18. So in every [like] instance.

185—186. Rules: two stanzas. But, if there be two squares of colours, with (or without[5]) absolute number, assume one of them at choice as (*pra-criti*) the affected square,[6] and let the residue be additive: and then proceed to investigate the root in the manner taught, provided there be more than one equation.

186. Or, if there be two squares of colours together with a factum

* The author employs the initials á, u and ga of the words ádya, uttara and gachcha, signifying. Initial term, Difference and Period (or number of terms) of a progression. See *Líl.* Ch. 4.

† By the rule in the *Lílávatí*, for the sum of a progression. *Líl.* § 119.

[3] One copy here inserts, 'L 3 G 5 A 2 and, making the additive ninefold, L 9 G 15 A 18.' This indication of the manner of finding the roots is, however, wanting in other collated copies of the text.

[4] Assume 9 for "least" root: its square 81, multiplied by the coefficient 3, is 243: from which subtract 18 for the negative additive; and the remainder 225 gives the square-root 15. SŪR.

[5] Collated copies exhibit "with:" but the commentator reads and interprets "without;" (*arúpacé* instead of *sarúpacé*). The author's own comment may countenance either reading.

[6] See note at the beginning of Ch. 3.

(bhávita); taking the square-root [of so much of it as constitutes a square] let the root be made equal to half the difference between the residue, divided by an assumed quantity, and the quantity assumed.

The root of the first side of equation having been taken, if there be on the second side two several squares of colours with or without unity (or any absolute number), make one square of a colour the subject *(pracriti)*,[1] and let the rest be the additive. Then, proceeding by the rule (§ 75) let a multiple (by one, or some other factor,) of the same colour which occurs in the additive, or such colour with a number (one, or another,) added to it, be put for the "least" root, selected by the calculator's own sagacity; and thence find the "greatest" root (§ 75). If the coefficient be an exact square, the roots are to be sought by the rule (§ 95) 'The additive divided by an assumed quantity, &c.'

If there be a *(bhávita)* product of colours, then by the above rule (§ 186) the root of so much of the expression as affords a root is to be taken;[2] and that root is to be made equal to the half of the difference between the quotient of the residue divided by an assumed quantity and that assumed quantity.[3]

But, if there be three or more squares or other terms of colours, then reserving two colours selected at pleasure, and putting arbitrary values for the rest, let the root [of the reserved] be investigated.

This is to be practised when there is more than one equation. But, if there be only one; then reserving a single colour, and putting arbitrary values for the rest, let the root be sought as before.

187. Example: Tell two numbers, the sum of whose squares multiplied by seven and eight respectively yields a square-root, and the difference does so being added to one.

Let the numbers be put *ya* 1, *ca* 1. The sum of these squares multiplied respectively by seven and eight, is *ya v* 7 *ca v* 8. It is a square. Making

[1] See note at the beginning of Ch. 3.

[2] The term consisting of the product of two factors may be thus exterminated, taking with it squares of both colours with proper coefficients to complete the square.—See SU'R.

[3] This is grounded on the rule of § 95. The compound square has unity for coefficient; and the residue is the additive; the "least" root, which is the root of that square, is deduced from the additive by the rule cited; and needs no division, the square-root of the coefficient being unity.

it equal to square of *ní*, and subtracting; the two sides of equation are *ya v* 7 *ca v* $\dot{8}$ *ní v* 1. Adding eight times the square of *ca*, the root of the second side of equation is *ní* 1; and roots of the first side, viz. *ya v* 7 *ca v* 8, are to be investigated by the method of the affected square. Here the number *(anca)*, which is joined with the square of *ya*, is *(pracrĭti)* the subject affecting it: the residue is additive: *C* 7, A *ca v* 8.* Roots found by the rule (§ 75), assuming *ca* 2, are L, *ca* 2; and G, *ca* 6. "Greatest" root is a value of *ní*; "least" is so of *ya*. Substituting with it for *ya*, the two numbers become *ca* 2, *ca* 1. Again the difference of the squares of these multiplied respectively by seven and eight, together with one added to it, is *ca v* 20 *ru* 1. It is a square.[2] Proceeding then as before, "least" root comes out 2 or 36. This is a value of *ca*. Substituting with it, the two numbers are obtained: viz. 4 and 2; or 72 and 36.

188. Example: Bring out quickly two numbers such, that the sum of the cube [of the one] and square [of the other] may be a square; and the sum of the numbers themselves be likewise a square.

Put the numbers *ya* 1 *ca* 1. The sum of the square and cube of these is *ya v* 1 *ca gh* 1. It is a square. Making it equal to the square of *ní* and adding cube of *ca*, the root of one side is *ní* 1; and, of the other (viz. *ya v* 1 *ca gh* 1) roots are to be sought by the method of the affected square. The number, which is joined with the square of *ya*, is the coefficient; the rest is the additive: C, *ya v* 1 A, *ca gh* 1. Then, by the rule (§ 95), taking *ca* for the assumed quantity under that rule, the two roots come out *ca v* $\frac{1}{2}$ *ca* $\dot{\frac{1}{2}}$ and *ca v* $\frac{1}{2}$ *ca* $\frac{1}{2}$. "Least" root is value of *ya*. Substituting with it for *ya*, the two numbers are *ca v* $\frac{1}{2}$ *ca* $\frac{1}{2}$ and *ca* 1. The sum of these is *ca v* $\frac{1}{2}$ *ca* $\frac{1}{2}$. It is a square. Making it then equal to *pí*; and multiplying both sides by

* If the "least" root be put *ca* 2; the "greatest," as inferred from it, (§ 75) is *ca* 6.—SU'R. Square of 2 multiplied by *pracrĭti* 7, is 28; and, with the additive, 36; the square-root of which is 6. RA'M.

[2] Put it equal to square of *pí*; and proceed to investigate the root of *ca v* 20 *ru* 1. Assume for "least" root 2. Its square 4, multiplied by 20 and added to 1, is 81: the square-root of which is 9. Then by composition of like (§ 77) C 20 L 2 G 9 A 1
L 2 G 9 A 1
other roots are deduced 1 36 [g 161]. SU'R. and RA'M.

four,[1] and adding unity, the root of the first side is *ca* 2 *ru* 1; and roots of the second (viz. *pív* 8 *ru* 1) investigated by the method of the affected square are 6 and 17,[2] or 35 and 99. Making "greatest" root equal to the root of the foregoing side of equation (*ca* 2 *ru* 1) the value of *cá* comes out 8 or 49. Substituting therewith, the two numbers are found 28 and 8, or 1176 and 49.

Or let two numbers be put *ya v* 2 *ya v* 7. The sum of these is of itself obviously a square, *ya v* 9. The sum of the cube and square of these is *ya v gh* 8 *ya v v* 49. It is a square. Make it equal to square of *ca*. Depressing the side of the equation by the biquadrate of *ya*, and proceeding as before taught,[3] the value of *ya* is obtained 2, or 7, or 3. Substituting therewith, the two numbers are found 8 and 28; or 98 and 343; or 18 and 63.[4]

189. Example: Tell directly two numbers such, that the sum of their squares, added to their product, may yield a square-root: and their sum, multiplied by that root and added to unity, may also be a square.

Let the numbers be put *ya* 1, *ca* 1. The sum of their squares, added to their product, is *ya v* 1 *ya. ca bh* 1 *ca v* 1. This has not a square-root. Therefore putting it equal to square of *ní*, and adding square of *ca*,[5] and multiplying by thirty-six, the root of the side involving *ní* is obtained, viz. *ní* 6; and the other side is *ya v* 36 *ya. ca bh* 36 *ca v* 36: in which the root of so much of it as affords a square-root is to be taken by the preceding rule (§ 186) viz. *ya* 6 *ca* 3, and the residue, namely *ca v* 27, being divided by *ca* as an assumed quantity [§ 95], and from the quotient the same assumed

[1] After reducing to a common denomination and dropping the denominator.—RÁM. Multiplying both sides by eight. SÚR.

[2] The commentators (SÚR. and RÁM.) direct 6 to be put; and proceeding by § 75, deduce G 17. But, if 1 were put tentatively, it would answer; G being in that case 3; and the further pair of roots is derived from composition of these sets by § 77, viz. C. 8 L 1 G 3 A 1 / *L* 6 *G* 17 *A* 1 whence by cross multiplication, &c. 1 35 g 99 a 1. The lower numbers seem to have been omitted by the author and commentators, because the numbers sought (*ca* being 1) would come out 0 and 1, which they consider to be unsatisfactory for an answer.

[3] See § 180.

[4] Put 2, 3 or 7 for "least" root: the "greatest" is 9, 11 or 21; which multiplied by the square of "least" (§ 180) give 36, 99, or 1029.

[5] That is, bringing it back, after subtraction, to the same side on which it first stood.

quantity being subtracted, and the remainder halved [ibid.], gives *ca* 13; which, made equal to that root, brings out the value of *ya*; viz. *ca* $\frac{5}{3}$. Substituting with this, the two numbers are found *ca* $\frac{5}{3}$ and *ca* 1. The sum of their squares *ca v* $\frac{34}{9}$, added to their product *ca v* $\frac{5}{3}$, is *ca v* $\frac{49}{9}$: the square-root of which is *ca* $\frac{7}{3}$. The sum of the numbers, *ca* $\frac{8}{3}$, multiplied by this, with unity added, is *ca v* $\frac{56}{9}$ *ru* $\frac{9}{9}$. Making this equal to the square of *pí*, "least" root[1] found by investigation, is 6 or 180. It is the value of *ca*. Substituting with it, the two numbers come out 10 and 6; or 300 and 180. In like manner a multiplicity of answers may be obtained.

190. Example of a certain ancient author.[2] Tell me quickly, algebraist, two numbers such, that the cube-root of half the sum of their product and least number, and the square-root of the sum of their squares, and those extracted from the sum and difference increased by two, and that extracted from the difference of their squares added to eight, being all five added together, may yield a square-root: excepting, however, six and eight.

The conditions of the problem being numerous, the solution, unless at once, does not succeed. The intelligent algebraist must therefore so put the quantities, as that all the conditions may be answered by one symbol.[3] Accordingly the two quantities are put *ya v* 1 *ru* 1 and *ya* 2. The cube-root of half the sum of their product and the least number is *ya* 1. The square-root of the sum of their squares is *ya v* 1 *ru* 1. The square-root of their sum [increased by two] is *ya* 1 *ru* 1. The square-root of their difference [increased by two] is *ya* 1 *ru* $\dot{1}$. The square-root of the difference of their squares [with eight added] is *ya v* 1 *ru* 3.

The sum of these [five] is *ya v* 2 *ya* 3 *ru* 2. It is a square. Make it equal to square of *ca*. Multiplying both sides of equation by eight, and adding the absolute number nine,[4] the root of the first side is *ya* 4 *ru* 3;

[1] By rule § 75, put 6; and proceeding as there indicated, its square 36, multiplied by the coefficient $\frac{56}{9}$, is $\frac{2016}{9}$; and, with the additive ($\frac{9}{9}$), $\frac{2025}{9}$: of which the root is $\frac{45}{3}$; or, abridged, 15. Therefore L 6 G 15, and by composition (§ 77) L 180 G 449. SU'R. RA'M.

[2] Introduced to exhibit facility of solution. RA'M.

[3] The two quantities must be put such, that the five roots, which are prescribed, may be possible. SU'R.

[4] See § 131.

and roots of the other (namely of *ca v* 8 *ru* 25) investigated by the method of the affected square are 5 and 15; or 30 and 85; or 175 and 495. Making an equation of "greatest" root with the former (*ya* 4 *ru* 3) the value of *ya* is obtained 3, or $\frac{41}{2}$, or 123. By substituting with the value so found, the two numbers come out 8 and 6: or $\frac{1677}{4}$ and 41; or 15128 and 246; and in like manner, many other ways.

Or else one quantity may be put square of *ya* added to twice *ya*; and the other twice *ya* less two absolute: viz. *ya v* 1 *ya* 2 and *ya* 2 *ru* 2̇.

Or one quantity may be put square of *ya*, less twice *ya*; and the other twice *ya* less absolute two: viz. *ya v* 1 *ya* 2̇ and *ya* 2 *ru* 2̇.

Or one quantity may be square of *ya* with four times *ya* and three absolute; and the other twice *ya* with four absolute: viz. *ya v* 1 *ya* 4 *ru* 3 and *ya* 2 *ru* 4.

"As supposition, which thus is a thousand-fold, is to the dull abstruse, the mode of putting suppositions is therefore unfolded in compassion to them."

191—192. Rule: two stanzas. Let the root of the difference be first put, an unknown number, with or without absolute number: that root of the difference, added to the square-root of the quotient of the additive of the difference of squares divided by the additive of the difference of the numbers, will be the root of the sum. The squares of these with their additives subtracted, are the difference and sum: from which the numbers are found by the rule of concurrence.[1]

193. Example: Tell me, gentle and ingenuous mathematician, two numbers, besides six and seven, such that their sum and their difference, with three added to each, may be squares; that the sum of their squares less four, and the difference of their squares with twelve added, may also be squares; and half the product less the smaller number may be a cube; and the sum of all their roots, with two added, may likewise be a square.

Put the symbol of the unknown less unity for the root of the difference: viz. *ya* 1 *ru* 1̇. Then by that analogy (and according to the last rule) the two numbers are put *ya v* 1 *ru* 2 and *ya* 2.* The roots are

[1] *Lil.* § 55.

* Put for the root of the difference with three added to render it square, *ya* 1 *ru* 1̇. Add the

ya 1 *ru* i; *ya* 1 *ru* i; *ya v* 1; *ya v* 1 *ru* 4; *ya* 1. The sum of these, with two added to it, is *ya v* 2 *ya* 3 *ru* 2. It is a square. Let it be equal to square of *ca*. The two sides of equation become *ya v* 2 *ya* 3 / *ca v* 1 *ru* 2 Multiplying by eight and adding nine, the root of the first side is *ya* 4 *ru* 3; and the roots of the second (*ca v* 8 *ru* 25) by the method of the affected square are L 5 G 15 or *L* 175 *G* 495.* "Greatest" root being equal to the former root (*ya* 4 *ru* 3), the value of *ya* is obtained 3 or 123; and, substituting with these values, the two numbers come out 7 and 6 or 15127 and 246.†

194. Example by an ancient author:[3] Calculate and tell, if you know, two numbers, the sum and difference of whose squares, with one added to each, are squares: or which are so, with the same subtracted.

In the first example, let the squares of the numbers be put *ya v* 4 and *ya v* 5 *ru* i. The sum and difference of these with unity added, afford each a square-root. The square-root of the first assumed quantity is one of the numbers, viz. *ya* 2. Roots of the second, namely *ya v* 5 *ru* i, investigated by the method of the affected square,[4] are 1 and 2, or 17 and 38. Of these, "greatest" root is the second number, and "least" is a value of *ya*; from which the first number is deducible. Substituting then with that value, the two numbers are 2 and 2, or 34 and 38.

square-root of the quotient of the additive of difference of squares by the additive of difference of numbers, viz. *ru* 2, the sum is *ya* 1 *ru* 1; the root of the sum with three added to render it square. Their squares are *ya v* 1 *ya* 2 *ru* 1 and *ya v* 1 *ya* 2 *ru* 1; and, subtracting the additives of the sum and difference, there remain the sum and difference of the numbers, *ya v* 1 *ya* 2 *ru* 2 and *ya v* 1 *ya* 2 *ru* 2. Half the sum and difference of these are the numbers themselves.

* By § 75, the first roots are had by position: the next by combination, under § 77.

† The same is found by the process of the foregoing rule. Let the root of the difference be put 122. Divide the additive of the difference of the squares, by the additive of the numbers, 12 by 3; the quotient is 4. Its square-root is 2. Add this to the root of the difference, the result is the root of the sum: (2 added to 122; making 124.) The squares of these, less the additives, give the sum and difference: 14881 and 15373. Whence, by the rule of concurrence (*Líl.* § 55) the two numbers are deduced, 15127 and 246. SÚR. and RÁM.

[3] It comprises two distinct examples. SÚR.

[4] Put tentatively 1 for "least" root; and the "greatest" by § 75 is found 2. Then combining like roots (§ 77), there result L 4 G 9. Combining these dissimilar roots (ibid.) others result adapted to the second example l 17 g 38; or, combining like, *l* 72 *g* 161.

In the second example, similarly, the first number is *ya* 2; and, for the second, roots are to be investigated from this *ya v* 5 *ru* 1, by the method of the affected square. They are 4 and 9; or 72 and 161. With "least" the first root (or number,) is raised; and "greatest" is the second. Thus the two numbers come out 8 and 9; or 144 and 161.

Here such number, as, with the least, whether added or subtracted, yields a square-root, must be the second coefficient.[1] The way to find it is as follows.

Let the least square quantity [that is, the coefficient] be put 4. The second, with this added or subtracted, must afford a square-root. Being doubled, it is 8. This is the difference of the squares of certain two numbers; and it is consequently equal to the product of the sum and difference. The difference of the numbers, therefore, is assumed 2: and by the rule (*Lil.* § 57) for finding two numbers from the difference of squares, and difference of the numbers, the roots of the difference of squares and of the sum of the squares are found 1 and 3. Adding the least square quantity to the square of the first, or subtracting it from the square of the second, there results the second [viz. 5]. Here the least square quantity must be so devised, as that the second may be an integer.

Or, in like manner, another is assumed 36. Doubled, it is 72. This is the difference of two squares: and six being put for the difference of the numbers, the second is brought out 45. Or, with four put, it comes out 85; or, with two, 325.[2]

Or else another ground of assumption may be shown, as follows. The sum of the squares with twice the product of the two quantities added or subtracted, must afford a root. That twice the products of two quantities may be an exact square, one should be put a square, and the other half a square; for the product of squares is square. Thus they are assumed, one a square, the other half a square: 1 and 2. Twice their product is 4. This is least square number [or coefficient]. The sum of their squares is 5. This is second quantity.

Or let the one square, and the other half square, be 9 and 2. Twice their product is 36. This is the least square number. The sum of the squares is 85. This is second quantity.

[1] *Távad-vyacta*, the known number annexed to *távat* (or *yávat-távat*) the unknown quantity. See the author's remark towards the close of his comment. SU'R.

[2] And similarly a multiplicity may be found.

These known numbers are multiplied by square of *yávat-távat:* and, in the first example, the second quantity has unity subtracted from it; in the second example, it must have the same added to it. So doing, those two square quantities are so contrived as to fulfil both conditions of the problem. But having extracted the square root of the first, the root of the second is to be found by the method of the affected square, as before observed.

Thus [the problem is solved] many ways.

195—196. Rule: two stanzas. In such instances, if there remain the [simple] unknown with absolute number, find its value by making it equal to the square or [other power] of another symbol with unity:[1] and substituting with this value in [the expression of] the quantity, proceed to the further operation,[2] making the root of the former equal to the other symbol and unity.

After the root of the first side of the equation has been taken, if there be, on the other side, the simple unknown with absolute number, or without it; in such case, making an equation with the square of another colour with unity, and thence bringing out the value of that unknown, and substituting with this value in the expression of the quantity, proceed again to the further operation; and, in so doing, make an equation of the root of the first side with the other symbol and unity. But, if there be no further operation, then the equation is to be made with a known square and so forth.

197. Example: If thou be expert in the extirpation of the middle term in analysis, tell the number, which being severally multiplied by three and five, and having one added to the product, is a square.

In this case put the number *ya* 1. This, tripled, with one added, is *ya* 3 *ru* 1. It is a square. Making it equal to square of *ca*, and adding unity on both sides, [to replace it on its original side,] the root of the side of equation containing *ca* is *ca* 1. Making the other side, namely *ya* 3 *ru* 1, equal to

[1] Since the root cannot in such case be sought by the rule of Chapter 3, as there is not an affected square: for the simple unknown only remains: but *(pracrĭti)* an affected square consists in a square of the unknown. Its root therefore can only be possible by equating it with the square of some quantity whatsoever. SÚR.

[2] Not, if no further operation depend: for the value would be an unknown. But make it equal to a known square, &c. and thus the value is absolute. RÁM.

the square of thrice *ní* joined with unity, viz. *ní v* 9 *ní* 6 *ru* 1, the value of *ya* is obtained; substituting with which the number comes out *ní v* 3 *ní* 2. Again, this multiplied by five, with one added to the product, is *ní v* 15 *ní* 10 *ru* 1. It is a square. Making it equal to square of *pí*, the equation after like subtraction is *ní v* 15 *ní* 10 Multiplying both sides by
pí v 1 *ru* 1
fifteen and adding twenty-five, the root of the first side is *ní* 15 *ru* 5. Roots of the second, viz. *pí v* 15 *ru* 10, investigated by the method of the affected square,[1] are 9 and 35, or 71 and 275. "Least" is the value of *pí*. "Greatest" being equal to the root of the first side of the equation *ní* 15 *ru* 5, the value of *ní* comes out 2 or 18. Substituting with its value for it, the number is found, 16 or 1008.

Or let the number be *ya* 1; and, as this tripled, with one added, is a square, make it equal to square of *ca;* and, after equal subtraction, find the value of *ya;* which, substituted accordingly, gives for the number *ca v* $\frac{1}{3}$ *ru* $\frac{\dot{1}}{3}$. Or let its value be so put at the first, that one of the conditions may be of itself fulfilled, *ca v* $\frac{1}{3}$ *ru* $\frac{\dot{1}}{3}$. This multiplied by five, with one added to the product, *ca v* $\frac{5}{3}$ *ru* $\frac{\dot{2}}{3}$, yields a square root. Making it equal to square of *ní*, the root of the side involving *ní* being extracted is *ní* 1; and the roots of the other side *ca v* $\frac{5}{3}$ *ru* $\frac{\dot{2}}{3}$, being investigated by the method of the affected square, are 7 and 9. "Least" is value of *ca*, and substituting with it (in *ca v* $\frac{1}{3}$ *ru* $\frac{\dot{1}}{3}$) the number is found 16; the same as before.

198. Example by an ancient author: What number, multiplied by three, and having one added to the product, becomes a cube; and the cube-root, squared and multiplied by three, and having one added, becomes a square?

Let the number be put *ya* 1. This tripled, with one added, is *ya* 3 *ru* 1. It is a cube. Making it equal to cube of *ca*, the value of *ya* is found

[1] Put 1 for "least" root: the "greatest" by rule § 75 is 5. Then by composition C 15 L 1 G 5 A 10 / L 1 G 5 A 10 other roots are found, *L* 10 *G* 40 *A* 100; whence, by § 79, l 1 g 4 a 1; and by composition C 15 L 1 G 5 A 10 / l 1 g 4 a 1 like roots are *l* 9 *g* 35 *a* 10; and by further combination C 15 *l* 9 *g* 35 *a* 10 / l 1 g 4 a 1 they come out *λ* 71 *γ* 275 *α* 10. That is, 1 and 5; or 9 and 35; or 71 and 275. The first pair is not noticed, apparently because the number thence deduced would be cipher.

$\frac{ca\ gh\ 1\quad ru\ 1}{3}$ The cube-root of three times that, with unity added to the product, being squared and tripled, and having one added to it, is *ca v* 3 *ru* 1. It is a square. Put it equal to square of *ní* 3 *ru* 1: and the equation is

ca v 3
ní v 9 *ní* 6

Adding unity to both sides, the root of the second side is *ní* 3 *ru* 1; and those of the other, investigated by the method of the affected square,[1] are L 4 G 7. Substituting as before with the value of *ca*,[2] the number comes out 21 (or $\frac{3374}{3}$).

199. Example: Say quickly what are two numbers, of which, as of six and five, the difference of the squares being severally multiplied by two and by three, and having three added to the products, shall in both instances be square?

200. Maxim: Intelligent calculators commence the work sometimes from the beginning [of the conditions], sometimes from the middle, sometimes from the end; so as the solution may be best effected.[3]

In this instance, let the difference of squares be put *ya* 1. This doubled, with three added, (*ya* 2 *ru* 3) is a square. Make it equal to square of *ca*; and with the value of *ya* thence deduced, substitute for the quantity, which thus becomes $\frac{ca\ v\ 1\quad ru\ 3}{2}$. This again tripled, with three added, is a square. Make it equal to square of *ní*; and, like subtraction taking place, the sides of equation are

ca v 3 *ru* 0
ní v 2 *ru* 3

Multiplying them by three, the root of the first is *ca* 3; and the roots of the second (*ní v* 6 *ru* 9) investigated by the

[1] Put 1 for "least" root, the greatest is 2 by § 75. Then by composition of like, another pair of roots is thence found (§ 77) L 4 G 7; and by combination of unalike, another pair l 15 g 26.—SU'R. The first pair is unnoticed as it would here also bring out the number required, a cipher.

[2] In the expression $\frac{ca\ gh\ 1\quad ru\ 1}{3}$. Cube of 4 is 64; less one, is 63; divided by 3, is 21.—SU'R. So cube of 15 is 3375; less one, is 3374; divided by 3, is $\frac{3374}{3}$.

[3] Sometimes assumption is commenced by intelligent persons from the beginning of the conditions as enunciated; sometimes from the middle; sometimes from the end, by inversion: so as the work of solution be accomplished. That is, in the instance, the difference only is put as unknown; without putting the numbers themselves so. SU'R. RA'M.

method of the affected square,[1] are 6 and 15, or 60 and 147. "Greatest" being equalled to the [root of the] first side, the value of *ca* is obtained 5, or 49. And substituting with this value, the difference of squares comes out 11, or 1199.[2] Then by the rule (*Líl.* § 57) for finding two numbers from difference of squares and difference of the simple quantities, putting unity for their difference, the two numbers are found, 5 and 6, or 599 and 600: or, putting eleven for their difference, the two numbers are 49 and 60.

201. Rule: a stanza and a half.[3] If the simple unknown be multiplied by the quantity which was divisor of the square, &c. [on the other side]; then, that its value may in such case be an integer, a square or like [term] of another symbol must be put equal to it: and the rest [of the operations] will be as before taught.

In the case of a square, &c. and in that of a pulverizer or the like, after the root of one side of the equation has been taken, if there be on the other side an unknown multiplied by the quantity which was divisor of the square, &c. the square and other term of another symbol together with absolute number added or subtracted, must be put equal to it; that so its value may come out integer. The rest [of the steps] are as taught in the preceding rules.

202. Example: What square, being lessened by four and divided by seven, yields no remainder? or what other square, lessened by thirty? If thou know, tell promptly.

Put the number *ya* 1. Its square, less four, and divided by seven, is exhausted. Let the quotient be *ca*. Making an equation of the divisor multiplied by that, with this *ya v* 1 *ru* 4̇, the root of the first side is *ya* 1. Since the other side, *ca* 7 *ru* 4, yields no root, put it equal to square of seven *ní* and two absolute. The value of *ca* is had without a fraction *ní v* 7 *ní* 4: and the quantity put is the root of the second side of equation, or *ní* 7 *ru* 2. This being equal to the root of the first side, or *ya* 1, the value

[1] The lowest number, which answers for "least" root, found by position (§ 75) is 6; and the corresponding "greatest" is 15. From which by § 79 are deduced L 2 G 5 A 1; and by combination of unalike (§ 77) C 6 L 6 G 15 A 9, another pair of roots is derived l 60 g 147 a 9.
L 2 G 5 A 1

[2] Square of 49 is 2401; which, less 3, is 2398; and halved, 1199.

[3] The unfinished stanza is completed at § 208.

of *ya* is *ní* 7 *ru* 2, with the additive. It comes out 9;[1] and the square of this will be the number sought, 81.

For the instruction of the dull, the way, which is to be be followed in the selection of another symbol, is set forth by ancient authors.

203—205. Rule: three stanzas. Choosing a number such that its square, divided by the divisor, may yield no residue, as also the same number, multiplied by twice the root of the absolute number; let another colour be put multiplied by that [as coefficient], and with the root of the absolute number added to it.

204. But, if the absolute number do not yield a square-root, then, after abrading the number by the divisor, add [to the residue] so many times the divisor as will make a square.[2] If still it do not answer, [the problem is] imperfect.

205. If by multiplication[3] or addition the first [side of equation] was made to afford a square root; in that case also, the divisor [is to be retained], as enunciated by the conditions; but the absolute number, as adjusted by subtraction and so forth, is right.[4]

Such a number, as that its square divided by the divisor shall be exhausted; that is, yield no residue; and the same number multiplied by two and by the square-root of the absolute number, being divided by the divisor, shall be in like manner exhausted, yielding no remainder; by such coefficient, let another colour be multiplied and so be put with the root of the absolute number. But, if there be not a root of the absolute number, then, the absolute number having been abraded by the divisor, superadd [to the residue] so many times the divisor as will make a square. Let its square-root be [used for] the absolute root. Even, with so doing, if a square be not produced, then that example must be deemed imperfect and wrong. If the first side of equation multiplied by some number, or with one added to it,

[1] Putting unity for *ní*.—SÚR. Supposing 2, it comes out 16; or with 3, it is 23.

[2] And then proceed according to the foregoing rule, using its root as root of the absolute number.

[3] The commentator SÚRYADÁSA reads *hitwá* and interprets it 'subtracting;' but collated copies of the text exhibit *hatwá*, multiplying: and this seems the preferable reading. See § 128.

[4] For the purpose of the preceding rule (§ 204).

afford a square-root; in such case the divisor should be taken as enunciated, and not as either multiplied or divided: but the absolute number is to be taken precisely as it stands when equal subtraction has been made.

The like is also to be understood in the case of a cube: as follows. Such number, as that its cube divided by the divisor may be exhausted, exhibiting no residue, and the same number multiplied by three and by the cube-root of the absolute number, being divided by the divisor, may also be exhausted; by such coefficient let another colour be multiplied and so be put together with the cube-root of the absolute number. If there be not a cube-root of the absolute number, then, after abrading the number by the divisor, add [to the residue] so many times the divisor, as may make a cube. Then the cube-root is treated as root of the absolute number. Even with so doing, if there be not a complete cube, the instance is wrong. This is to be applied further on.[1]

To proceed to the second example (§ 202). Let the number be put *ya* 1. Its square is *ya v* 1. Doing with it as directed, the root of the first side is *ya* 1; and treating the second side, *ca* 7 *ru* 30, as prescribed by the rule (§ 204), after abrading the absolute number by the divisor, superadding twice the divisor, viz. 14, the root is *ru* 4. By making an equation of the square of seven *ní* with this added (*ní* 7 *ru* 4) the value of *ca* is obtained *ní v* 7 *ní* 8 *ru* 2̇. But the assumed quantity *ní* 7 *ru* 4 is the root of the second side of equation, and equal to the root of the preceding one *ya* 1. Framing an equation with them, the number is found by the former process *ní* 7 *ru* 4, with the additive.[2] It comes out 11.[3]

[1] See § 206.

[2] Put *ya* 1. Its square, less thirty, divided by seven, yields no remainder (§ 202). Let the quotient be *ca*. This* multiplied by seven (*ca* 7) is equal to that (*ya v* 1 *ru* 30). (Statement for equal subtraction *ya v* 1 *ca* 0 *ru* 30̇ / *ya v* 0 *ca* 7 *ru* 0). After subtraction there remains *ya v* 1 / *ca* 7 *ru* 30 Root of the first side is *ya* 1. In the other side, by rule § 204, abrading the absolute 30 by the divisor, the residue is 2; to which add a multiple of the divisor (§ 204), viz. twice the divisor, the sum is 16; and its square-root, 4.—SU'R. and RA'M. The square of this added to seven *ní* (*ní* 7 *ru* 4) is *ní v* 49 *ní* 56 *ru* 16; equal to *ca* 7 *ru* 30. Whence the value of *ca* is deduced *ní v* 7 *ní* 8 *ru* 2.—RA'M. The assumed quantity *ní* 7 *ru* 4 is the root of the second side of equation, and is equal to the root of the first *ya* 1. Whence the value of *ya* is found *ní* 7 *ru* 4. SU'R. and RA'M.

[3] Putting unity for *ní*. RA'M.

* Both commentaries have 'square of this:' but erroneously.

If seven *ní* were put with a negative absolute number, a different result would be obtained.

206. Example:[1] Tell me what is the number, the cube of which, less six, being divided by five, yields no residue? if thou be sufficiently versed in the algebra[2] of cubes.

Here put the number *ya* 1. Doing with it as directed,[3] the cube-root of the first side is *ya* 1; and the other side is *ca* 5 *ru* 6; from which, by the foregoing rule (§ 203—5) adapted to cubes (choosing a number such that its cube may be exactly divisible by the divisor, as well as its multiple into thrice the root of the absolute number;) or by analogy, making it equal to the cube of five *ní* with six absolute, and proceeding as before, the number with its additive is found *ní* 5 *ru* 6.[4]

207. Example:[5] If thou be skilled in computation, tell me the number, the square of which being multiplied by five, having three added, and being divided by sixteen, is exhausted.

Let the quantity be put *ya* 1. Doing with this as said, and multiplying both sides of equation by five, the square-root of the first side is *ya* 5. In the other side *ca* 80 *ru* 1̇5, retaining the divisor as enunciated, and taking the

[1] An instance of the rule (§ 203—5) applied to cubes.

[2] *Cuṭṭaca.*

[3] Put *ya* 1. Its cube less six, *ya gh* 1 *ru* 6̇, being divided by five, is exhausted. Let the quotient be *ca*. Multiplied by five it* is equal to that. Statement for equal subtraction
ya gh 1 *ca* 0 *ru* 6̇
ya gh 0 *ca* 5 *ru* 0
After subtraction, the root of the first side is *ya* 1.—SÚR. and RÁM. In the other side, *ca* 5 *ru* 6, by rule § 204, abrading the absolute number 6 by the divisor 5, the residue is 1; to which add a multiple of the divisor (§ 204): forty-three times the divisor added to 1 is 216. Its cube-root is 6. Added to five *ní*, is *ní* 5 *ru* 6. The cube *ní gh* 125 *ní v* 450 *ní* 540 *ru* 216 is equal to the second side of equation *ca* 5 *ru* 6. Whence the value of *ca* is found without fraction *ní gh* 25 *ní v* 90 *ní* 108 *ru* 42.—RÁM. The assumed quantity *ní* 5 *ru* 6 is cube-root of the second side, of equation *ca* 5 *ru* 6; and equal to the root of the first side, or *ya* 1. The value of *ya* is hence deduced *ní* 5 *ru* 6.—SÚR. and RÁM.

[4] By substitution of 1 for *ní*, the number comes out 11.—SÚR. Putting nought, it is 6; or supposing two, it is 16.

[5] An instance of the rule § 205. SÚR.

* Both the commentaries here also exhibit "cube of this." Whether by error of the authors or transcribers may be doubted.

absolute number as it is adjusted by subtraction (§ 205), the result is *ca* 16 *ru* $\dot{1}5$. Making an equation of this with eight *ní* and unity, the value of *ca* is obtained without fraction, *ní v* 4 *ní* 1 *ru* 1.[1] Equating the assumed root *ní* 8 *ru* 1 with the root of the first side *ya* 5, the value of *ya* is found by means of the pulverizer, *pí* 8 *ru* 5. If the root were supposed eight *ní* with negative unity, the result would be *pí* 8 *ru* 3.[2]

[1] It is five times too great. The augmented divisor 80 should be used to find the true value of the quotient *ca*.

[2] Put *ya* 1. Its square, multiplied by five and having three added, is *ya v* 5 *ru* 3; and is exactly divisible by sixteen. Be the product *ca*. Multiplied by sixteen, it* is equal to that. After equal subtraction, the remainder of equation is *ya v* 5 / *ca* 16 *ru* $\dot{3}$. Multiplying both sides by five, (*ya v* 25 / *ca* 80 *ru* $\dot{1}5$) the root of the first side is *ya* 5. Of the other side (*ca* 80 *ru* 15) putting the enunciated divisor sixteen for the [coefficient of] colour by rule § 205, or making the absolute number, as it is altered by subtraction and other operations, the correct absolute (§ 205); the statement is *ca* 16 *ru* $\dot{1}5$. Put it equal to the square of eight *ní* with unity (*ní* 8 *ru* 1)† the statement is *ca* 0 *ní v* 64 *ní* 16 *ru* 1 / *ca* 16 *ní v* 0 *ní* 0 *ru* $\dot{1}5$. Having made the subtraction, the remainder of equation is *ní v* 64 *ní* 16 *ru* 16 / *ca* 16; and divided by the divisor (16) *ní v* 4 *ní* 1 *ru* 1. SU'R. and RA'M.

The assumed root *ní* 8 *ru* 1 is equal to the root of the first side *ya* 5. Statement for equal subtraction *ya* 0 *ní* 8 *ru* 1 / *ya* 5 *ní* 0 *ru* 0. After subtraction, the remainder of equation is *ní* 8 *ru* 1 / *ya* 5. Proceeding by the rule (§ 101) there results Dividend *ní* 8 Additive *ru* 1. / Divisor *ya* 5. Then, by the rule (§ 55), there arises an uneven series ⊣ ⊣ ⊣ ⊣ ○. Multiplying by penult and so on (§ 55) the pair of numbers deduced is ∞ ⊶. The series being uneven, the quotient and multiplier are subtracted from their abraders (§ 57), viz. 8 and 5. Whence the quotient and multiplier with their additives are obtained *pí* 8 *ru* 5 / *pí* 5 *ru* 3. The quotient is the value of *ya* the colour of the divisor (§ 154), and the multiplier is the value of *ní* the colour of the dividend. Statement in their order, *pí* 8 *ru* 5 value of *ya*. / *pí* 5 *ru* 3 —— of *ní*

Or let the assumption be eight *ní* with unity negative. Statement of the two sides of equation *ya* 0 *ní* 8 *ru* $\dot{1}$ / *ya* 5 *ní* 8 *ru* 0; and after subtraction *ní* 8 *ru* $\dot{1}$ / *ya* 5; whence Dividend *ní* 8 Additive $\dot{1}$. / Divisor *ya* 5. Then by

* The same error again occurs in both commentaries, which here put the "square." It occasionally reappears in all three instances in course of the operations which follow: still however leaving it doubtful whether it be not imputable to transcribers.

† The root *ní* 8 *ru* $\dot{1}$ is rightly assumed conformably with the rule § 205. For 15, abraded by the original divisor 16, gives a residue 15, to which adding a multiple of divisor by one, the sum (the signs being contrary) is 1; and its square-root 1 is to be used as root of the absolute. The coefficient 8 of the new symbol *ní* is duly selected such that its square and its multiple by twice that root of the absolute, shall both be divisible by 16. But the square of this assumed root is not equal to *ca* 16 *ru* $\dot{1}5$, but to *ca* 80 *ru* $\dot{1}5$ and to *ya v* 25.

The scope of the precept 'many different ways are to be devised' (§ 173) has been thus exhibited in a multiplicity of instances. Something too has been shown concerning the solution of quadratics by the pulverizer. Other devices, as practicable, are to be applied by intelligent algebraists.

the rule (§ 55) is deduced the uneven series १ १ १ १ ०, and from this the pair of numbers ८ ३. The series being uneven, but the additive being negative (§ 59), they are quotient and multiplier: *pí* 8 *ru* 3 value of *ya.* Substituting with *pí* 8 *ru* 5 for value of *ya* 1 (putting unity for *pí*) it is
pí 5 *ru* 2 —— of *ní*
13; or substituting with *pí* 8 *ru* 3 it is 11. RÁM.

CHAPTER VIII.

EQUATION INVOLVING A FACTUM OF UNKNOWN QUANTITIES.

NEXT, the product of unknown[1] is propounded.

208. Rule: two half stanzas.[2] Reserving one colour selected, let values chosen at pleasure be put for the rest by the intelligent algebraist. So will the factum be resolved. The required solution may be then completed by the first method of analysis.[3]

In an instance where a factum arises from the multiplication of two or more colours together, reserving one colour at choice, put arbitrary numeral values for the rest, whether there be one, two, or more. Substituting with those assumed values for the colours as contained in the sides of equation, and adding them to absolute number, and having thus broken the factum, find the value of the [reserved] colour by the first method of analytic solution.

209. Example: Tell me, if thou know, two numbers such, that the sum of them, multiplied severally by four and by three, may, when added to two, be equal to the product of the same numbers.

Let the numbers be *ya* 1 *ca* 1. Dealing with them as expressed, the two sides of equation are *ya* 4 *ca* 3 *ru* 2. Thus a factum being raised, let an
ya. ca bh 1

[1] *Bhávita.* See § 21 and comment upon § 100.

[2] Completing a stanza begun in a preceding rule (§ 201) and beginning another which is completed in the following (§ 212).

[3] By that taught under the head of simple or uniliteral equation. Ch. 4.

arbitrary value be put for *ca*, under the rule (§ 208): as, for instance, five: Substituting with it for *ca* in the first side of equation, and adding the term to the absolute number, it becomes *ya* 4 *ru* 17; and the other side becomes *ya* 5: whence by like subtraction, as before, the value of *ya* is obtained 17. Thus the two numbers are 17 and 5. Or substituting six for *ca*, the two numbers come out 10 and 6. In like manner, by means of various suppositions, an infinity of answers may be obtained.

210. Example: What four numbers are such, that the product of them all is equal to twenty times their sum? say, learned algebraist, who art conversant with the topic of product of unknown quantities.

Here let the first number be *ya* 1; and the rest be arbitrarily put 5, 4 and 2. Their sum is *ya* 1 *ru* 11. Multiplied by twenty, *ya* 20 *ru* 220. Product of all the quantities *ya* 40. Statement for equation *ya* 40 *ru* 0
ya 20 *ru* 220
Hence by the first analysis, the value of *ya* is found 11; and the numbers are 11, 5, 4 and 2. Or [with a different supposition] they are 55, 6, 4 and 1; or 60, 8, 3 and 1; or 28, 10, 2 and 1. In like manner a multiplicity may be found.

211. Example: Say what is the pair of numbers, of which the sum, the product and both squares being added together, the square-root of the aggregate, together with the pair of numbers, may amount to twenty-three? or else to fifty-three? Tell them severally; and in whole numbers. If thou know this, thou hast not thy equal upon earth for a good mathematician.

In this case, let the numbers be put *ya* 1 *ru* 2. The aggregate of their product, sum and squares, is *ya v* 1 *ya* 3 *ru* 6. It is equal to the square of twenty-three less the sum of the numbers (*ya* $\dot{1}$ *ru* 21), viz. *ya v* 1 *ya* 42 *ru* 441. From this equation the value of *ya* is obtained $\frac{29}{3}$; and thus the two numbers are $\frac{29}{3}$ and 2.

Or else let the numbers be supposed *ya* 1, *ru* 3. Proceeding as before, the two numbers are thence found $\frac{27}{11}$ and 3. In like manner putting five for the assumed quantity: the two come out in whole numbers 7 and 5.

In the second example, put the quantities *ya* 1 *ru* 2. The aggregate of their product, sum and squares is *ya v* 1 *ya* 3 *ru* 6. It is equal to the square of fifty-three, less the sum of the numbers (*ya* $\dot{1}$ *ru* 51) viz. *ya v* 1 *ya* 102 *ru* 2601. From the equation of these, by the foregoing process, the two numbers are $\frac{173}{7}$ and $\frac{2}{1}$. Or integers they are 11 and 17.

Thus, one quantity being put an absolute number, the other is brought out an integer with much trouble. How it may be done with little labour, is next shown.

212—214. Rule: two and a half stanzas. Removing the factum from one side, and the simple colours and absolute number from the other, as optionally selected, and dividing both sides of the equation by the coefficient of the factum, divide the sum of the product of the coefficients of the colours added to the absolute number by any assumed number; the quotient and the number assumed must be added to the coefficients of the colours, at choice; or be subtracted from them: the sums, or the differences, will be the values of the colours: and they must be understood to be so reciprocally.[1]

Removing by subtraction the factum from one of the equal sides, and the simple colours and absolute number from the other, and then reducing the two sides to the lowest denomination, by the coefficient of the factum as common measure, and dividing by some arbitrarily assumed number the product of the coefficients of the colours on the second side added to the absolute number, the assumed quantity and the quotient, having the coefficients of the two colours added to them respectively, as selected at pleasure, are values of the colours; and to be so understood reciprocally: that is, the one, to which the coefficient of *ca* is annexed, is the value of *ya;* and that, to which the coefficient of *ya* is added, is the value of *ca*. But, if, owing to the magnitude, the condition be not answered, when that has been done, the coefficients, less the quotient and assumed number, are the values reciprocally.

First Example: "Tell two numbers such, that the sum of them, multiplied by four and three, may, added to two, be equal to the product" (§ 209).

Here, that which is directed, being done, the two sides of equation are *ya* 4 *ca* 3 *ru* 2.[2] The sum of the product of the coefficients with absolute
ya. ca bh 1
number is 14. This, divided by one put as the assumed number, gives 1 and 14 for assumed number and quotient. These, with the two coefficients re-

[1] See BRAHMEGUPTA 18, § 36; which appears from a subsequent passage (ibid. § 38) and the scholiast's remark on it, to be a rule borrowed from a still earlier writer.

[2] The subtraction (or transposition) and division by the coefficient (which, in the instance, is unity,) leaves the equation unaltered. SÚR.

spectively added, taking them at choice, furnish the values of *ya* and *ca*, either 4 and 18, or 17 and 5. By the supposition of two, they come out 5 and 11; or 10 and 6.

The demonstration follows. It is twofold in every case: one geometrical, the other algebraic.[1] The geometric demonstration is here delivered. The second side of the equation is equal to the factum of the quantities. But that factum is the area of an oblong quadrangular figure.[2] The two colours are its side and upright. *ca* 1 *ya* 1 Within that plain figure is contained four times *ya* with thrice *ca* and twice unity. Within this figure, then, four times *ya* being taken away, as also *ca* less four, multiplied by its own coefficient, it becomes 4 *ya* 3 *ca* And the second side of the equation being so treated, there results *ru* 14. This is the area of the remaining rectangle at the corner, within the rectangle representing the factum of the quantities. It is a product arising from the multiplication of a side and upright. But they are here unknown. Therefore an assumed number is put for the side; and if the area be divided by that, the quotient is the upright. Either of the two (side, or upright,) with the addition of a number equal to the coefficient of *ya*, is the upright of the rectangle representing the factum; because that upright was lessened by it when four times *ya* was taken from the rectangle representing the factum. In like manner, the other, with the addition of a number equal to the coefficient of *ca*, is the side. These precisely are values of *ya* and *ca*.

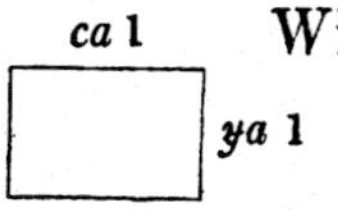

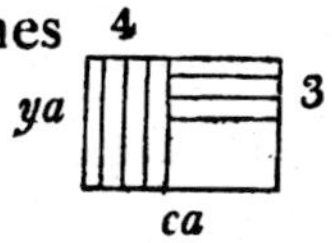

The algebraic demonstration is next set forth. That also is grounded on figure. Let other colours, *ní* 1 and *pí* 1, be put for the length of the side and upright in the smaller rectangle within the larger one, which consists of a side and upright represented by *ya* and *ca*. Then either of them, added to a number equal to the coefficient of *ya*, is the value of *ya* the side of that rectangle: viz. *ní* 1 *ru* 4 and *pí* 1 *ru* 3. Substituting with these for *ca* and *ya* in both sides of the equation, the upper side of it becomes *pí* 4 *ní* 3 *ru* 26;

[1] *Cshétra-gatá*, geometric: *Ráśi-gatá*, algebraic or arithmetical. (*Varńa-gatá*, algebraic exclusively.)

[2] *Áyata-chaturasra*. See *Líl.* Ch. 6.

and that containing the factum is transformed into *ní. pí bh* 1 *ní* 3 *pí* 4 *ru* 12. Like subtraction being made, the lower side of equation is *ní. pí bh* 1 ;. and the upper side is *ru* 14. It is the area of that inner rectangle; and it is equal to the product of the coefficient added to the absolute number.* How values of the colours are thence deduced, has been already shown.

This very operation has been delivered, in a compendious form, by ancient teachers. The algebraic demonstration must be exhibited to those who do not comprehend the geometric one.

'Mathematicians have declared algebra to be computation joined with demonstration: else there would be no difference between arithmetic and algebra.'

Therefore this explanation of the principle of the resolution has been shown in two several ways..

It has been said above, that the product of the coefficient, added to the absolute number, is the area of another small rectangle within that which represents the factum of the unknown, and situated at its corner. Sometimes, however, it is otherwise. When the coefficients are negative, the rectangle representing the factum will be within the other at its corner. When the coefficients are greater than the side and upright of the rectangle representing the factum, and are affirmative, the new rectangle will stand without that which represents the factum, and at its corner. See

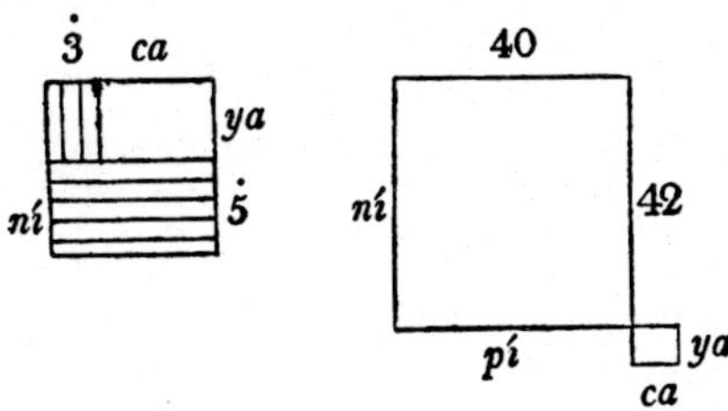

When it is so, the coefficients, lessened by subtraction of the assumed number and quotient, are values of *ya* and *ca*.

215. Example: What two numbers are there, twice the product of which is equal to fifty-eight less than the sum of their multiples by ten and fourteen?

Let the two numbers be put *ya* 1, *ca* 1. What is directed being done

* $\overline{4\times3}+2=14$.

with them, and the equation being divided by the coefficient of the factum, the result is *ya* 5 *ca* 7 *ru* 29 The sum of the product of coefficients with
ya. ca bh 1
the absolute number, viz. 6* is divided by two; and the assumed number and quotient are 2 and 3. The coefficients with these added are either 10 and 7, or 9 and 8; and, with the same subtracted, are 4 and 3, or 5 and 2: the numbers required.

216. Example: What two numbers are there, the product of which, added to triple and quintuple the numbers themselves, amounts to sixty-two? Tell them, if thou know.

Here also, what is expressed being done, there results *ya* $\dot{3}$ *ca* $\dot{5}$ *ru* 62
ya. ca bha 1
The sum of the product of coefficients with absolute number is 77.† The assumed number and quotient, 7 and 11. The coefficients, with these added, make the numbers 6 and 4, or 2 and 8. They should be added only, as the numbers come out negative, if they be subtracted.[3]

The foregoing third and fourth examples: "What is the pair of numbers, &c." (§ 211.)

Put the two numbers *ya* 1, *ca* 1. The aggregate of their product, sum and squares, is *ya v* 1 *ca v* 1 *ya. ca bh* 1 *ya* 1 *ca* 1. Since this does not afford a square-root, equal it with the square of twenty-three less the two quantities (*ya* $\dot{1}$ *ca* $\dot{1}$ *ru* 23) viz. *yav* 1 *cav* 1 *ya. ca bh* 2 *ya* $\dot{46}$ *ca* $\dot{46}$ *ru* 529. Dropping the equal squares, and subtraction being made, the remaining equation divided by the coefficient of the factum of the unknown (viz. unity[4]) gives *ya* 47 *ca* 47 *ru* $\dot{529}$. The product of the coefficients added to the absolute number is 1680;[5] and this, being divided by forty as assumed number, gives quotient and arbitrary number 42 and 40. Here the quotient and arbitrary assumed number must only be subtracted from the coefficients;

* (5×7)—29=6.

† (—3×—5)+62=77.

[3] The coefficients, with the arbitrary assumed number and quotient subtracted, make $\dot{12}$ and $\dot{14}$, or $\dot{16}$ and $\dot{20}$. RÁM.

[4] The equal squares being dropped, the statement for subtraction is *ya. ca bh* 1 *ya* 1 *ca* 1
ya. ca bh 2 *ya* $\dot{46}$ *ca* $\dot{46}$ *ru* 529
After subtraction *ya* 47 *ca* 47 *ru* 529
ya. ca bh 1 RÁM.

[5] 2209—529.

and the numbers will thus come out 7 and 5. If they were added, the condition, that they shall amount to twenty-three (§ 211), would not be fulfilled.[1]

"Or else amount to fifty-three" (§ 211). In this example, that which has been directed being done, there arises *ya* 107 *ca* 107 *ru* 2809. Here the sum of the product of coefficients with the absolute number is 8640. The arbitrary number and quotient 90 and 96. The coefficients less these quantities are the numbers required, 11 and 17.

So, likewise, in other instances.

In some cases, where the equations are numerous, finding various values of the factum of unknown quantities, and with those values equated and reduced to a common denomination, the two quantities may be discovered from the equation, by the former process of analytic solution.

From the mention of quantities in the dual number, it is evident of course, that arbitrary values are to be put for the rest of the colours, in the cases of three or more.

[1] By addition, the numbers are 87 and 89. The square-root of the aggregate (23402) is 153. The pair of numbers added together, 176. If the root be taken negative, the amount is 23. SU'R.

CHAPTER IX.

CONCLUSION.

217. ON earth was one named MAHÉŚWARA, who followed the eminent path of a holy teacher among the learned. His son, BHÁSCARA, having from him derived the bud of knowledge, has composed this brief treatise of elemental computation.[1]

218. As the treatises of algebra by BRAHMEGUPTA,[2] ŚRÍD'HARA and PADMANÁBHA are too diffusive, he has compressed the substance of them in a well reasoned compendium, for the gratification of learners.

219—223. For the volume contains a thousand lines[3] including precept and example. Sometimes exemplified to explain the sense and bearing of a rule; sometimes to illustrate its scope and adaptation: one while to show variety of inferences; another while to manifest the principle. For there

[1] *Laghu Víja-gańita.*

[2] The text expresses *Brahmáhwaya-víja*, algebra named from *Brahma;* alluding to the name of BRAHMEGUPTA, or to the title of his work *Brahmesidd'hánta*, of which the 18th chapter treats of algebra. The commentator accordingly premises 'Since there are treatises on algebra by BRAHMEGUPTA and the rest, what occasion is there for this?' The author replies "As the treatises, &c." RÁM.

[3] *Anushtubh.* Lines of thirty-two syllables, like the metre termed *anushtubh.* This intimation of the size of the volume regards both the prose and metrical part. The number of stanzas including rules and examples is 210; or, with the peroration, 219. Some of the rules, being divided by intervening examples in a different metre, have in the translation separate numbers affixed to the divisions. On the other hand a few maxims, and some quotations in verse, have been left unnumbered.

is no end of instances: and therefore a few only are exhibited. Since the wide ocean of science is difficultly traversed by men of little understanding: and, on the other hand, the intelligent have no occasion for copious instruction. A particle of tuition conveys science to a comprehensive mind; and having reached it, expands of its own impulse. As oil poured upon water, as a secret entrusted to the vile, as alms bestowed upon the worthy, however little, so does science infused into a wise mind spread by intrinsic force.

It is apparent to men of clear understanding, that the rule of three terms constitutes arithmetic; and sagacity, algebra. Accordingly I have said in the chapter on Spherics:[1]

224. 'The rule of three terms is arithmetic; spotless understanding is algebra.[2] What is there unknown to the intelligent? Therefore, for the dull alone, it[3] is set forth.'

225. To augment wisdom and strengthen confidence, read, do read, mathematician, this abridgment elegant in stile, easily understood by youth, comprising the whole essence of computation, and containing the demonstration of its principles, replete with excellence and void of defect.

[1] *Gólád'hyáya.* Sect. II. § 3.

[2] *Víja.*

[3] The solution of certain problems set forth in the section. The preceding stanza, a part of which is cited by the scholiast of the *Lílávatí*, (Ch. 12), premises, 'I deliver for the instruction of youth a few answers of problems found by arithmetic, algebra, the pulverizer, the affected square, the sphere, and [astronomical] instruments.' *Gól.* Sect. II. § 2.

GAṄITÁD'HYÁYA, ON ARITHMETIC;

THE TWELFTH CHAPTER OF THE

BRAHME-SPHUṬA-SIDD'HÁNTA,

BY BRAHMEGUPTA;

WITH SELECTIONS FROM THE COMMENTARY ENTITLED

VÁSANÁ-BHÁSHYA,

BY CHATURVÉDA-PRIT'HÚDACA-SWÁMÍ.

CHAPTER XII.

ARITHMETIC.

SECTION I.

1. He, who distinctly and severally knows addition and the rest of the twenty logistics, and the eight determinations including measurement by shadow,[1] is a mathematician.[2]

2. Quantities, as well numerators as denominators, being multiplied by

[1] Addition, subtraction, multiplication, division, square, square-root, cube, cube-root, five [should be, six] rules of reduction of fractions, rule of three terms [direct and inverse,] of five terms, seven terms, nine terms, eleven terms, and barter, are twenty *(paricarman)* arithmetical operations. Mixture, progression, plane figure, excavation, stack, saw, mound, and shadow, are eight determinations *(vyavahára)*. CH.

For topics of Algebra, see note on § 66.

[2] *Gańaca*, a calculator; a proficient competent to the study of the sphere. CH.

the opposite denominator, are reduced to a common denomination. In addition, the numerators are to be united.[1] In subtraction, their difference is to be taken.[2]

3. Integers are multiplied by the denominators and have the numerators added. The product of the numerators, divided by the product of the denominators, is multiplication[3] of two or of many terms.[4]

4. Both terms being rendered homogeneous,[5] the denominator and nu-

[1] SCANDA-SÉN-ÁCHÁRYA, who has exhibited addition by a rule for the summation of series of the arithmeticals, has done so to show the figure of sums; and he has separately treated of figurate quantity *(cshétra-ráśi)*, to show the area of such figure in an oblong. But, in this work, addition being the subject, sum is taught; and the author will teach its figure by a rule for the summation of series (§ 19). In this place, however, sum and difference of quantities having like denominators are shown: and that is fit. CH.

[2] Example of addition:* What is the sum of one and a third, one and a half, one and a sixth part, and the integer three, added together?

Statement: $1\frac{1}{3}$ $1\frac{1}{2}$ $1\frac{1}{6}$ 3. Or reduced $\frac{4}{3}$ $\frac{3}{2}$ $\frac{7}{6}$ $\frac{3}{1}$.

The numerator and denominator of the first term being multiplied by the denominator of the second, 2, and those of the second by that of the first, 3 they are reduced to the same denominator ($\frac{8}{6}$ $\frac{9}{6}$; and, uniting the numerators, $\frac{17}{6}$). With the third term no such operation can be, since the denominator is the same: union of the numerators is alone to be made; $\frac{24}{6}$, which abridged is) 4. So with the fourth term: and the addition being completed, the sum is 7.

Subtraction is to be performed in a similar manner; and the converse of the same example may serve. CH.

[3] *Pratyutpanna*, product of two proposed quantities.—CH. See a rule of long multiplication, § 55.

[4] Example: Say quickly what is the area of an oblong, in which the side is ten and a half, and the upright seventy sixths.

Statement: $10\frac{1}{2}$ $11\frac{2}{3}$. Multiplying the integers by the denominators, adding the numerators, and abridging, the two quantities become $\frac{21}{2}$ and $\frac{35}{3}$. From the product of the numerators 735, divided by the product of the denominators 6, the quotient obtained is 122 $\frac{1}{2}$. It is the area of the oblong.

Others here exhibit an example of the rule of three terms, making unity stand for the argument or first term. For instance, if one *pala* of pepper be bought for six and a half *pańas*, what is the price of twenty-six *palas?* Answer: 169 *pańas.*] CH.

[5] The method of rendering homogeneous has been delivered in the foregoing rule (§ 3) "Integers are multiplied by the denominators," &c.—CH. It is reduction to the form of an improper fraction.

* It is not quite clear whether the examples are the author's or the commentator's. The metre of them is different from that of the rules; and they are not comprehended, either in this or in the chapter on Algebra, in the summed contents at the close of each. They are probably the commentator's; and consigned therefore to the notes.

merator of the divisor are transposed: and then the denominator of the dividend is multiplied by the [new] denominator; and its numerator, by the [new] numerator. Thus division[1] [is performed.]

5. The quantity being made homogeneous,[2] the square of the numerator, divided by the square of the denominator, is the square.[3] The root of the homogeneous numerator, divided by the root of the denominator, is the square-root.[4]

6. The cube of the last term is to be set down; and, at the first remove from it, thrice the square of the last multiplied by the preceding; then thrice the square of this preceding term taken into that last one; and finally the cube of the preceding term. The sum is the cube.[5]

[1] Example: In a rectangle, the area of which is given, a hundred and twenty-two and a half; and the side, ten and a half; tell the upright.

Statement: $122\frac{1}{2}$ $10\frac{1}{2}$. Reduced to homogeneous form $\frac{245}{2}$ $\frac{21}{2}$.

Here the side is divisor. Its denominator and numerator are transposed $\frac{2}{21}$. The numerator of the dividend, multiplied by this numerator, becomes 490; and the denominator of the dividend, taken into the denominator, makes 42. The one, divided by the other, gives the quotient $11\frac{2}{3}$. It is the upright.

Some in this place also introduce an example of the rule of three terms. Thus "A king gave to ten principal priests a hundred thousand pieces of money, together with a third of one piece. What was the wealth that accrued to one?" CH.

[2] As before.—CH. [That is, reduced to fractional form.]

Put unity as the denominator of an integer; and proceed as directed. CH.

[3] A square is the product of two like quantities multiplied together. § 62. The present rule is introduced to show how the square of a fraction is found. CH.

Example: Tell the area of an equilateral tetragon, the side and upright of which are alike seven halves.

Statement: Side $\frac{7}{2}$ Upright $\frac{7}{2}$. Product of the numerators 49. Product of the denominators 4. These products are squares, since the side and upright are equal.

The square of the numerator 49 being divided by the square of the denominator 4, the quotient $12\frac{1}{4}$ is the area of the tetragon. *Ib.*

[4] Example: Tell the equal side and upright of an equilateral tetragon, the area of which is determined to be twelve and a quarter.

Statement, after rendering homogeneous: $\frac{49}{4}$. The root of the homogeneous numerator 49, is 7: that of the denominator 4, is 2. Dividing by this the root of the numerator, the quotient is the square-root $\frac{7}{2}$. It is the length of the upright and of the side. CH.

[5] Continued multiplication of three like quantities is a cube. § 62. As 1, 8, 27, 64, 125, 216, 343, 512, 729, cubes of numbers from 1 to 9. The rule is introduced for finding the cube of ten

7. The divisor for the second non-cubic [digit] is thrice the square of the cubic-root. The square of the quotient, multiplied by three and by the preceding, must be subtracted from the next [non-cubic]; and the cube from the cubic [digit]: the root [is found].[1]

and so forth. The cube of any given quantity comprising two or more digits or terms is required. The cube of the last digit, found by continued multiplication, is to be set down. Then the square of that last digit, tripled and multiplied by the term next before the last, is to be set down, at one remove or place of figures from that of the cube previously noted; and to be added to it. [So the square of this term tripled and taken into the last digit.] Then the cube of the term so preceding is set down in the next place of figures; and added. Thus the cube of two terms or digits is found. For a number comprising three or more terms, put two of them [previously finding the cube of this binomial by the rule] for last term; and proceed in every other respect conformably with the directions; and then, in like manner, put the trinomial* for last term; and so on, to find the cube of a quantity containing any number of terms. CH.

Example: Tell the cubic content of a quadrangular equilateral well (or cistern) measured by three cubits cubed and the same in depth.

Statement: 27, 27, 27. The product of these three equal quantities is 19683. It is the content in cubits of a solid having twelve corners:† for "the multiplication of three like quantities is a twelve-angled solid."

The rule furnishes another method. The cube of twenty-seven is required. The cube of the last digit 2 is set down 8. The square of the last 8, tripled, is 12, and multiplied by the preceding is 84: set down at the first remove, and added to the cube previously noted, it makes 164. [Thrice the square of 7 multiplied by 2 is 294; put at the next place of digits and added, makes 1934.] Cube of the preceding digit 7 is 343. Added as before, it gives 19683. It is the solid content in cubits; that is, it contains so many twelve-angled excavations measured by a cubit.

The same is to be understood of a pile or stack; putting height instead of depth. CH.

[1] The first digit of the proposed cube is termed cubic; and proceeding inversely, the two next places of figures are denominated non-cubic; then one cubic, and two non-cubic; and so on alternately, until the end of the number. With this preparation, the rule takes effect. . . . . . The meaning is as follows: In the first place, the cube of some number is to be subtracted from the last of all the digits termed cubic; and that number is reserved, and set down apart with the designation of cube-root. Take its square and multiply this by three; and with the tripled square divide the digit standing next before that of which the cube-root was taken; and note the quotient in the second place contiguous in direct order to the reserved cube-root. Square the quotient, and multiply by three and by the cube-root first found; and subtract the product from the first non-cubic standing before that of which the division was made. Then taking the cube of the quotient subtract it from the next preceding cubic digit. Thus a binomial root is found. If more be requisite, put the binomial root for first term; and proceed in every respect according to the rule, using it as first cube-root: and then put the trinomial, and afterwards the tetranomial, for first radical term; until the proposed number be exhausted.

* *Dwipada,* binomial; *tripada,* trinomial; *chatushpada,* tetranomial.

† *Dwádasásri,* lit. dodecagon; but intending a cube or a parallelopipedon. See *Lílávati,* § 7.

8. The sum of numerators which have like denominators, being divided by the [common] denominator, is the result in the first reduction to homogeneousness:[1] in the second, multiply numerators by numerators, and denominators by denominators.[2]

Example: Tell the cubic-root of a stack, of which the flanks* and elevation are alike, and the solid content is equal to twelve thousand, one hundred and sixty-seven.

Statement: 12167. Here the digit 7 is named cubic; 6 and 1 non-cubic; 2 cubic. From that subtract the cube of two, the remainder is 4167. Cube-root 2; its square 4; tripled 12; this is the divisor. Dividing by that the second non-cubic digit, the quotient is 3 and remainder 567. The square of the quotient 9; multiplied by three, 27, and by the preceding, 54. Subtracted from the first non-cubic, the residue is 27. Cube of the quotient, 27, subtracted from the cubic place of figures, leaves no remainder. Thus the root is this binomial 23. So much is the height; as much the length; and as much the breadth of the pile.

For trinomials and the rest, proceed as directed.

Such is the method of finding the cube and cube-root of integers. For the cube of fractions, let the cube of the numerator, after the quantity has been rendered homogeneous [§ 3], and the cube of the denominator, be separately computed: and divide the one by the other, the quotient is the cube sought. For the cube-root, let the roots be separately extracted, and then divide the cube-root of the numerator by that of the denominator, the quotient is the cube-root of the fraction. CH.

[1] The author here teaches the method of finding the result of the first assimilation *(játi)* consisting in addition. The sum of numerators which have dissimilar denominators is never taken. All the quantities must be reduced to like denominators; and then the addition of numerators is made; and the sum is divided by a single common denominator.

Example: Half of unity, a sixth part of the same, a twelfth part of it, and a quarter, being added together, what is the amount?

Statement: $\frac{1}{2}$ $\frac{1}{6}$ $\frac{1}{12}$ $\frac{1}{4}$. Reduced to like denominators the numerators become $\frac{6}{12}$ $\frac{2}{12}$ $\frac{1}{12}$ $\frac{3}{12}$. Added together and divided by the numerator, the result is unity.

Example: Twenty-two, sixty-six, thirty-eight, thirty-nine, thirteen, a hundred and fourteen are put in the denominator's place, and five, seven, nine, one, four and eleven are their numerators. When they are added together what is the whole sum?

Statement: $\frac{5}{22}$ $\frac{7}{66}$ $\frac{9}{38}$ $\frac{1}{39}$ $\frac{4}{13}$ $\frac{11}{114}$. Answer: one.

But when the similar denomination is not obvious, the denominators being very large, divide both denominators by the remainder [or last result] of the reciprocal division of the two, and multiply by the two quotients the reversed denominators together with their quotients. Other methods may be similarly devised by one's own ingenuity.

Subtraction also takes place between like quantities: and the rule must be therefore applied to difference. CH.

[2] The author now teaches the method of finding the result of the second assimilation consisting in multiplication.

First multiply separately numerators by numerators, and denominators by denominators. Then proceed with the former part of the rule.

* *Párśwa*, flank or side.

9. In the third, the upper numerator is multiplied by the denominator.[1] In the two next, severally, the denominators are multiplied by the denominators; and the upper numerators by the same increased or diminished by their own numerators.[2]

Example: Half a quarter, a sixth part of a quarter, a twelfth part of a quarter, an eighth part of ten quarters, a fifth part of seven quarters: summing these and adding three twentieths, let us quickly declare the amount. It is a sum, which we must constantly pay to a learned astronomer.

Statement: $\frac{1}{4}\,\frac{1}{2}\quad \frac{1}{4}\,\frac{1}{6}\quad \frac{1}{4}\,\frac{1}{12}\quad \frac{10}{4}\,\frac{1}{8}\quad \frac{7}{4}\,\frac{1}{5}\quad \frac{3}{20}$:

Or, $\frac{1}{8}\quad \frac{1}{24}\quad \frac{1}{48}\quad \frac{10}{32}\quad \frac{7}{20}\quad \frac{3}{20}$. Answer: the sum is one.

[1] The author next shows the method of finding the result of the third assimilation consisting in division.

The dividend is intended by the term upper numerator: and the middle quantity together with its denominator is the divisor. Then the rule for transposition of numerator and denominator (§ 4) takes effect.

Example: In what time will [four] fountains, being let loose together, fill a cistern, which they would severally fill in a day; in half a one; in a quarter; and in a fifth part?*

Statement: $\begin{matrix}1 & 1 & 1 & 1\\ 1 & \frac{1}{2} & \frac{1}{4} & \frac{1}{5}\end{matrix}$ The rule being observed; $\frac{1}{1}\ \frac{2}{1}\ \frac{4}{1}\ \frac{5}{1}$. The sum is 12.

So many are the measures in a day with all the fountains. Then by the rule of three, if so many fillings take place in one day, in what time will one? Statement: $\frac{12}{1}\ |\ \frac{1}{1}\ |\ \frac{1}{1}$. Answer: $\frac{1}{12}$. In this portion of a day, all the fountains, loose together, fill the cistern.

Example: One bestows an unit on holy men, in the third part of a day; another gives the same alms in half a day; and a third distributes three in five days. In what time, persevering in those rates, will they have given a hundred?

Statement: $\begin{matrix}1 & 1 & 3\\ \frac{1}{3} & \frac{1}{2} & \frac{5}{1}\end{matrix}$ And, the rule being observed, $\frac{3}{1}\ \frac{2}{1}\ \frac{3}{5}$. Reducing these to a common denominator, and summing them, the result is $\frac{28}{5}$; the total amount, which all bestow in alms in a day. Then by the rule of three, if so many fifths of an unit be given in one day, in how many will a hundred units be given?

Statement: $\frac{28}{5}\ |\ \frac{1}{1}\ |\ \frac{100}{1}$. Answer: $17\frac{6}{7}$. CH.

[2] The author adds this rule to exhibit reduction of fractional increase and decrease *(bhágánuband'ha* and *bhagápaváha-játi)*; the two assimilations *(játi)* which follow next after the first, second and third; that is, the fourth and fifth.

In fractional increase the numerators standing above are multiplied by the denominators augmented by their own numerators; in fractional decrease by the same diminished by their own numerators. The remainder of the process consists in reduction to homogeneous form as before.

Example of fractional increase: A little boy, receiving from a merchant a quarter of an unit, dealt with commodities for gain, during six days, and obtained for his goods, on the respective days, a price with both profit and principal equal to the original money added to its half, its third, its quarter, its fifth part, its sixth, and its seventh: what was the amount? Another did the same with

* *Lílávatí*, § 94—95.

10. In the rule of three, argument, fruit and requisition [are names of the terms]: the first and last terms must be similar.[1] Requisition, multiplied by the fruit, and divided by the argument, is the produce.[2]

an unit: and a third did so, with six. Tell the amount of their dealings also, if thou be conversant with fractional increase.

Statement:

$\frac{1}{4}$	$\frac{1}{1}$	$\frac{6}{1}$
$\frac{1}{2}$	$\frac{1}{2}$	$\frac{1}{2}$
$\frac{1}{3}$	$\frac{1}{3}$	$\frac{1}{3}$
$\frac{1}{4}$	$\frac{1}{4}$	$\frac{1}{4}$
$\frac{1}{5}$	$\frac{1}{5}$	$\frac{1}{5}$
$\frac{1}{6}$	$\frac{1}{6}$	$\frac{1}{6}$
$\frac{1}{7}$	$\frac{1}{7}$	$\frac{1}{7}$

The denominator four, multiplied by the denominator two, makes 8. The upper numerator 1, multiplied by the denominator 2 added to its own numerator 1, viz. 3, gives 3; and the result is $\frac{3}{8}$. Proceeding in like manner with three and the rest of the denominators, the amount for the first boy is 1; for the second, 4; for the third, 24.

Example of fractional decrease: Eight *palas* of white sandal wood were carried by a merchant from *Canyacubja* to the northern mountain; and at five places offerings were made by him of a moiety, a third part, a fifth, a ninth, and an eighth part of his stock. What was the residue?*

Statement: 8
$\frac{1}{2}$
$\frac{1}{3}$
$\frac{1}{5}$
$\frac{1}{9}$
$\frac{1}{8}$

Multiply denominators by denominators; and the upper numerators by denominators lessened by their own numerators. This being done, the answer is 1 $\frac{89}{135}$ [should be $\frac{112}{135}$].

The author has delivered but five rules of reduction or assimilation *(játi)*; and has omitted the sixth, as it consists of the rest and is therefore virtually taught. It has been given by SCANDA-SÉNA and others under the name of *Bhága-mátá*. CH.

See *Bhága-mátrí-játi* in ŚRÍD'HARA's abridgment: § 56—57.

[1] The middle term is dissimilar. CH.

[2] The rule concerns integers. If there be fractions among the terms, reduce all to the same denominator. CH.

Example: A person gives away a hundred and eight cows in three days; how many kine does he bestow in a year and a month?

Statement: Days 3. Cows 108. Days 390.

Answer: 14040.

Example: A white ant advances eight barley corns less one fifth part of that amount in a day; and returns the twentieth part of a finger in three days. In what space of time will one, whose progress is governed by these rates of advancing and receding, proceed one hundred *yójanas*?

Statement: Daily advance 8 less $\frac{1}{5}$. Triduan retrogradation $\frac{1}{20}$ fing. Distance 100 y.

* The text of this example, its statement and the answer are very corrupt.

11. In the inverse rule of three terms, the product of argument and fruit, being divided by the demand, is the answer.[1]

11—12. In the case of three or more[2] uneven terms, up to eleven,[3] transition of the fruit takes place on both sides.

The product of the numerous terms on one side, divided by that of the fewer on the other, must be taken as the answer. In all the fractions, transition of the denominators, in like manner, takes place on both sides.[4]

Here this maxim applies "eight breadths of a barley corn are one finger; twenty-four fingers, one cubit; four cubits, one staff; eight thousand staves, one *yójana*."

The daily advance, in a homogeneous form, is $\frac{32}{5}$ of a barley-corn. Retrogradation in three days, $\frac{1}{20}$ of a finger; in one day, by the proportion, 'as three to that, so is one to how much?' $\frac{1}{60}$. The daily advance, divided by eight, is reduced to fingers, viz. $\frac{32}{40}$ or $\frac{4}{5}$. Reduced to the same denominator as the retrogradation, $\frac{48}{60}$. Subtracting the retrogradation, the neat progress is $\frac{47}{60}$. A hundred *yójanas*, turned into sixtieths of fingers, are 4608000000. Then $\frac{47}{60} \mid \frac{1}{1} \mid \frac{4608000000}{60} \mid$

Answer: Days 98042553. CH.

[1] Example: The load *(bhára)* was before weighed with a *tulá* of six *suverńas*, tell me, promptly, how much will it be, if weighed out with one of five?

Statement: *su.* 6; *bhá* 1; *su* 5.

Answer: 24 hundred *palas*.

Here this maxim serves "sixteen grains of barley are one *másha;* sixteen of these, a *suverńa;* four of which, make one *pala;* and two thousand *palas*, a *bhára*."

Example: Tell me, quickly, how many ten *c'háris*, which were meted with a measure of three and a half to the *prast'ha*, will be when meted with one of five and a half?

Statement: *cu* $\frac{7}{2}$; *c'há* 10; *cu* $\frac{11}{2}$.

Answer: *c'há* 6, *má* 1, *dró* 1, *á* 3, *pra* 1, *cu* $\frac{4}{11}$.

Maxim applicable to the instance. "Four *cúdabas* make one *prast'ha;* four of these, one *ád'haca;* four *ád'hacas*, a hollow *purátana;** four of these, a *máńicá;* four *máńicás*, one *c'hárí*, a measure familiar to the people of *Magad'ha*." CH.

[2] The case of three terms must be excluded, being already provided for (§ 10); and the rule concerns five, seven, nine and eleven terms. CH.

[3] Uneven; not even, as four, &c. would be. CH.

[4] Example: The interest is settled at ten in the hundred for three months: let the interest of sixty lent for five months be told.

Statement:			Answer: 10.
	3	5	
	100	60	
	10		

Transferring the term ten to the second side, the product of this becomes the more numerous one, viz. 3000; which, divided by the product of the fewer, three and a hundred, viz. 300, gives 10; the interest for five months.

Example: If the interest of thirty and a half, for a month and one third, be one and a half: be it here told what is the interest of sixty and a half for a year?

* *C'hátá purátana.* It is the *dróńa* of SRÍD'HARA and BHÁSCARA. See *Lil.* § 8, and *Gań. sár.* § 5.

13. In the barter of commodities, transposition of prices being first

Statement of homogeneous terms: $\frac{4}{3}$ $\frac{12}{1}$
$\frac{61}{2}$ $\frac{121}{2}$
$\frac{3}{2}$

Transposition of the fruit and of the denominators having been made, the statement is
4 12
1 3
61 121
2 2
2 3

Whence the answer is found as before $26\frac{95}{122}$.

Example: Forty is the interest of a hundred for ten months. A hundred has been gained in eight months. Of what sum is it the interest?

Statement: 10 8
100
40 100

Mutually transferring the fruits, forty on one side and a hundred on the other, the statement is
10 8
100
100 40

Whence proceeding as above, the answer comes out $\frac{625}{2}$.

The same answer may be found by two proportions or sets of three terms.

Example of seven terms: If three cloths, five [cubits] long and two wide, cost six *pańas*, and ten have been purchased three wide and six long, tell the price.

Statement: 2 3
5 6
3 10
6

Transposing, and proceeding as in the rule of five, the answer is 36.

Example: If three cloths, two wide and five long, cost six *pańas*; tell me how many cloths, three wide and six long, should be had for six times six?

Statement: 2 3
5 6
3
6 36

Making a mutual transfer, and in other respects proceeding as above, the answer is 10.

The answer may be proved by three proportions or sets of three terms.

Example of nine terms: The price of a hundred bricks, of which the length, thickness and breadth, respectively, are sixteen, eight and ten, is settled at six *dinàras*: we have received a hundred thousand of other bricks a quarter less in every dimension: say what we ought to pay.

Statement: 16 12
8 6
10 $\frac{30}{4}$
100 100000
6

Transposition of fruit and of denominator being made, the answer comes out $2531\frac{1}{4}$.

The answer may be proved by four proportions or sets of three terms.

Example of eleven terms: Two elephants, which are ten in length, nine in breadth, thirty-six in girt, and seven in height, consume one *dróńa* of grain. How much will be the rations of ten other elephants, which are a quarter more in height and other dimensions?

terms takes place; and the rest of the process is the same as above directed.[1]

Operations,[2] subservient to the eight investigations,[3] have been thus explained.

Statement:

2	10
10	$\frac{50}{4}$
9	$\frac{45}{4}$
36	45
7	$\frac{35}{4}$
1	

The fruit and denominators being transposed, and proceeding as above, the answer comes out 12 *drónas*, 3 *prast'has*, 1¼ *cúdaba*.

[1] Example: If a hundred of mangoes be purchased for ten *panas*; and of pomegranates for eight; how many pomegranates [should be exchanged] for twenty mangoes?

Statement:

10	8
100	100
20	

and after transposition of prices and transition of fruit;

8	10
100	100
	20

Answer: 25 pomegranates.

[2] *Paricarman:* algorithm, or logistics. See § 1.

These operations, as affecting surd roots, unknown quantities, affirmative and negative terms, and cipher, the author will teach in the chapter on (*cuttaca*) the pulverizer; and we shall there explain them under the relative rules. CH.

[3] *Vyavahára*, ascertainment, or determination. § 1.

SECTION II.

MIXTURE.

14. The argument taken into its time and divided by the fruit, being multiplied by the factor less one, is the time.[1] The sum of principal and interest, being divided by unity added to its fruit, is the principal.[2]

15. The product of the time and principal, divided by the further time, is twice set down.[3] From the product of the one by the mixt amount,

[1] The principal sum, multiplied by the time, reckoned in months, which regulates the interest, is divided by the interest: and the quotient is multiplied by one less than the factor; (if the double be inquired, by one; if the triple, by two; if the sesquilateral, by half;) the result is the number of months, in which the sum lent is raised to that multiple. Ch.

Example: If the interest of two hundred for a month be six *drammas*, in what time will the same sum lent be tripled?

Answer: $66\frac{2}{3}$ months.

Example: If the interest of twenty *pańas* for two months be five, say in what time will my principal be raised to the sesquialterate amount?

Answer: 4 months. *Ib.*

[2] Subtracting this from the amount given, the remainder is the interest. Or multiply the amount of principal and interest by the interest of unity and divide by unity added to its interest, the quotient is the interest. Ch.

Example: A sum lent at five in the hundred by the month amounted to six times six in ten months; what was the sum in this case lent?

Answer: Principal 24. Interest 12.

Example: Eight hundred *suvernas* were delivered to a goldsmith with these directions: "make vessels for the priests, and take five in the hundred for the making." He did as directed. Tell me the amount of wrought gold.

Answer: Wrought gold $761\frac{19}{21}$. Fashion $38\frac{2}{21}$.

The rule is applicable to analogous instances. *Ib.*

[3] The rate of interest by the hundred, at which the money was lent by the creditor, is not known. All that is known is, that the interest for a given number of months has been received

added to the square of half the other, extract the square-root: that root, less half the second, is the interest of principal.[1]

16. The contributions, taken into the profit divided by the sum of the contributions, are the several gains:[2] or, if there be subtractive or additive differences, into the profit increased or diminished by the differences; and the product has the corresponding difference subtracted or added.[3]

and lent out again at the same rate, and has amounted in a given number of months to a certain sum, principal and interest. The rate of interest is required; and the rule is propounded to find it. CH.

[1] Example: Five hundred *drammas* were a loan at a rate of interest not known. The interest of that money for four months was lent to another person at the same rate; and it accumulated in ten months to seventy-eight. Tell the rate of interest on the principal.

Answer: 60.

Here the demonstration is to be shown algebraically by solution of a quadratic equation, as follows. If the interest of five hundred for four months be *yávaca*; what is the interest of *yávaca* for ten months. Here, transition of the fruit taking place (§ 12), the principal taken into the time is the product of the fewer terms; and the product of the numerous terms is the square of *yávaca* multiplied by the further time. Those products are reduced to least terms by a common divisor equal to the further time: as is directed (§ 15). Thus, by the rule of three terms, the answer comes our *ya v* $\frac{1}{200}$; the interest of *yávaca*. Adding *yávaca*, it is the mixed amount; that is, *ya v* $\frac{1}{200}$ *ya* $\frac{1}{1}$. This is equal to seventy-eight. Reducing to uniformity and dropping the common denominator, the two sides of the equation become: 1st side *ya v* 1 *ya* 200 *ru* 0; 2d side *ya v* 0 *ya* 0 *ru* 1560. By the rule in the chapter on *cuttaca*, "of the coefficient of the square, &c."* the value of *yávaca* comes out 60; which is equal to that above found. CH.

[2] *Pracshépaca:* what is thrown or cast together: the proposed quantities, of which an union is made. CH.

Labd'hi, profit. *Lábha*, gain. *Uttara*, difference.

[3] Example of the first rule: A horse was purchased, with the principal sums, one, &c. up to nine, by dealers in partnership; and was sold [by them] for five less than five hundred. Tell me what was each man's share of the mixt amount.

Statement: Contributions 1, 2, 3, 4, 5, 6, 7, 8, 9. Their sum 45. The profit 495, divided by that, gives the quotient 11; by which the contributions being multiplied, become 11, 22, 33, 44, 55, 66, 77, 88, 99. These are the several gains of the dealers.

Example of the second rule: Four colleges, containing an equal number of pupils, were invited to partake of a sacrificial feast. A fifth, a half, a third, and a quarter came from the respective colleges to the feast; and, added to one, two, three and four, they were found to amount to eighty-seven; or, with those differences deducted, they were sixty-seven.

* *Varg'-áhata-rúpáńám*, &c. See Algebra of *Brahm.* § 34.

Statement: $\begin{matrix} 1 & 2 & 3 & 4 \\ \frac{1}{5} & \frac{1}{2} & \frac{1}{3} & \frac{1}{4} \end{matrix}$ Reduced to a common denomination and the denominator being dropped, they are $\begin{matrix} 1 & 2 & 3 & 4 \\ 12 & 30 & 20 & 15 \end{matrix}$ The number given is 87. It is the profit (§ 16). Deducting the sum of the differences (1, 2, 3, 4) viz. 10, the remainder is 77: which, divided by the sum of the contributions, 77, gives 1; and the contributions, multiplied by this quotient, and having their differences added, become 13, 32, 23, 19; or, added together, 87. The number of disciples in each college is 60. Or, subtracting the differences, the number of pupils that came from the four colleges to the feast is 11, 28, 17, 11; total 67.

Example: Three jars of liquid butter, water, and honey, contained thirty-two, sixty, and twenty-four *palas* respectively: the whole was mixed together, and the jars again filled; but I know not the several numbers. Tell me the quantity of butter, of water and of honey, in each jar.

Statement: Butter 32; water 60; honey 24: these are the contributions (§ 16). Their sum, 116; by which divide the profit, viz. butter 32, the quotient is $\frac{8}{29}$. The contributions, severally multiplied by this, give the gains, viz. butter in the butter-jar $8\frac{24}{29}$; in the water-jar, $16\frac{16}{29}$; in the honey-jar, $6\frac{18}{29}$. So water in the water-jar $31\frac{1}{29}$; in the honey-jar, $12\frac{12}{29}$; in the butter-jar, $16\frac{16}{29}$: honey in the honey-jar $4\frac{28}{29}$; in the butter-jar, $6\frac{18}{29}$; in the water-jar, $12\frac{12}{29}$. CH.

Remark.—In this chapter of arithmetic, the computation of gold [or alligation] is omitted. On that account, the following stanza is here subjoined. "Add together the products of the weight into the fineness of the gold; and divide by the given touch: the quotient is the quantity. Or divide by the sum of the gold, the quotient is the touch."

Thus five *suvernas* of the touch of twelve, six of that of thirteen, and seven of that of fourteen, ($\begin{matrix} 5 & 6 & 7 \\ 12 & 13 & 14 \end{matrix}$ or, multiplying weight into fineness, 60, 78, 98;) being added together, are 236. By whatever touch this mass is divided, the quotient is the quantity of gold of that fineness. For instance, if the touch be sixteen, dividing by 16, the quotient is 14 *su.* 12 *má.* Dividing by fifteen, it is $16\frac{6}{15}$.* The number of *suvernas* in the mass is of one fineness. The mass of gold, therefore, is to be divided by the sum of the weights: the quotient is the touch of that number of *suvernas*. Thus, dividing the aggregate of products of weight into fineness, 236, by the sum of the weights 18, the quotient $13\frac{1}{9}$ is the touch. CH.

* So the MS. But should be $15\frac{11}{15}$.

SECTION III.

PROGRESSION.

17. The period less one, multiplied by the common difference, being added to the first term, is the amount of the last. Half the sum of last and first terms is the mean amount: which, multiplied by the period, is the sum of the whole.[1]

[1] To find the contents of a pile in the form of half the *méru-yantra* [or spindle]. CH.

Example: A stack of bricks is seen, containing five layers, having two bricks at the top, and increasing by three in each layer: tell the whole number of bricks.

Statement: Init. 2; Diff. 3; Per. 5. Answer: 40.

Example: The king bestowed gold continually on venerable priests, during three days and a ninth part, giving one and a half [*bháras*] with a daily increase of a quarter: what were the mean and last terms, and the total?

Statement: Init. $1\frac{1}{2}$; Diff. $\frac{1}{4}$; Per. $3\frac{1}{9}$.

Period $\frac{28}{9}$, less one, is $\frac{19}{9}$; multiplied by the difference, it is $\frac{19}{36}$; and added to the first term, becomes $\frac{73}{36}$. This is the last term. Added to the first term and halved, it gives $\frac{127}{72}$. This is the mean amount: multiplied by the period, it yields the total $\frac{889}{162}$; or 5 *bháras*, 9 hundred [*palas*] and $\frac{60}{80}$ [of a hundred].

Example: Tell the price of the seventh conch; the first being worth six *pańas*, and the rest increasing by a *pańa*?

Statement: Init. 6; Diff. 1; Per. 7. Answer: 12.

Example: A man gave his son-in-law sixteen *pańas* the first day; and diminished the present by two a day. If thou be conversant with progression, say how many had he bestowed when the ninth day was past?

Statement: Init. 16; Diff. 2; Per. 9. Answer: 72; received by the son-in-law: or 72 the father-in-law's; being his disbursement.

Example: [The first term being five; the difference three; and the period eight; what is the sum? the last term? and the mean amount?*]

Statement: Init. 5; Diff. 3; Per. 8. Answer: Last term 26. Mean $\frac{31}{2}$. Sum 124.

Here one side is to be put equal to the period of the progression; and a second, equal to its

* The terms of the question are wanting in the original.

18. Add the square of the difference between twice the initial term and the common increase, to the product of the sum of the progression by eight times the increase: the square-root, less the foregoing remainder divided by twice the common increase, is the period.[1]

mean term: and the figure of a rectangle is to be thus exhibited. Then so many little areas, in the figure of the progression, [formed] by its area, as are excluded [on the one part,] are gathered in front within that oblong. Therefore the finding of the area is congruous. CH.

To show the rule for finding the sum of a series increasing twofold, or threefold, &c. three stanzas of my own [the commentator PRIT'HU'DACA'S] are here inserted: 'At half the given period put "square;" and at unity [subtracted] put "multiplier;" and so on, until the period be exhausted. Then square and multiply the common multiplier inversely in the order of the notes. Let the product less one be divided by the multiplier less one, and multiplied by the amount of the initial term; and call the result area [or sum], the progression being [geometrical] twofold, &c. This method is here shown from the combination of metre in prosody.' The meaning is this: if the period be an even number, halve it, and note "square" in another place; when the number is uneven, subtract unity, and note "multiplier" in that other place and contiguous. Proceed in the same manner, halving when the number is even, and subtracting one when it is uneven, and noting the marks "square" and "multiplier," one under the other, in order as they are found, until the period be exhausted. The lowermost mark must of course be "multiplier." It is equal to the [common] multiplier [of the progression]. Setting down that on the working ground, square the quantity when "square" is noted, and multiply it where "multiplier" is marked: proceeding thus in the inverse order, to the uppermost note. From the quantity which is thus obtained, subtract unity; divide the remainder by the amount of the [common] multiplier less one; and multiply the quotient by the number of the initial term. This being done, the product is the sum of a progression, where the difference is twofold or the like. *Ib.*

Example: How much is given in ten days, by one who bestows six with a threefold increase daily?

Statement: Init. 6; Com. mult. 3; Per. 10. Answer: 177144.

Example: Say how much is given by one, who bestows for three days, three and a half [daily] with increase measured by the [common] multiplier five moieties?

Statement: Init. $\frac{7}{2}$; Diff. mult. $\frac{5}{2}$; Per. 3.

Put "mult." for subtraction of unity; "square" for the half; and again "mult." for unity subtracted: mult. The multiplier is two and a half or $\frac{5}{2}$, at the first place. Squared at the
sq.
mult.

second, it is $\frac{25}{4}$; and again multiplied at the third, $\frac{125}{8}$. Unity being subtracted, it is $\frac{117}{8}$. Divided by multiplier less one ($\frac{3}{2}$) it becomes $\frac{117}{12}$. This multiplied by the initial term, and abridged, yields $\frac{273}{8}$.

[1] The first term, common increase, and total amount, being known, to find the period. CH.

Example: Say how many are the layers in a stack containing a hundred bricks, and having at the summit ten, and increasing by five.

Statement: Init. 10; Com. diff. 5; Per.? Sum 100.

19. One, &c. increasing by one, [being added together] are the sum of a

Operation: Twice the initial, 20, less the increase 5, is 15; the square of which is 225. The sum 100, eight (8) and increase 5, multiplied together, make 4000. Add to this the square of the remainder, 225, the total is 4225. Its square-root 65, less the foregoing remainder 15, gives 50; which divided by twice the common increase, 10, yields the period 5.

So in other cases likewise.

Here the principle is the resolution of a quadratic equation. For instance: Init. 10; Com. diff. 5; Per. *ya* $\dot{1}$. This less one [§ 17] becomes *ya* 1 *ru* $\dot{1}$; which, multiplied by the common increase 5, makes *ya* 5 *ru* $\dot{5}$; and added to the initial term 10, affords *ya* 5 *ru* 5, the last term. Added to first term, it is *ya* 5 *ru* 15; which halved gives *yá* $\frac{5}{2}$ *ru* $\frac{15}{2}$. It is the mean amount; and, multiplied by the period, yields *ya v* $\frac{5}{2}$ *ya* $\frac{15}{2}$, the sum of the whole: which is equal to a hundred. Making an equation, two is multiplier of a hundred, being the [denominator, or] divisor standing beneath,* as before shown. The quantity being so treated, and the rule for preparing the equation† observed, the first side of the equation is *ya v* 5 *ya* 15; and the second side is *ru* 200. Then, proceeding by the rule "Multiply by four times [the coefficient of] the square" and so forth,‡ the absolute number becomes 4000. It is the product of the multiplication of the sum, common increase and eight. For the multiplier being two, the quantity must be multiplied by that and by four: wherefore multiplication by eight is specified. The unity, which is subtracted from *yávaca*, becomes negative: it is multiplied by the common increase; and thus a number equal to the common increase becomes negative: this being added to the initial term, and the result again added to the initial term, an affirmative quantity equal to twice the initial is introduced: taken together, the difference is the sum of the negative and affirmative quantities and is fitly called the remainder. It is here the coefficient of *yávaca*. Then, observing the rule for adding "the square of [the coefficient of] the middle term,"‖ the absolute number is as here shown: viz. 4225. Its root, 65, less the [coefficient of the] middle term, is 50: which, divided by twice the [coefficient of the] square, is the middle [term of the equation], that is to say the period of the progression: viz. 5. For *yávaca* is here the period.

If the initial term be unknown, but the common increase, period and sum be given, divide the sum of the progression by the period: the quotient is the mean amount. Double it; and subtract the product of the period less one taken into the common increase: half the remainder is the initial term. For instance: Init.? Diff. 3; Per. 5; Sum 40. This, divided by the period, gives 8, the mean amount; which doubled is 16. The period less one is 4; and the common increase 3: their product 12. Subtracting this from the foregoing, the remainder is 4: its half 2 is the initial term. This is to be applied in other cases also.

Where the common increase is unknown; divide in like manner the sum by the period, the quotient is the mean amount. Double it; and subtract twice the initial term: the quotient of the re-

* *Ad'hast'ha-ch'héda.*

† Algebra of *Brahm.* § 33.

‡ See Algebra of *Brahm.* § 32. A rule of the same import with that of SRÍD'HARA cited by BHÁSCARA. *Víj.-gan.* § 131.

‖ *Ibid.*

given period. That sum being multiplied by the period added to two, and being divided by three, is the sum of the sums.[1]

20. The same,[2] being multiplied by twice the period added to one, and being divided by three, is the sum of the squares.[3] The sum of the cubes

mainder by the period less one, is the common increase. For instance; Init. 2; Diff.? Per. 7. Sum 77. Deducing the mean amount from the sum by the period, doubling it [and proceeding in other respects as directed, the common difference comes out 3.*]

[If the first term and common difference be both unknown, deduce the mean amount from the sum by its period; and doubling it*] set down the result as a reserved quantity. Then put an arbitrary common increase; and by that multiply the period less one. Subtract the product from the reserved quantity: the moiety of the residue is the initial term; and the common increase, as assumed. For instance: Init.? Diff.? Per. 9. Sum 576. The quotient of this by the period is the mean amount 64: the double of which is called the reserved quantity, 129. Putting one for the common increase, the period less one, multiplied by that, is 8: which being subtracted from the reserved quantity, and the remainder being halved, yield Initial term 60; Diff. 1; Per. 9. Or, putting two for the ammon difference, the result is Init. 56; Diff. 2; Per. 9; Sum 576. Or, assuming two and a half, it comes out Init. 54; Diff. $\frac{5}{2}$; Per. 9; Sum 576. This is applicable in all cases and in whole numbers.

But, if the first term, common difference and period be all three unknown; put an arbitrary number for the period, and proceed as just shown.

If the difference, period, mean amount and sum total of a progression be required in square numbers, put any square quantity for the period of the progression. The period multiplied by sixteen serves for the common difference; and the square of two less than the period for the initial term. With these, the mean amount and sum total are found as before. For instance: let the square number 9 be the period. Multiplied by sixteen, it gives 144. The period less two is 7; the square of which, 49, is the initial term. Init. 49; Diff. 144; Per. 9; Mean amount 625. Sum 5625. All five are square numbers.

In like manner a variety of examples may be devised for the illustration of the subject. For fear of rendering the book voluminous, they are not here instanced. as we have undertaken to interpret the whole astronomical system *(sidd'hánta)*. CH.

[1] A rule to find the content of a pile of sums. CH.

Example:† Per. 5. The sum of this, consisting of the arithmeticals one, &c. increasing by one, is 15, which, multiplied by the period added to two, viz. 7, is 105. Divided by three, the quotient is 35, the content, in bricks, of a pile of sums, the period of which is five. CH.

[2] To find the content of a pile of quadrates; and one of cubics. CH.

[3] Example:† Per. 5. This doubled, and having one added to it, is 11. The sum of the period, viz. 15, beine multiplied by that, is 165: which divided by three, gives 55. It is the content, in bricks, of a pile of quadrates, the period of which is five. CH.

* The manuscript is here deficient: but the context renders it easy to supply the defect.

† The questions are not proposed in words at length: or else the manuscript is in this respect deficient.

is the square of the same.[1] Piles [may be exhibited*] with equal balls [or cubes;[2] as a practical illustration*] of these [methods.*].[3]

[1] Example:† Per. 5. The sum of this is 15. Its square is 225; the content, in bricks, of a pile of cubics, the period of which is five.

[2] Bricks in the form of regular dodecagons.—CH. Meaning cubes. See *Lílávatí*, § 7, note.

[3] As the author has mentioned a pile of balls, the method of finding the content is here shown. Let the area of the circle be found by the method subsequently taught [§ 40] and be reserved. The square-root of it is to be extracted; and by that root multiply the reserved area. This being done, the area of the globe is found. But in the circle the area is an irrational quantity. This again then is to be multiplied by the square of the surd: and the square-root of the product is the content of the globe and is a surd. CH.

* CHATURVÉDA.

† The questions are not proposed in words at length: or else the manuscript is in this respect deficient.

SECTION IV.

PLANE FIGURE.[1]

TRIANGLE and QUADRILATERAL.

21. THE product of half the sides and countersides[2] is the gross area of a triangle and tetragon.[3] Half the sum of the sides set down four times, and

[1] Triangles are three; tetragons five, and the circle is the ninth plane figure. Thus triangles are *(sama-tribhuja)* equilateral, *(dwi-sama-tribhuja)* isosceles, and *(vishama-tribhuja)* scalene. Tetragons are *(sama-chaturasra)* equilateral; *(áyata-sama-chaturasra)* oblong with equal sides [two and two]; *(dwi-sama-chaturasra)* having two equal sides; *(tri-sama-chaturasra)* having three sides equal; *(vishama-chaturasra)* having all unequal. CH.

[2] *Báhu-pratibáhu*, or *bhuja-pratibhuja* (§23): opposite sides.

[3] Example: What is the area of an equilateral triangle, the side of which is twelve?

Statement: 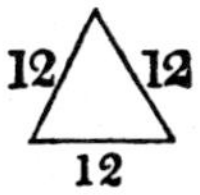 12, 12, 12 (triangle). The sum of sides and of countersides, 12 and 24; their moieties 6 and 12; the product of which is 72, the gross area.

Example: What is the area of an isosceles triangle the base of which is ten and the sides thirteen?

Statement: 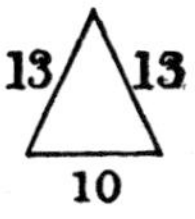 13, 13, 10 (triangle). The moieties of the sums of opposite sides, 5 and 13; their product 65, the gross area.

Example: What is the area of a scalene triangle, the base of which is fourteen and the sides thirteen and fifteen?

Statement: 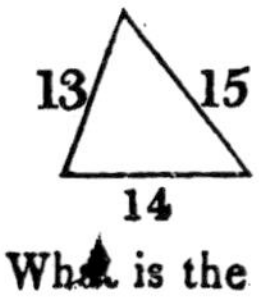 13, 15, 14 (triangle). Answer: 98 the gross area.

Example: What is the area of an equilateral tetragon, the side of which is ten?

Statement: 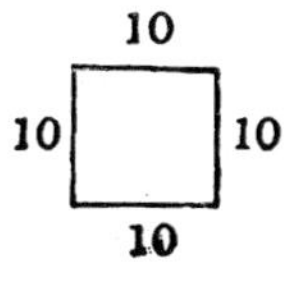 10, 10, 10, 10 (square). Answer: 100, the gross as well as exact area.

severally lessened by the sides,[1] being multiplied together, the square-root of the product is the exact area.[2]

Example: What is the area of an oblong, two sides of which are twelve; and two, five?

Statement: (oblong, sides 12, 5, 5, 12) Answer: 60, the gross and exact area.

12

5 5

12

Example: What is the area of a quadrilateral having two equal sides thirteen, the base fourteen, and the summit four?

4

Statement: 13 13 Answer: 117 the gross area.

14

Example: Tell the area of a quadrilateral having three equal sides twenty-five, and base thirty-nine?

25

Statement: 25 25 Answer: 800 the gross area.

39

Example: Tell the gross area of a trapezium, of which the base is sixty, the summit twenty-five, and the sides fifty-two and thirty-nine?

25

Statement: 52 39 Answer: $1933\frac{3}{4}$ the gross area. CH.

60

[1] The sides of the quadrilateral are severally subtracted from the half of the sum in all four places; but the sides of the triangle are subtracted in three, and the fourth remains as it stood. CH.

[2] Examples as above. Sides of the equilateral triangle 12; the sum 36; its half set down four times 18, 18, 18, 18; which severally lessened by the sides gives 6, 6, 6, 18. The product of those numbers is 3888, the surd root of which is the exact area.

Sides of the isosceles triangle 10, 13, 13; the sum 36. Its half 18, lessened severally by the sides, gives 5, 5, 8, 18. The product whereof is 3600. The square-root of this is the exact area, 60.

Sides of the scalene triangle 14, 13, 15. Half the sum 21, less the sides, gives 7, 8, 6, 21. Product 7056; the root of which is the exact area 84.

The gross area of the equilateral tetragon, as of the oblong, is the same with the exact area.

Sides of the tetragon with two equal sides, 14, 13, 13, 4. The exact area, as found by the rule, is 108.

Sides of the tetragon having three equal sides, 39, 25, 25, 25. Exact area 768.

Sides of the trapezium 60, 52, 39, 25. Exact area 1764. CH.

Putting a side of a tetragon equal to the segment of the base, and an upright equal to the per-

22. The difference of the squares of the sides being divided by the base, the quotient is added to and subtracted from the base:[1] the sum and the remainder, divided by two, are the segments. The square-root, extracted from the difference of the square of the side and square of its corresponding segment of the base, is the perpendicular.[2]

23. In any tetragon but a trapezium, the square-root of the sum of the products of the sides and countersides,[3] is the diagonal. Subtracting from the square of the diagonal the square of half the sum of the base and summit, the square-root of the remainder is the perpendicular.[4]

pendicular, the area of a figure is represented by little square compartments formed by as many lines as are the numbers of the upright and side. CH.

[1] The bottom *(adhas)* or lower line of every triangle is the base *(bhú)*, literally ground. The flanks *(párśwa)* are termed the sides *(bhuja)*. In an equilateral triangle, or in an equicrural one, the two segments of the base are equal. In a scalene triangle, the greater segment answers to the greater side; and the least segment to the least side. The perpendicular is the same, computed from either side. CH.

[2] Example: An isosceles triangle, the base of which is ten, and the sides thirteen.

Statement:  13 13 10 Answer: Segments 5 and 5. Perpendicular 12.

Example: A scalene triangle, the base of which is fourteen, and the sides thirteen and fifteen.

Statement: 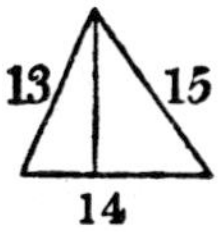 13 15 14 Answer: Segments 5 and 9. Perpendicular 12. CH.

The segments are found by halving the sum and difference: for it is directed in a subsequent rule (§ 23) to subtract the square of the upright from that of the diagonal; and the two segments are thence deduced by the rule of concurrence. The perpendicular is found by extracting the square-root of the remainder when the square of the side has been subtracted from the square of the diagonal; that remainder being the square of the upright: for the perpendicular is the upright. *Ib.*

[3] The opposite sides; (See § 21) the flanks, and the base and summit. CH.

[4] Example: An equilateral tetragon, the side of which is twelve.

Statement: 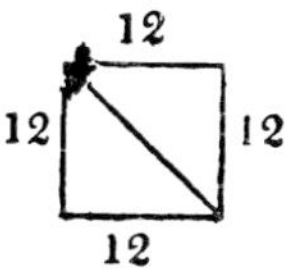 12 12 12 12 Answer: Diagonal, the surd root of 288. Perpendicular 12.

The example of an oblong is similar.

24. Subtracting the square of the upright[1] from the square of the diagonal, the square-root of the remainder is the side; or subtracting the square of the side, the root of the remainder is the upright: the root of the sum of the squares of the upright and side is the diagonal.

25.[2] At the intersection of the diagonals, or the junction of a diagonal and a perpendicular, the upper and lower portions of the diagonal, or of the perpendicular and diagonal, are the quotients of those lines taken into the corresponding segment of the base and divided by the complement[3] of the segments.[4]

Example: A tetragon having two equal sides, thirteen; and the base fourteen, and summit four.

Statement: 4; 13, 13; 14. Answer: Diagonal 15. Perpendicular 12.

In like manner, a tetragon with three equal sides: (See § 26.) CH.

The product of the base and summit is equal to the square of the greater segment less the square of the least. The square of the flanks is equal to the square of the perpendicular added to the square of the least segment. Their sum is the sum of the squares of the perpendicular and greater segments, and is the sum of the squares of the upright and side: and its square-root consequently is the diagonal. Half the sum of the base and summit is the greater segment: it is the side. Subtracting the square of it from the square of the diagonal, the remainder is the square of the upright. Its square-root is the upright termed the perpendicular. *Ib.*

[1] One side being so termed (*báhu* or *bhuja*), the other is called upright. It matters not which. CH.

[2] In tetragons having two or three equal sides, as above noticed, to show the method by which the upper and lower portions of the diagonals, as divided by the intersection of the diagonals, may be found: and the upper and lower portions of both diagonal and perpendicular, as divided by the intersection of the perpendicular and diagonal. CH.

[3] *Swayuti*, the line which joins the extremities of the perpendicular and diagonal. It is the greater segment of the base or complement of the less: and answers to BHÁSCARA's *pít'ha*. *Lil.* § 195.

[4] Example: The tetragon with two equal sides as last mentioned.

Statement: 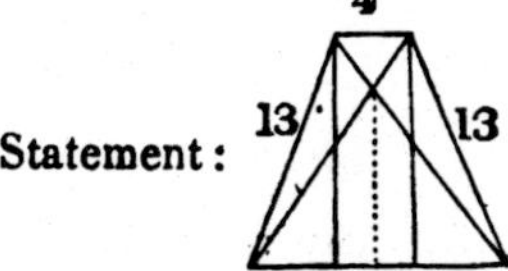

The segment of the base 7, multiplied by the diagonal 15, makes 105. Divided by the complement 9, the quotient is 11$\frac{2}{3}$. It is the lower portion of the diagonal; and subtracted from 15, leaves the upper portion 3$\frac{1}{3}$. So for the second diagonal.

26. The diagonal of a tetragon other than a trapezium, being multiplied by the flank, and divided by twice the perpendicular, is the central line;[1] and so is, in a trapezium, half the square-root of the sum of the squares of opposite sides.[2]

27. The product of the two sides of a triangle, divided by twice the per-

In like manner, at the intersection of the perpendicular. The segment 5, multiplied by the diagonal 15, and divided by the complement 9, gives the lower portion $8\frac{1}{3}$; which, subtracted from the diagonal, leaves the upper portion $6\frac{2}{3}$. The perpendicular likewise, 12, taken into the corresponding segment 5, makes 60; which, divided by the complement 9, yields $6\frac{2}{3}$ the lower portion of the perpendicular: and this, subtracted from 12, leaves $5\frac{1}{3}$ the upper portion of it. CH.

Put the proportion 'If the entire diagonal be hypotenuse answering to a side equal to the complement, what will be the hypotenuse answering to a side equal to the given segment of the base?' The result gives the portion of the diagonal below the intersection. A similar proportion gives the segment of the perpendicular. Thus the lower portions are found: and, subtracting them from the whole length, the remainder is the upper portion of the diagonal or of the perpendicular. *Ib.*

[1] *Hrĭdaya-rajju*, the central line, is the semidiameter of a circle in contact with the angles. CH. In an equilateral or an oblong tetragon it is equal to the semidiagonal.—*Ib.*

Cóna-sprĭg-vrĭtta, or *bahir-vrĭtta;* a circle in contact with the angles; an exterior circle: one circumscribed.

[2] Example: The tetragon with two equal sides, as last noticed.

Statement: 4, 13, 13, 14

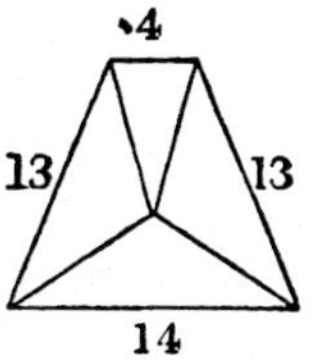

The diagonal 15, multiplied by the side 13, is 195: divided by twice the perpendicular, the quotient is $8\frac{1}{8}$; the length of the central line.

Example: The tetragon with three equal sides before exhibited (§ 21).

Statement: 25, 25, 25, 39

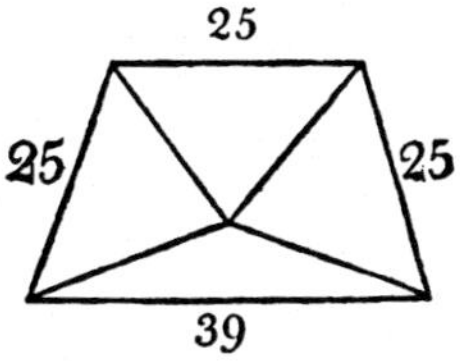

Diagonal 40, multiplied by the side 25, makes 1000; which, divided by the perpendicular doubled, gives the central line $20\frac{5}{6}$.

Example: The trapezium of which the base is sixty, the summit twenty-five, and the sides fifty-two and thirty-nine.

Statement: 25, 52, 39, 60

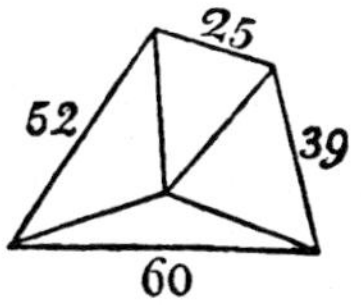

The squares of the base and summit 60 and 25 are 3600 and 625. The sum is 4225; its root 65: the half of which $\frac{65}{2}$ is the central line. Or the squares of the flanks 52 and 39 are 2704 and 1521; the sum of which is 4225; and half the root, $\frac{65}{2}$. CH.

pendicular, is the central line: and the double of this is the diameter of the exterior circle.[1]

28.[2] The sums of the products of the sides about both the diagonals being divided by each other, multiply the quotients by the sum of the products of opposite sides; the square-roots of the results are the diagonals in a trapezium.[3]

[1] Example: An isosceles triangle, the sides of which are thirteen, the base ten, and the perpendicular twelve.

Statement: 13 13 10 Product of the sides 169; divided by twice the perpendicular, gives the central line $7\frac{1}{24}$.* CH.

Let twice the perpendicular be a chord in a circle, the semidiameter of which is equal to the diagonal. Then this proportion is put: If the semidiameter be equal to the diagonal in a circle in which twice the perpendicular is a chord, what is the semidiameter in one wherein the like chord is equal to the flank? The result is the semidiameter of the circumscribed circle, provided the flanks be equal. But, if they be unequal, the central line is equal to half the diagonal of an oblong the sides of which are equal to the base and summit; or half the diagonal of one, the sides of which are equal to the flanks. It is alike both ways. *Ib.*

For the triangle the demonstration is similar; since here the diagonal is the side. *Ib.*

[2] This passage is cited in BHÁSCARA's *Lílávatí*, § 190.

[3] Example: A tetragon of which the base is sixty, the summit twenty-five, and the sides fifty-two and thirty-nine.

Statement:

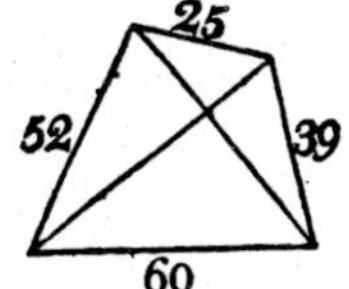

The upper sides about the greater diagonal are 39 and 25; the product of which is 975. The lower sides about the same are 60 and 52; and the product 3120. The sum of both products 4095. The upper sides about the less diagonal are 25 and 52; the product of which is 1300. The lower sides about the same, 60 and 39; and the product 2340. The sum of both 3640. These sums divided by each other are $\frac{4095}{3640}$ and $\frac{3640}{4095}$, or abridged $\frac{9}{8}$ and $\frac{8}{9}$. The product of opposite sides 60 and 25 is 1500; and of the two others 52 and 39 is 2028: the sum of both, 3528. The two foregoing fractions, multiplied by this quantity, make 3969 and 3136; the square-roots of which are 63 and 56, the two diagonals of the trapezium. CH.

This method of finding the diagonals is founded on four oblongs. *Ib.*

The brief hint of a demonstration here given is explained by GAṄÉSA on *Lílávatí*, § 191. Two triangles being assumed, the product of their uprights is one portion of a diagonal, and the pro-

* The manuscript here exhibits $8\frac{1}{8}$: but is manifestly corrupt: as is the text of the rule and in part the comment on it

duct of their sides is the other; as before shown. (Dem. of § 191—2.) The two sides on the one part of the diagonal are deduced from the reciprocal multiplication of the hypotenuses of the assumed triangles by their uprights: and the product of the sides is consequently equal to the product of the uprights taken into the product of the hypotenuses. So the product of the two sides on the other part of the diagonal, resulting from the reciprocal multiplication of the hypotenuses by the sides of the assumed triangles, is equal to the product of the sides of the triangles taken into the product of their hypotenuses. Therefore the sum of those products of the sides of the trapezium is equal to the diagonal multiplied by the product of the hypotenuses. The sides about the other diagonal are formed by the upright of one triangle and side of the other reciprocally multiplied by the hypotenuses. Their product is equal to the product of the reciprocal upright and side taken into the product of both hypotenuses. Hence the sum of the products is equal to the diagonal multiplied by the product of the hypotenuses. Therefore dividing one by the other, and rejecting like dividend and divisor (i. e. the product of the hypotenuses), there remain the diagonals divided by each other. Now the sum of the products of the multiplication of opposite sides is equal to the product of the diagonals [as will be shown]. Multiplying this by the fractions above found, and rejecting equal dividends and divisors, there remain the squares of the diagonals: and by extraction of the roots the diagonals are found. Now to show, that the sum of the products of opposite sides is equal to the product of the diagonals: the three sides of each of the assumed triangles being multiplied by the hypotenuse of the other, two other rectangular triangles are formed: and duly adapting together the halves of these, a figure is constituted, the sides of which are equal to the uprights and sides of the two triangles. It is the very trapezium; and its area is the sum of the areas of the triangles.

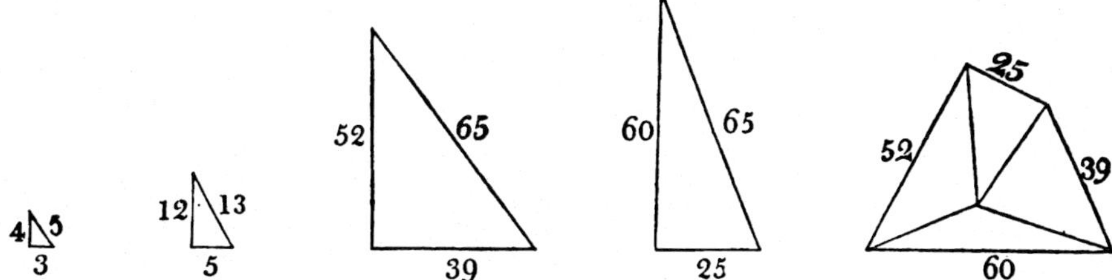

Here half the product of the base and summit is the area of one triangle; and half the product of the flanks is so of the other. Therefore half the sum of the products of opposite sides is the area of the quadrangle. Now the four triangles before mentioned, with four others equal to them, being duly adapted together, these eight compose an oblong quadrilateral with sides equal to the diagonals of the trapezium. See

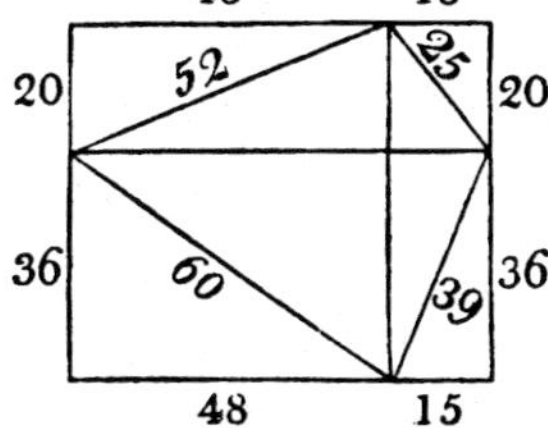

Half the area of the oblong or product of the diagonals, as is apparent, will be the area of the trapezium. It is half the sum of the products of opposite sides. Therefore the sum of the products of the opposite sides is equal to the product of the diagonals.

GAŃ.

29. Assuming two scalene triangles[1] within the trapezium, let the segments for both diagonals be separately found as before taught; and then the perpendiculars.[2]

30—31. Assuming two triangles within the trapezium, let the diagonals be the bases of them.[3] Then the segments, separately found, are the upper and lower portions formed by the intersection of the diagonals.[4] The lower

[1] The first with the greater diagonal for one side, and the least flank for the other side; and the second having the least diagonal for one side, and the greater flank for the other: and the base of the tetragon being base of both triangles. Then the segments are to be separately found in both triangular figures, by the rule before taught (§ 22); and then find the two perpendiculars by the sequel of the rule. CH.

[2] In the unequal tetragon just mentioned, one triangle will be this

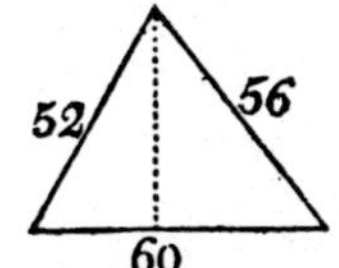

and the other this

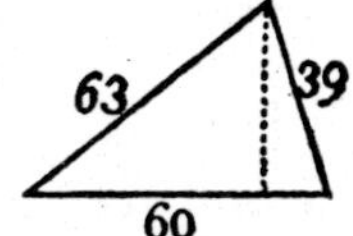

In the one the difference of the squares of the sides 432, divided by the base, gives $7\frac{1}{5}$, which subtracted from, and added to, the base, makes $52\frac{4}{5}$ and $67\frac{1}{5}$. These divided by two are $26\frac{2}{5}$ and $33\frac{3}{5}$ the two segments. Whence, taking the root of the difference of the squares of the side and its segment, the greater perpendicular is deduced $44\frac{4}{5}$. In the other triangle, the two segments found by the rule are $9\frac{3}{5}$ and $50\frac{2}{5}$; whence the least perpendicular comes out $37\frac{4}{5}$. CH.

[3] The greater diagonal is the base of one; and the summit and greater flank are its sides. The least diagonal is the base of the other; and the summit and least flank are the sides. CH.

[4] In the tetragon just now instanced, the scalene triangle with the greater diagonal for base is this

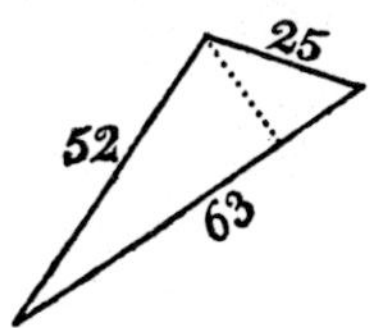

The segments of its base as found by the rule (§ 22) are 48 and 15. These are respectively the lower and upper portions of the greater diagonal.

The scalene triangle with the less diagonal for base is

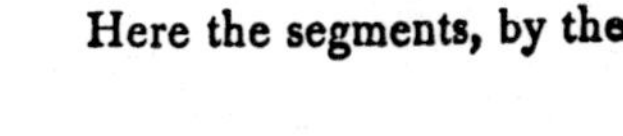

Here the segments, by the same rule (§ 22), are 36 and 20. They are the lower and upper portions of the least diagonal.

Or find the segments of one only: the perpendicular, found by the rule (§ 22) is the upper portion of the second diagonal: and subtracting that from the entire length, the remainder is the

portions of the two diagonals are taken for the sides of a triangle; and the base [of the tetragon] for its base. Its perpendicular is the lower portion of the [middle] perpendicular of the tetragon: the upper portion of it is the moiety of the sum of the [extreme] perpendiculars less the lower portion.[1]

32.[2] At the intersection of the diagonals and perpendiculars, the lower segments of the diagonal and of the perpendicular are found by proportion: those lines less these segments are the upper segments of the same. So in the needle[3] as well as in the *(páta)* intersection [of prolonged sides and perpendiculars].[4]

lower portion of it. Thus, in the foregoing example, the least segment in the first triangle is 15. Its square 225, subtracted from the square of the least side 625, leaves 400, the root of which is 20. It is the upper portion of the smaller diagonal, and subtracted from the whole length 56, leaves the lower portion 36.

CH.

[1] In the same figure, the scalene triangle composed of the two lower segments of the diagonals together with the base is this

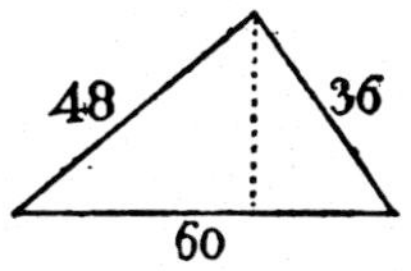

48 36

60

Here the perpendicular found by the rule (§ 22) is 28 $\frac{4}{5}$. It is the lower portion of the mean perpendicular. The greatest and least perpendiculars being 44 $\frac{4}{5}$ and 37 $\frac{4}{5}$, the moiety of their sum is 41 $\frac{3}{10}$. This is the length of the entire mean perpendicular. Subtracting from it its lower segment the residue is its upper segment 12 $\frac{1}{2}$.

CH.

[2] A rule to find the upper and lower portions of the diagonals and perpendiculars cut by the intersection of diagonals and perpendiculars, within a trapezium; also the lines of the needle and a figure of intersection.

[3] *Súchí*, the needle; the triangle formed by the produced flanks of the tetragon. The section of a cone or pyramid.

Páta, sampáta, tripáta, intersection; of a prolonged side and perpendicular. The figure formed by such intersection.

[4] Example: In a trapezium the base of which is sixty; one side fifty-two; the other thirty-nine; and the summit twenty-five: the greater diagonal sixty-three; the less, fifty-six: the greater perpendicular forty-five less one fifth; its segments of the base, the greatest thirty-three and three-fifths, the least twenty-six and two-fifths: the least perpendicular thirty-seven and four-fifths; its segments of the base, greatest fifty and two-fifths, least nine and three-fifths: the perpendicular passing through the intersection of the diagonals, forty-one and three-tenths; its segments of the base, greatest thirty-eight and two-fifths, least twenty-one and three-fifths; tell the upper and lower portions of the perpendiculars, the intersections [of prolonged sides and perpendiculars] and the needle.

Here at the intersection of the diagonals, the segments of the greater diagonal, found as before (§ 30), are 48 and 15; those of the less are 36 and 20.

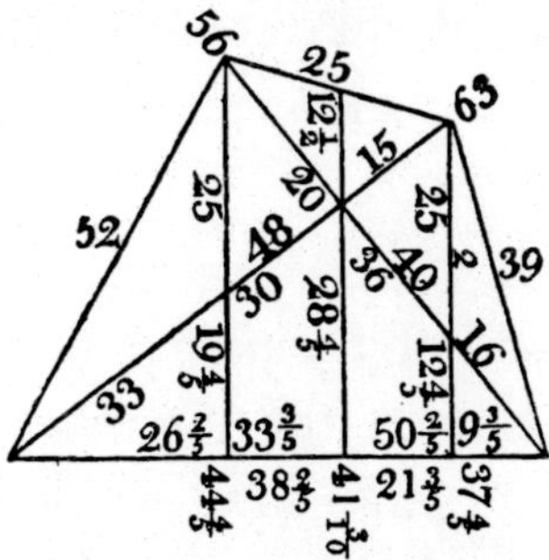

At the junction of the greater diagonal and greater perpendicular; the proportion is as diagonal sixty-three to the complement* fifty and two fifths, so, to the segment twenty-six and two-fifths, what? or rendered homogeneous, $\frac{252}{5}$ | 63 | $\frac{135}{5}$ |. Answer: 33. It is the lower portion of the diagonal. Again, as the same complement is to the least perpendicular, so is the above mentioned segment to what? Statement: $\frac{252}{5}$ | $\frac{189}{5}$ | $\frac{132}{5}$ | Answer: $19\frac{4}{5}$. It is the lower portion of the perpendicular. Subtracting these from the whole diagonal 63 and entire perpendicular $44\frac{4}{5}$, the remainders are the upper segments of the diagonal and perpendicular; 30 and 25.

Next, at the junction of the less diagonal and less perpendicular: as the complement thirty-three and three-fifths is to the diagonal fifty-six, so is the segment nine and three-fifths to what? Statement: $33\frac{3}{5}$ | 56 | $9\frac{3}{5}$ |. Answer: 16, the lower portion of the diagonal. So, putting the perpendicular for the middle term, the lower portion of the less perpendicular comes out $12\frac{4}{5}$. By subtraction from the entire diagonal and perpendicular, their upper segments are obtained 40 and 25.

In like manner, for any given question, the solution may be variously devised with the segment of the base for side, the segment of the perpendicular for upright, and the segment of the diagonal for hypotenuse.

The operation on the needle is next exhibited

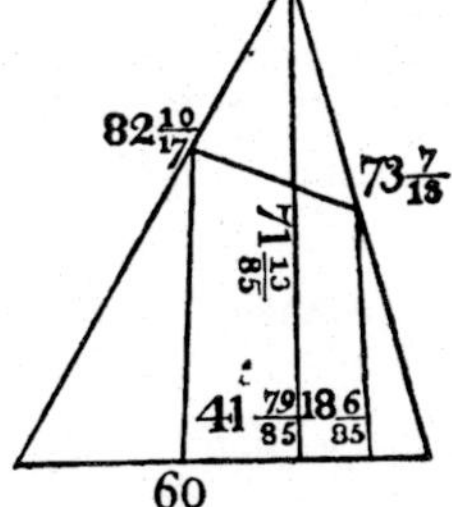

The segments of the base on either side of the perpendicular let fall from the top of the needle come out $41\frac{79}{85}$ and $18\frac{6}{85}$.† With either of these segments the mean perpendicular is found by proportion: if the least segment $9\frac{3}{5}$ give the least perpendicular $37\frac{4}{5}$, what does the segment $18\frac{6}{85}$ give? Answer: $71\frac{13}{85}$. It is the perpendicular let fall from the summit of the needle. In the same manner, with the greater segment, the same length of the perpendicular is deduced.

Next, to find the sides of the needle: As the least perpendicular is to the side thirty-nine, so is the middle perpendicular to what? Statement: $37\frac{4}{5}$ | 39 | $71\frac{13}{85}$. Answer: $73\frac{3}{13}$. Or the side may be found from the segments: thus $9\frac{3}{5}$ | 39 | $18\frac{6}{85}$. Answer: $73\frac{3}{13}$ as before. To find the

* *Swa-yuti.* See note to § 25.

† The text relative to the method of finding these segments is irretrievably corrupt; and has been therefore omitted in the version.

greater side: As the greater perpendicular is to the side fifty-two, so is the perpendicular of the needle to what? 44 $\frac{4}{5}$ | 52 | 71 $\frac{13}{85}$ |. Answer: 82 $\frac{10}{17}$. Or proportion may be taken with the segments of the base: 26 $\frac{2}{5}$ | 52 | 41 $\frac{72}{85}$ |. Answer: 82 $\frac{10}{17}$, as before. See figure [as above].

Now to find the intersections [of the prolonged sides and perpendiculars]. If the segment of the base belonging to the greater perpendicular, or 26 $\frac{2}{5}$, answer to that perpendicular, 44 $\frac{4}{5}$, what will the segment 50 $\frac{2}{5}$ answer to? Answer: 85 $\frac{29}{35}$, the perpendicular prolonged to the intersection. Again: As the greater perpendicular 44 $\frac{4}{5}$ is to the side 52, so is the perpendicular of the intersection 85 $\frac{29}{35}$ to what? Answer: 99 $\frac{3}{11}$, the side of the figure. In like manner to find the perpendicular of the second figure of intersection: If the segment of the base appertaining to the less perpendicular answer to this perpendicular, what does the segment thirty-three and three-fifths correspond to? Answer: 132 $\frac{3}{10}$, the perpendicular of second figure. To find the side of the same: As the least perpendicular 37 $\frac{4}{5}$ is to the side 39, so is the perpendicular just found 132 $\frac{3}{10}$ to what? Answer: 136½, the side. Or it may be found from the segments. Thus, as the segment answering to the least perpendicular, 9 $\frac{3}{5}$, is to the side 39, so is the segment 33 $\frac{3}{5}$ to what? Answer: the greater side 136½ as before. See figure of the needle with the intersections and perpendiculars.

136½
99 $\frac{3}{11}$
132 $\frac{3}{10}$
85 $\frac{29}{35}$
52
44 $\frac{4}{5}$
37 $\frac{4}{5}$
39
26 $\frac{2}{5}$
9 $\frac{3}{5}$
60

In like manner, the flank intersections* are computed. If the segment appertaining to the greater perpendicular 26 $\frac{2}{5}$ answer to that perpendicular, what will the segment 60 correspond to? Answer: 101 $\frac{9}{11}$. To find the side of the same: As the segment of the base for the greater perpendicular is to the side fifty-two, so is the segment sixty to what? 26 $\frac{2}{5}$ | 52 | 60 |. Answer: 118 $\frac{2}{11}$. So, on the other part: If the segment of the base for the less perpendicular answer to that perpendicular, what will the segment sixty correspond to? 9 $\frac{3}{5}$ | 37 $\frac{4}{5}$ | 60 |. Answer: 236 ¼. To find the side: As the least segment is to the side thirty-nine, so is the segment sixty to what? 9 $\frac{3}{5}$ | 39 | 60. Answer: 242 $\frac{3}{4}$. See

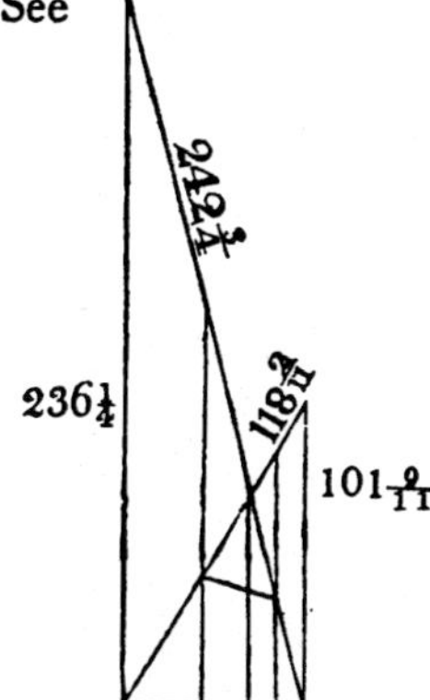

* Párśwa-páta, the intersection of the prolonged flank and perpendicular raised at the extremity of the base.

33.[1] The sum of the squares of two unalike quantities are the sides of an isosceles triangle; twice the product of the same two quantities is the perpendicular; and twice the difference of their squares is the base.[2]

34. The square of an assumed quantity being twice set down, and divided by two other assumed quantities, and the quotients being severally added to the quantity first put, the moieties of the sums are the sides of a scalene triangle: from the same quotients the two assumed quantities being subtracted, the sum of the moieties of the differences is the base.[3]

35.[4] The square of the side assumed at pleasure, being divided and then

In the top above the summit of the trapezium a distribution of the figure is to

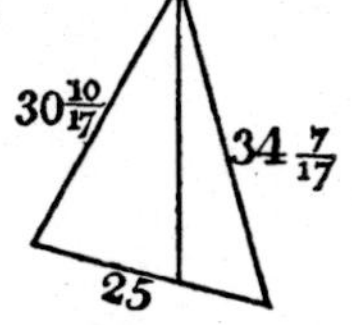

be in like manner made by proportions selected at choice.

Since every where the segment of the base is a side, the corresponding perpendicular an upright, and the flank an hypotenuse, the several lines above-stated may be found in various ways by the rule, that subtracting the square of the upright from the square of the hypotenuse, the square-root of the residue will be the side; or that subtracting the square of the side, the root of the remainder will be the upright.

In the same manner, in tetragons with two or three equal sides, the perpendicular of the needle and its segments of the base are to be found. But there can be no needle to an equilateral tetragon, nor to an oblong. CH.

[1] To find an equicrural triangle; preparatory to showing a rectangular one. The next following rule (§ 34) is for finding a scalene triangle. An equilateral one may consist of any quantity assumed at pleasure for the side; since all the sides are equal. CH.

[2] Example: Let the unalike quantities be put 2 and 3. Their squares are 4 and 9; the sum of which is 13; and the sides are of this length. Twice the product is 12; and is the perpendicular. Again, the squares of the same number are 4 and 9: the difference is 5; which multiplied by two makes 10, the base. [See § 22.] CH.

[3] Example: Let 12 be assumed. Its square is 144. Put the two numbers 6 and 8; and severally divide: the quotients are 24 and 18: which, added to the number originally put, make 36 and 30, the moieties whereof are 15 and 13, the two sides. The same quotients, 24 and 18, less the assumed numbers 6 and 8, make 18 and 10; the moieties of which are 9 and 5: and the sum of these, 14, is the base.—CH. [See § 22.]

[4] To find an oblong tetragon. The equilateral tetragon may be assumed with any quantity: since all the sides are alike.—CH. The subsequent rules, § 36—38, deduce tetragons with two and three equal sides or with all unequal.

lessened by an assumed quantity, the half of the remainder is the upright of an oblong tetragon; and this, added to the same assumed quantity, is the diagonal.[1]

36. Let the diagonals of an oblong be the flanks of a tetragon having two equal sides. The square of the side of the oblong, being divided by an assumed quantity and then lessened by it, and divided by two, the quotient increased by the upright of the oblong is the base; and lessened by it is the summit.[2]

37. The three equal sides of a tetragon, that has three sides equal, are the squares of the diagonal [of the oblong]. The fourth is found by subtracting the square of the upright from thrice the square of the [oblong's] side. If it be greatest, it is the base; if least, it is the summit.[3]

38. The uprights and sides of two rectangular triangles reciprocally multiplied by the diagonals are four dissimilar sides of a trapezium. The greatest is the base; the least is the summit; and the two others are the flanks.[4]

[1] Example: Let the side be put 5. Its square is 25, which divided by the assumed quantity one makes 25; and subtracting from this the same assumed quantity, half the remainder is 12, and is the upright. This added to the assumed divisor is the diagonal 13.—CH. [See § 21 and 23.]

[2] Example: If the diagonal of the oblong be thirteen, the side twelve and the upright five; what tetragon with two equal sides may be deduced from it? The diagonals 13 and 13 are the flanks. The square of the side 12 is 144. Divided by an assumed number 6, it gives 24; from which subtracting the number put 6, remains 18; the half whereof is 9. This with the upright 5 added, makes 14, the base. Again, the same moiety 9, with the upright 5 subtracted, leaves 4, the summit.—CH. [See § 21 and 23.]

[3] Example: Find a tetragon with three equal sides from an oblong the diagonal of which is five, the side four, and upright three. Square of the diagonal 25; the length of the sides. Square of the side 16, tripled, is 48: from which subtracting 9, the square of the upright 3, the remainder 39 is the base. Or let the side be three and upright four. Square of the side tripled is 27; and subtracting from this the square 16 of the upright 4, the remainder 11 is the summit.—CH. [See § 21 and 26.]

[4] Example: In one oblong the diagonal is five, the upright three, and the side four. In the second the diagonal is thirteen, the upright twelve, and the side five. The uprights and sides of each of the two rectangular triangles, viz. 12, 5, 3, and 4, being multiplied by the diagonal (hypotenuse) of the other, give 60, 25, 39 and 52. Here the greater number 60 is the base; the least 25 is the summit; the remaining two, 39 and 52, are the flanks.—CH. [See § 21 and 28.]

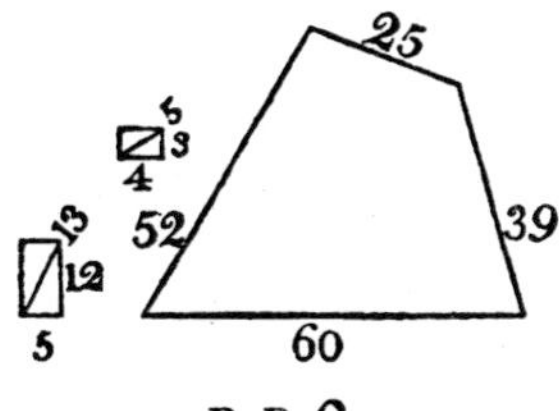

39.[1] The height of the mountain, taken into a multiplier arbitrarily put, is the distance of the town. That result being reserved, and divided by the multiplier added to two, is the height of the leap. The journey is equal.[2]

40. The diameter and the square of the semidiameter, being severally multiplied by three, are the practical circumference and area. The square-roots extracted from ten times the squares of the same are the neat values.[3]

[1] Within an oblong tetragon, to describe a figure such, that the sum of the side and one portion of the upright may be equal to the diagonal and remaining portion of the upright: so as the journeys may be equal.—CH. See *Lílávatí*, § 154, and *Víja-gańita*, § 126; where the same problem is introduced: substituting, however, in the example, a tree, an ape and a pond, for a hill, a wizard and a town.

[2] Example: On the top of a certain hill live two ascetics. One of them, being a wizard, travels through the air. Springing from the summit of the mountain, he ascends to a certain elevation, and proceeds by an oblique descent, diagonally, to a neighbouring town. The other, walking down the hill, goes by land to the same town. Their journeys are equal. I desire to know the distance of the town from the hill, and how high the wizard rose.

This being proposed, the rule applies; and its interpretation is this: any elevation of the mountain is put; and is multiplied by an arbitrarily assumed multiplier: the product is the distance of the town from the mountain. Then divide this reserved quantity by the multiplier added to two, the quotient is the number of *yójanas* of the wizard's ascent. The sum of the hill's elevation and wizard's ascent is the upright; the distance of the town from the mountain is the side: the square-root of the sum of their squares is the diagonal (hypotenuse): it is the oblique interval between the town and the summit of the rise.

Thus, let the height of the mountain be twelve. This, multiplied by an arbitrarily assumed multiplier four, 12 by 4, makes 48. It is the distance of the town from the hill. This divided by the multiplier added to two, 48 by 6, gives 8. It is the ascent. Here the upright is 20: its square is 400. The side is 48; the square of which is 2304. The sum of these squares is 2704; and its square-root 52. The semirectangle* is thus found. Here also the sum

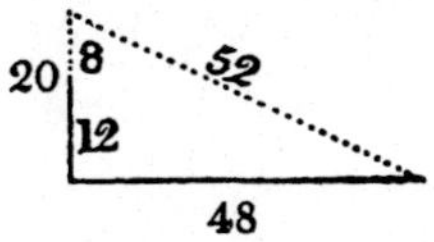

20 | 8 | 52
12
48

of the side and lower portion of the upright is 60, the journey of one of the ascetics: and the upper portion added to the hypotenuse is that of the other, likewise 60.

The author will treat of rectangular triangles and surd roots, in the chapter on Algebra (*cuttacá-d'hyáya*†) under the rule, which begins, "Be a surd the perpendicular. Its square, &c." We also shall there expound it. CH.

[3] Example: Of a circle, the diameter whereof is ten, what is the circumference? and how much the area?

* *Áyatárd'ha*, half an oblong.

† See *Brahm.* Alg. § 26.

41. In a circle the chord is the square-root of the diameter less the arrow taken into the arrow and multiplied by four.[1] The square of the chord divided by four times the arrow, and added to the arrow, is the diameter.[2]

Statement: 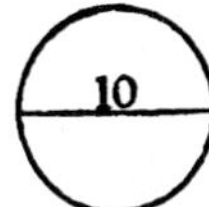

Diameter 10, multiplied by three, 30; this is the gross circumference. Semidiameter 5: its square 25; tripled, 75; the gross area for practice.

Diameter 10: its square 100, multiplied by ten, 1000. The surd root of this is the circumference of a circle the diameter whereof is ten. Square of the semidiameter 25: This again squared and decupled is 6250. Its surd-root is the area of the circle. CH.

[1] Example: Within a circle, the diameter of which is ten, in the place where the arrow is two, what is the chord?

Diameter 10: less the arrow 2; remains 8. This multiplied by the arrow makes 16; which multiplied by 4. gives 64: the square-root of which is 8. See figure

The principle of the rule for finding the square of the chord (in the construction of tabular sines) is here to be applied. But the square is in this place multiplied by four, because the entire chord is required. CH.

[2] Example: Chord 8. Its square 64, divided by four times the arrow 2, viz. 8; gives the quotient 8: to which adding the arrow, the sum is 10.

Example 2d: A bambu, eighteen cubits high, was broken by the wind. Its tip touched the ground at six cubits from the root. Tell the length of the segments of the bambu.

Statement: Length of the bambu 18. It is the diameter less the least arrow.* The ground from the root, to the point where the tip fell, is 6: it is the semichord. Its square is 36. This is equal to the diameter less the arrow multiplied by the arrow.* Dividing it by the diameter less the arrow, viz. 18, the quotient is 2. It is the arrow. Adding this to the diameter less the arrow, the sum is the diameter, 20. Half of this, 10, is the semidiameter. It is the upper portion of the bambu and is the hypotenuse. Subtracted from eighteen, it leaves the upright, or lower portion of the bambu, 8. The side is the interval between the root and tip, 6. The point of fracture of the bambu is the centre of the circle. See figure

Example 3d: In limpid water the stalk of a lotus eight fingers long was to be seen.† ....
That visible [portion of] stalk is the smaller arrow. The place of submersion, 24, is the semi-

* What is termed by us "diameter less the arrow," is by ÁRYA-BHAṬṬA denominated the greater arrow. For he says, 'In a circle the product of the arrows is equal to the square of the semichord of both arcs.' CH.

† The remainder of the passage, in which the question was proposed, is wanting.

42. Half the difference of the diameter and the root extracted from the difference of the squares of the diameter and the chord is the smaller arrow.[2]

chord. From the square of this [semi-]chord 576 divided by the smaller arrow 8, the quotient 72 is obtained, which is the greater arrow. The sum of both arrows, viz. 80, is the diameter of the circle. Its half is 40, the semidiameter. It is the hypotenuse, and is the length of the stalk of lotus. Subtracting the smaller arrow, the remainder is the depth of water and is the upright 32. The side is the space to the place of submersion and is the semichord. See

Example 4th: A cat, sitting on a wall four cubits high, saw a rat prowling eight cubits from the foot of the wall. The rat too perceived the puss and hastened towards its abode at the foot of the wall; but was caught by the cat proceeding diagonally an equal distance. In what point within the eight cubits was the rat caught; and what was the distance they went? Tell me, if thou be conversant with computation concerning circles.

Statement: Height of the wall 4. Distance to which the rat had gone forth 8. These are semichord and greater arrow. The square of the semichord, 16, being divided by the greater arrow 8, the quotient is the smaller arrow 2. The sum of both arrows is the diameter 10. Its half is the semidiameter 5. It is the rat's return. Subtracting it from the eight cubits, the remainder is the interval between the foot of the wall and point of capture, or upright 3. The side is 4. The root of the sum of their squares is the hypotenuse: it is the cat's progress, and is equal to the rat's progress homewards.*

Let the figure be exhibited as before. In the centre of it is the place of capture.

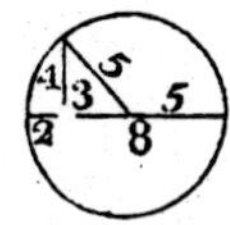

In like manner other examples may be shown for the instruction of youth. Else all this is obvious, when the relation of side, upright and hypotenuse is understood. CH.

[1] The chord and diameter being given, to find the smaller arrow. And, when two circles, the diameters of which are known, cut each other, to find the two arrows. CH.

[2] Example: Chord 8. Its square 64. Diameter 10. Its square 100. Difference 36. Its root 6. Subtracting this from the diameter 10, the moiety of the remainder is 2 and is the smaller arrow.

The same figure is here contemplated. Within it let an oblong be inscribed, with the chord for its side, the difference between the diameter and twice the arrow for its upright, and the entire diameter for its diagonal. It is this Here, the square-root of the difference between the squares of the diameter and chord is equal to the root of the residue of subtracting the square of the side from the square of the diagonal, and is the upright: and, that being taken from the diameter, two portions remain equal to the smaller arrow, one at either extremity. Hence the rule § 42. Let all this be shown on the figure. CH.

* The three last instances are imitated in BHASCARA'S *Lílávatí*, § 148—153, and *Víj.-gaṅ.* § 124—125 and 139.

The erosion[1] being subtracted from both diameters, the remainders, multiplied by the erosion and divided by the sum of the remainders, are the arrows.[2]

43.[3] The square of the semichord being divided severally by the given arrows, the quotients, added to the arrows respectively, are the diameters.[4] The sum of the arrows is the erosion: and that of the quotients is the residue of subtracting the erosion.[5]

[1] *Gràsa,* the erosion, the morcel bitten; the quantity eclipsed.

Samparca, intersection.

[2] Example: The measure of *Ráhu* is fifty-two; that of the moon, twenty-five: the erosion is seven.

Diameters 52 and 25. Remainders after subtracting the erosion 45 and 18. These multiplied by the erosion, make 315 and 126: which, divided by the sum of the residues 63, give 5 and 2, for the segments cut by a chord passing through the points of intersection of the circles. The arrow of *Ráhu* is two; that of the moon five. See

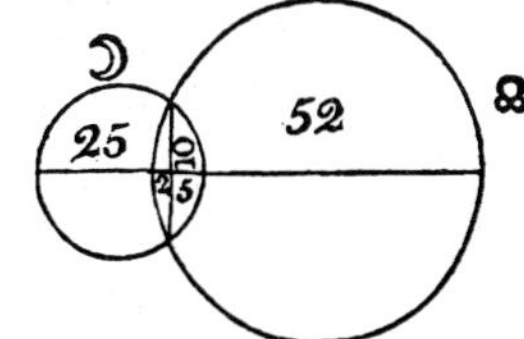

Here the erosion is the profit; and the diameters less the erosion, are the contributions; and the segments are found by the rule, § 16. The greater quotient belongs to the least circle; and the less quotient, to the greater circle. CH.

[3] In the like case of the intersection of two circles, the chord and arrows being known, to find the diameters: And, the diameter and arrows being given, to deduce the quantity eclipsed and the residue. CH.

[4] Example: The intersection of the circles last mentioned.

Chord 20. Its half 10: the square of which is 100. Divided by the two arrows severally, viz. 5 and 2; the quotients are 20 and 50: which, with the arrows respectively added, make 25 and 52. They are the diameters. See foregoing diagram.

Demonstration: So much as is the square of the semichord, is the square of greater and less arrows multiplied together. The quotient of the division thereof by the less arrow is the greater arrow; and the sum of the greater and less arrows is the diameter, as even the ignorant know.

CH.

[5] Example: The arrows just found, 5 and 2. Their sum is 7. It is the erosion or quantity eclipsed. The quotients 20 and 50. Their sum 70. It is the residue, subtracting the erosion [from the sum of the diameters].

The principle is here obvious. CH.

SECTION V.

EXCAVATIONS.

44. The area of the plane figure, multiplied by the depth, gives the content of the equal [or regular] excavation; and that, divided by three, is the content of the needle.[1]

In an excavation having like sides [length and depth] at top and bottom, [but varying in depth,] the aggregates[2] [or products of length and depth of the portions] being divided by the common length, [and added together,] give the mean depth.[3]

45—46. The area, deduced from the moieties of the sums of the sides at top and at bottom, being multiplied by the depth, is the practical measure

[1] Example: Tell the content of a well, in which the sides are ten and twelve, alike above and below, and the depth five.

Statement: Here the area is 120: which, multiplied by the depth 5, gives the content in cubic cubits, 600.

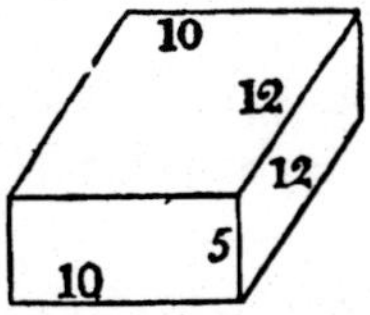

In the like instance, if the well terminate in a point, the foregoing divided by three gives 200, the content of the needle or pyramid.

[2] *Aicya*, lit. aggregate: explained by the commentator the product of the length and depth of the portions or little excavations differing in depth.

Ecágra, the whole of the long side which is subdivided.

Sama-rajju, equal or mean string: the mean or equated depth (*sama-béd'ha*).

[3] Example: A well thirty cubits in length, and eight in breadth, comprizes within it five portions of excavation, by which the side is subdivided into parts measuring four, &c. [up to eight]. The depth severally measures nine, seven, seven, three and two. Say quickly what is the mean string [mean depth] of the excavations.

of the content.[1] Half the sum of the areas at top and at bottom, multiplied by the depth, gives the gross content. Subtracting the practical content from the other, divide the difference by three, and add the quotient to the practical content, the sum is the neat content.[2]

Statement:

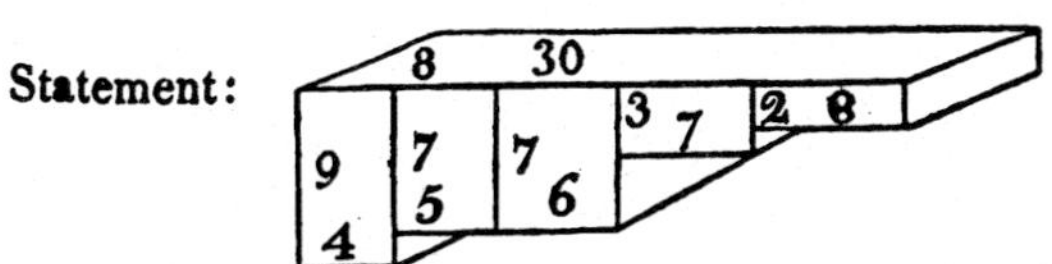

Here the aggregates in their order are 36, 35, 42, 21, 16. These, divided by the whole length 30, give $\frac{36}{30}$ $\frac{35}{30}$ $\frac{42}{30}$ $\frac{21}{30}$ $\frac{16}{30}$; which added together make $\frac{150}{30}$; the quotient is the mean depth 5. The area of the plane figure 240, multiplied by that, is 1200. It is the solid content of the entire excavation. It may be proved by adding together the several contents of the parts: viz. of the 1st, 288; of the 2d, 280; of the 3d, 336; of the 4th, 168; of the 5th, 128: total 1200.

[1] *Vyavahárica,* designed for practical use.

Autra, gross. [The etymology and proper sense of the term are not obvious; and are unexplained.]

Súcshma, neat, or correct.

[2] Example: A square well, measured by ten cubits at the top and by six at the bottom, is dug thirty cubits deep. Tell me the practical, the gross, and the neat contents.

Here the side at the top is 10; that at the bottom is 6. The sum of these is 16; its moiety 8. The same in the other directions, 8. The area with these sides is 64; which, multiplied by the depth 30, makes 1920. It is the practical content.

Sides at the top 10, 10. Area deduced from them 100. Sides at the bottom 6, 6. Area deduced from these 36. Sum of the areas 136. Its half 68; multiplied by the depth 30, makes 2040. It is the gross content.

Subtracting the practical content from this, the difference is 120. Divided by three, it gives 40. Adding this to the practical content 1920, the sum is 1960 the neat content.

SECTION VI.

STACKS.[1]

47. The area of the form [or section][2] is half the sum of the breadth at bottom and at top multiplied by the height: and that multiplied by the length is the cubic content: which divided by the solid content of one brick, is the content in bricks.[3]

[1] There is no difference in principle between the measure of excavations and of stacks; unless that what is there depth is here height. Every thing else is alike in both. Ch.

[2] *Acríti:* the form or shape of the wall, as it appears in one cubit's length, according to its height and the thickness at bottom and top.—Ch. Section of the wall.

[3] Example 1st: Tell the content of a stack which is a hundred cubits in length; five in thickness at bottom, and three at top; and seven high.

Statement:

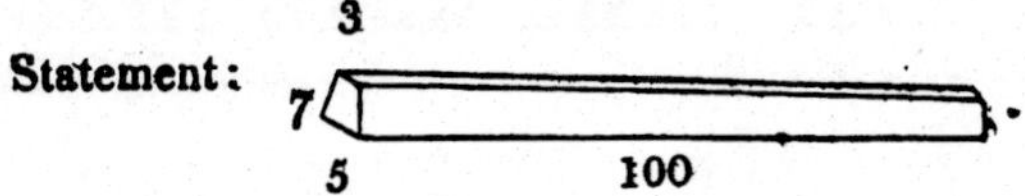

Breadth at top 3; at bottom 5. Sum 8. Its half 4, multiplied by 7, is 28: which, multiplied by the length 100, makes 2800. So many are the cubic contents in the wall. The dimensions of a brick may be arbitrarily assumed. Say a cubit long; half of one broad; and a sixth part thick. Statement: $\frac{1}{1}$ $\frac{1}{2}$ $\frac{1}{6}$. Product $\frac{1}{12}$. The whole cubic amount 2800, divided by that, gives 33600 for the number of bricks.

Example 2d: A sovereign piously caused a quadrangle to be built for a college, the wall measuring a hundred cubits without and ninety-six within, and seven high, with a gate four by three, and wickets half as big on the sides. How many bricks did it contain?

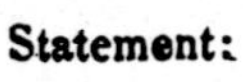

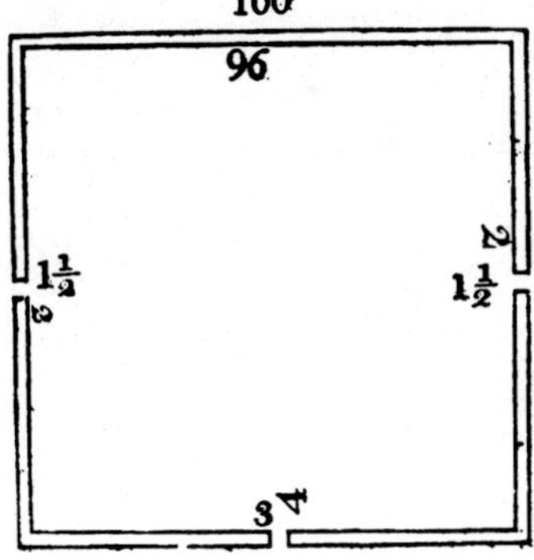

Here the area of the exterior figure is 10,000: that of the interior one 9216. The difference is 784. It is the area of the figure covered by the walls. Multiplied by the height 7, it gives the content 5488; from which subtracting the gates 36, the remainder is the exact cubic content 5452. Dividing this by the content of a brick $\frac{1}{12}$, the quotient is the number of bricks 65424.

SECTION VII.

SAW.

48—49. The product of the length and thickness in fingers, being multiplied by the number of sections and divided by forty-two, is the measure in *cishcangulas.*[1] That quotient, divided by ninety-six, gives the work,[1] if the timber be *śáca* or the like;[2] but, if it be *śálmalí,* the divisor is two hundred; if *víjaca,* a hundred and twenty; if *sála, saraṅa* and the rest, one hundred; if *sapta-vidáru,* sixty-four.[3]

[1] *Áyáma,* breadth, or rather *(dairghya)* length. *Vistára,* width, or rather *(ghanatwa)* thickness. *Márga,* the way or path of the saw; the section. *Cishcangula,* a technical term in use with artisans. *Carman,* the work; that is, the rate of the workman's pay: a technical use of the term. Ch.

[2] *Sáca,* Tectona grandis. *Śálmalí,* Bombax heptaphyllum: it is the softest wood used for timber. *Víjaca,* Citrus medica. *Sála,* Shorea robusta. *Saraṅa,* same with *Sarala?* Pinus longifolia. *Vidáru,* not known. *C'hadira,* Mimosa catechu; the hardest wood employed as timber.

The following passage of Árya-bhaṫṫa is cited by Gaṅéśa in his commentary on the *Lílávatí.* 'The product of the breadth [or length] and thickness, in fingers, being multiplied by the intended sections, and divided by five hundred and seventy-six, the quotient is the *(p'hala)* superficial measure of the cutting, provided the timber be *C'hadira* (Mimosa catechu). If the wood be *Śríparṅí* (   ), *Śáonca* (Tectona grandis), &c. the divisor should be put three hundred and fifty; if the wood be *Jambu* (Eugenia Jamboo), *Víja* (Citrus medica), *Cadamba* (Nauclea orientalis and Cadamb), or *Amlí* (Tamarindus indica), it should be twenty less than four hundred. The divisor should be two hundred and fifty, if the timber be *Śála, Amra* and *Sarala* (Shorea robusta, Mangifera indica and Pinus longifolia). If it be *Śálmalí* (Bombax heptaphyllum), &c. the divisor is two hundred. Money is to be paid according to the divisor.'

[3] Example: A seasoned timber of *(Víjáca)* citron wood, ten cubits in length and six fingers in width [thickness], is sawed in seven sections. Say what is the price of the labour, if the rate of work be eight *paṅas.*

Statement: 

Product of the thickness 6, by the length 240, is 1440.

Multiplied by 7, it makes 10080; which, divided by forty-two (42), gives 240. These are *cishcangulas.* The timber being wood of the *Víja* tree, that is divided by one hundred and twenty. The quotient is the quantity of the work, 2. Multiplied by the rate of the pay, viz. 8, the product is the number of *paṅas* 16. This amount is to be paid to the artisan. Ch.

SECTION VIII.

MOUNDS OF GRAIN.

50. The ninth part of the circumference is the depth [height] in the case of bearded corn; the tenth part, in that of coarse grain; and the eleventh, in that of fine grain.[1] The height, multiplied by the square of the sixth part of the circumference, is the content.[2]

51. The circumference of a mound resting against the side of a wall, or within or without a corner, is multiplied by two, by four, or by one and a third; and, proceeding as before, the content is found; and that is divided by the multiplier which was employed.

[1] *Śúcin*, bearded corn: viz. rice, as *shashticá*[*] and the rest.

St'húla, coarse grain: barley, &c.

Anu, fine grain: mustard and the like. CH.

[2] The content, as thus found, is the number of solid cubits; (the circumference having been taken with the cubit:) and thence the number of *prast'has* is to be deduced by the rule of three, according to the proportion of the cubit to the particular *prast'ha* in use.—CH. It is the content in solid cubits or *c'háris* of *Magad'ha*.—*Gań. sár.* Ch. 12.

[3] Example: What is the content of a mound of rice upon level ground the circumference being thirty-six?

Statement: Circum. 36. Its ninth part 4. This is the height of the mound. The sixth part of the circumference of the mound is 6; its square is 36: multiplied by the height, it makes 144, the content of the mound in cubits.

Example 2d: A mound of barley, the circumference of which is thirty? Answer: 75.

Example 3d: A mound of mustard-seed, sixty-six cubits in circumference? Answer: 726.

Example 4th: A mound of rice resting against a wall, and measuring eighteen?

18 doubled is 36. With this circumference the content found as before is 144; which, divided by the particular multiplier 2, gives 72, the solid content in cubits of the portion of a mound.

Example 5th: A mound of rice resting against the outer angle of a wall and measuring twenty-seven?

27 multiplied by one and a third makes 36, the circumference. Hence the content 144; which divided by the particular multiplier $\frac{4}{3}$, gives 108, the content of a mound that is a quarter less than a full one.

[*] *Shashtica* or *Shasti*; vulg. *Sáti* (Hind): so named because it is sown and reaped in sixty days. Oryza sativa; var.

SECTION IX.

MEASURE BY SHADOW.

52.[1] THE half day being divided by the shadow (measured in lengths of the gnomon) added to one, the quotient is the elapsed or the remaining portion of day, morning or evening. The half day divided by the elapsed or remaining portion of the day, being lessened by subtraction of one, the residue is the number of gnomons contained in the shadow.[2]

53.[3] The distance between the foot of the light and the bottom of the gnomon, multiplied by the gnomon of given length, and divided by the difference between the height of the light and the gnomon, is the shadow.[4]

[1] To find the time from the shadow; and the shadow from the time. CH.

[2] This rule being useless, no example is given. It does not answer for finding either the shadow or the time, in a position even equatorial; but has been noticed by the author in this place, copying earlier writers of treatises on computation. CH.

See the concluding chapter of ŚRÍD'HARA'S *Gańita-sára*, where the same rule is given, and examples of it subjoined.

[3] Given the length of the gnomon standing at a known distance from the foot of a light in a known situation, to find the shadow. CH.

[4] Example: The height of the light to the tip of the flame is a hundred fingers. [The distance a hundred and ten. The gnomon twelve.*]

110, multiplied by the gnomon 12, is 1320. Subtracting the gnomon 12 from the height 100, the remainder is 88. Dividing by this, the quotient is 15, the shadow of a gnomon twelve fingers high.

Here the rule of three terms is applicable: if an upright equal to the difference of the two heights answer to a side equal to the interval of ground between the foot of the light and the gnomon, what will answer to the given gnomon? See

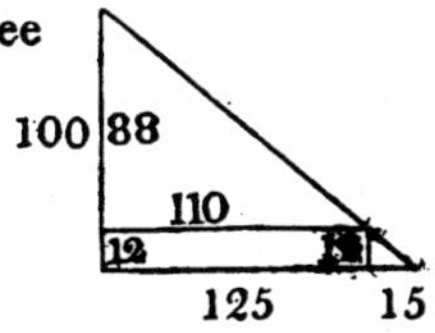

100 88
110
12 12
125 15

* The text is deficient: but is supplied by the operation in the sequel.

54.[1] The shadow multiplied by the distance between the tips of the shadows and divided by the difference of the shadows, is the base. The base, multiplied by the gnomon, and divided by the shadow, is the height of the flame of the light.[2]

[1] The difference between two positions of the gnomon being known, to find the distance between the foot of the light and gnomon; and the elevation of the light: CH.

[2] The shadow of a gnomon twelve fingers high is in one place fifteen fingers. The gnomon being removed twenty-two fingers further, its shadow is eighteen. The distance between the tips of the shadows is twenty-five. The difference of the length of the shadows is three.

Distance between the tips of the shadows 25. By this multiply the shadows 15 and 18: the products are 375 and 450; which, divided by the difference of the shadows 3, give the several quotients 125 and 150. They are the bases; that is, the distances of the tips of the shadows from the foot of the light.

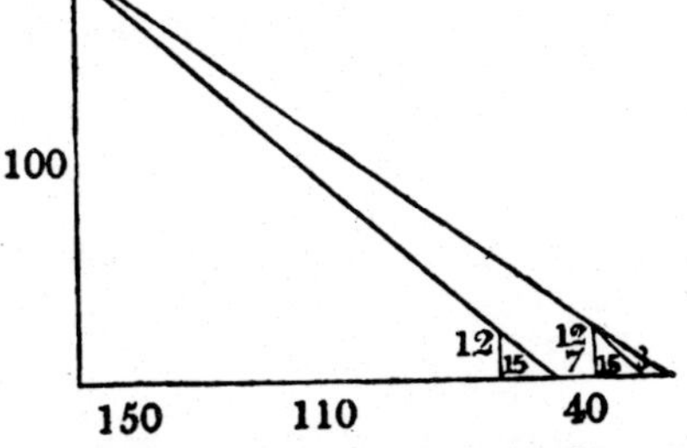

The grounds or bases 125 and 150, multiplied by the gnomon 12, make 1500 and 1800; which, divided by the respective shadows, give the quotients 100 and 100; or the elevation of the light, alike both ways.

Here also the operation of the rule of three is applicable: 'If to the difference of the shadows answers a side equal to the distance between the tips of the shadows, what will answer to the length of the shadow?' The answer is a side, which is the distance of the foot of the light to the tip of the shadow.

So to find the upright, the proportion is: 'If an upright equal to the gnomon answer to a side equal to the shadow, what will answer to a side equal to the base?' The answer gives the height of the flame of the light.

SECTION X.

SUPPLEMENT.

55.[1] THE multiplicand is repeated like a string for cattle,[2] as often as there are integrant portions[3] in the multiplier, and is severally multiplied by them, and the products are added together: it is multiplication. Or the multiplicand is repeated as many times as there are component parts in the multiplier.[4]

[1] In the rule of multiplication (§ 3) it is said "The product of the numerators divided by the products of the denominators is multiplication." But how the product is obtained was not explained. On that account the author here adds a couplet to show the method of multiplication. CH.

[2] *Gó-sútricá;* a rope piqueted at both ends; with separate halters made fast to it for each ox or cow.

[3] *C'handa;* portions of the quantity as they stand; contrasted with *bhéda*, segments or divisions; being component parts, which, added together, make the whole; or aliquot parts, which, multiplied together, make the entire quantity.

[4] Example: Multiplicand two hundred and thirty-five. Multiplicator two hundred and eighty-eight.

The Multiplicator is repeated as often as there are portions in the multiplicator:

235	2
235	8
235	8

Multiplied by the portions of the multiplier in their order, there results 470 : which, added
1880
1880
together according to their places, make 67680.

Or the multiplicand is repeated as often as the parts 9, 8, 151, 120; and multiplied by them

235	9	2115
235	8	1880
235	151	35485
235	120	28200

The sum is the quantity resulting from multiplication, as before, 67680.

Or the parts of the multiplier are taken otherwise: as thus 9, 8, 4; the continued multiplication of which is equal to the multiplier 288. So with others. And the multiplicand is successively multiplied by those divisors, which taken into each other equal the multiplicator. Thus the multiplicand 235, multiplied by 9, makes 2115; which, again, taken into 8, gives 16920; and this, multiplied by 4, yields 67680.

This method by parts is taught by SCANDA-SÉNA and others. In like manner the other methods of multiplication, as *tat-st'ha* and *capáta-sand'hi*, taught by the same authors, may be inferred by the student's own ingenuity. CH.

56. If the multiplicator be too great or too small,[1] the multiplicand is to be multiplied by the excess or defect as put; and the product of the multiplicand by the quantity so put is added or subtracted.[2]

57. The quotient of a dividend by a divisor increased or diminished by an assumed quantity,[3] is reserved; and is multiplied by the assumed quantity, and divided by the original divisor; and the quotient of this division, added to, or subtracted from, the reserved quantity, is the correct quotient.[4]

58. The product of quotient and divisor,[5] being divided by the multiplicator, is the multiplicand; or divided by the multiplicand, is the multi-

ŚRÍDHARA's rule is as follows: 'Placing the multiplicand under the multiplying quantity in the order of the foldings *(capáta-sand'hi crama)*, multiply successively, in the direct or in the inverse order, repeating the multiplier each time. This method is termed *capáta-sand'hi.** The next is termed *tat-st'ha*, because the multiplier stands still therein *(tasmin tisht'hati)*. By division of the form or separation of the digits *(rúpa-st'hána-vibhága)* that named from parts *(c'han̄da)* becomes two-fold. These are four methods for the operation of multiplication *(pratyutpanna)*.'—*Gan.-sár.* § 15—17.

[1] When the quantity to be multiplied has by mistake been multiplied by a multiplicator too great or too small; to correct the error in such case, the author adds a couplet. CH.

[2] Example: Multiplicand 15; multiplicator 20. This multiplicand has, by mistake, been multiplied by four more, viz. by 24. The product is 360. Here the number put is 4; and multiplicand 15: their product 60. It is subtracted from the number as multiplied: and, with a reproof to the blundering calculator, he is told "the true product is 300."

Or the multiplicand has been multiplied by four less; viz. 16; and the product stated is 240. Here the product of the multiplicand and number put is 60; which is added, as the multiplication was short; and the correct result is 300. CH.

[3] When the dividend has been divided by a divisor increased or diminished by an assumed quantity; to correct the quotient. CH.

[4] Example: Dividend 300. Original divisor 20.

The division being made with that increased by four, viz. 24, the quotient was 12½. This is reserved, and is multiplied by the assumed number 4: product 50: whence, by the original divisor, the quotient is had 2½. This, added to the reserved quantity 12½, makes 15.

Or the same dividend 300, being divided by four less than the right divisor, viz. by 16, the quotient was 18¾. This multiplied by the assumed number 4, makes 75; which divided by the original divisor 20, yields 3¾: and this quotient, subtracted from the reserved quantity 18¾, leaves 15. CH.

[5] Of multiplicand, multiplicator, divisor and quotient, to find any one, the rest being known. CH.

* From *capáta*, a folding door, and *sand'hi*, junction.

plicator: the product of multiplicand and multiplier, divided by the divisor, is the quotient; or divided by the quotient, is the divisor.[1]

59. If two of the quantities, whether multiplicand and multiplicator, or divisor and quotient, be wanting,[2] [the given quantities are to be changed for the others, and arbitrary quantities to be put in their places.[3]][4]

60. Multiply the multiplicand or the multiplicator by the denominator of the divisor: and the divisor is to be multiplied by the denominator of the multiplicand, and by that of the multiplicator.[5]

61. Making unity denominator of an integer, let all the rest of the pro-

[1] Example: Divisor 20; multiplicand 32; multiplicator 5; quotient 8.

First to find the multiplicand. The product of divisor and quotient, 20 and 8, is 160: which, divided by multiplicator 5, gives 32.

Next for the multiplicator. The product of divisor and quotient is 160; which, divided by the multiplicand 32, yields 5.

Then for the quotient. The product of the multiplicand and multiplier, 32 and 5, is 160: which, divided by the divisor 20, affords 8.

Lastly, for the divisor. The product of the multiplicand and multiplier is 160: which, divided by the quotient 8, produces 20. CH.

[2] If a couple of the quantities be wanting [that is, unknown], to find them. CH.

[3] The text is deficient in the manuscript; but is here supplied from the commentator's gloss.

[4] Example: Divisor 20; multiplicand 32; multiplier 5; quotient 8.

The multiplicator and multiplicand being wanting; the divisor and quotient are 20 and 8. These are put for multiplicand and multiplicator. Their product is 160. Hence, putting four for the quotient, the divisor is found 40; or putting eight, it is 20: and so on arbitrarily. Or the arbitrary number may be the divisor; whence the quotient is to be deduced: and so on variously.

Or, the divisor and quotient being wanting, the multiplicand and multiplicator are 32 and 5. These are converted into divisor and quotient, or quotient and divisor. Their product is 160. Putting ten, an assumed quantity, either for the multiplicand or for the multiplicator; the other, namely multiplier or multiplicand, is deduced 16: or, putting five, the number deduced is 32. So, a hundred different ways. CH.

[5] To make the terms homogeneous in the rule of three.—CH. It it the same in effect with that before delivered and expounded. § 4. *Ib.*

cess be as above described.[1] The divisor and multiplicator, or divisor and multiplicand,[2] are to be abridged by a common measure.[3]

62. The integer, multiplied by the sexagesimal parts of the fraction belonging thereto, and divided by thirty, is *the square of the fractional portion*,[4] to be added to the square of the whole degrees.[5] A square and a cube are the products of two, and of three, like quantities multiplied together.[6]

63.[7] Twice the less portion[8] of a quantity [added to the greater[9]] being multiplied by the greater and added to the square of the less, is the entire square.[10] Or, an arbitrary number being added to, and subtracted from, the

[1] It has been so shown by us in preceding examples.—CH. See note on § 5.

[2] Never the multiplicand and multiplicator. CH.

[3] They are to be reduced to least terms by a common divisor, if the case comport it; to abbreviate the work.

Example: Divisor 20; multiplicand 40.

These, being abridged by the common measure twenty, become 1, 2.

So, divisor 20; multiplicator 4.

Reduced by the common measure four, they become 5, 1.

[4] *Vicala-varga*, square of the minutes; the multiple of the fraction to be added to the square of the integer, to complete the square of the compound quantity. See § 64.

[5] To find the square of a quantity, that includes minutes of a degree. CH.

The rule may be stated otherwise [and more generally]. The integer, multiplied by the numerator of its attendant fraction, which has a given denominator, being divided by [half] its denominator, is to be added to the square of the integer portion.—*Ibid.* This method gives the square grossly: being less than the truth by the product of the minutes by minutes, expressed in sexagesimal seconds. *Ib.*

Example: What is the square of fifteen degrees and a half? Statement: 15° 30′.

The integer 15, multiplied by the sexagesimal parts or minutes, 30, is 450: which, divided by thirty, gives 15, to be added to the square of the whole degrees, or 225; making in all 240.

So square of twelve and a twelfth part? Statement: 12° 05. Answer: 146.

[6] Definition of square and cube.—CH. The continued multiplication of four or more like quantities is termed *tadgata*, as the author afterwards notices in the chapter on Algebra *(cuttacád'hyáya)*. *Ib.*

[7] To find the square of a quantity. CH.

[8] Or the greater may be taken; or any two portions of the proposed quantity may be employed; or a greater number of portions. CH.

[9] The text is obscure, and the comment deficient: but either it must be thus supplied, or the sense must be 'the quantity added to its least portion': or else the square of the greater portion, as well as of the less, must be added after the multiplication.

[10] Example: Square of twenty-five.

Here five is the less portion, and twenty the greater. The less portion of the quantity doubled

quantity, the product of the sum and difference, added to the square of the assumed number, is the square required.[1]

64—65.[2] To the square of the given least quantity add the *square of the fractional portion*[3] of the other, and from it subtract the same:[4] the sum and difference are divided by twice the *other number*,[5] and in the second place by the same divisor together with the first quotient added and subtracted: the [last corrected] divisor with the same quotient [again] added and subtracted, being halved, is the root[6] of the sum and of the difference of squares. Or the *other number*, with the quotient added and subtracted, is so.[7]

[and added to the greater] is 30: which, being multiplied by the greater, makes 600. The square of the less 25. Their sum is 625, the square of twenty-five.

Or the greater portion [added to the quantity] is 45: which, multiplied by the less is 225. Added to the square of the greater, viz. 400, the sum is 625.

Or one portion of the quantity 20 [doubled and] multiplied by the second, makes 200: and this, added to the squares of the portions, 400 and 25, gives 625.

Or one portion of the quantity 5, doubled, and multiplied by the other, makes 200; and this, added to the squares of the portions, produces 625.

Or let there be three portions of twenty-five: as 5, 7 and 13. One portion of the quantity, 5, doubled, is 10: which, multiplied by the second 7, makes 70: and added to the squares of the portions, viz. 25 and 49, produces 144. Its root is 12: with which and with thirteen the operation proceeds. CH.

[1] Example 25.

Adding and subtracting the arbitrarily assumed number five, it becomes 30 and 20. The product of these is 600: which, added to the square of the assumed number 5, viz. 25, makes 625. CH.

[2] To find a quantity such that its square shall be equal to the sum of the squares, or to the difference of the squares, of two quantities, of which the greater does not exceed the *square of the fractional portion*, nor the square of the less number. CH.

[3] Square of the sexagesimal minutes; that is, the multiple of the fraction. See § 62.

[4] The rule serves for finding both quatities at once; the additions being every where adapted to bring out the root of the sum of the squares; and the subtractions, to give the root of the difference of the squares.

[5] *Itara*, the other; other than the least; that is, the greater number.

[6] Approximately.

[7] Example for the sum: Let the greater number, termed the *other* quantity, together with its minutes, be 15° 40′; and the least be 14. The square of the latter is 196. The *square of the fractional portion* is 20. Added they make 216; which, divided by twice the *other* quantity 15, viz. 30, gives in the first place 6; and this, added to the divisor 30, makes 36; by which, in the second place, the correct quotient comes out 6. This again is added to the correct divisor; and the sum is 42: which halved yields 21, the number sought. For the square of the number thus found is

66. This is a portion only of the subject.[1] The rest will be delivered under the construction of sines,[2] and under the pulverizer.[3] [End of] chapter twelfth [comprising] sixty-six couplets on addition, &c.

441 : from which subtracting the square of the least number 196, the remainder is 245, the square* of the greater number 15° 40′. Subtracting this square, the remainder is nought.

Or, adding the quotient 6 to the *other* quantity 15, the sum 21 is a number equal to the square-root of the sum of the squares.

Example of the difference: The *other* quantity or greater number is 12° 50′. The least 10. The square of this, 100. The *square of the fractional portion* is 20; which, subtracted, leaves 80. This, divided by the *other* quantity doubled, 24, yields in the first place 4; which, subtracted from the [first] divisor, leaves 20. The corrected quotient 4, subtracted from the corrected divisor 20, affords the remainder 16, the half of which 8 is equal to the difference of squares.

Or, subtracting the quotient 4 from the *other* quantity 12, the residue 8 is a number equal to the square root of the difference of squares. Thus, its square is 64; and so much is the difference between the squares of the greater and least quantities 164† and 100. CH.

[1] A portion only has been here shewn; and a portion only has been by us expounded. Else a hundred volumes would be requisite under a single head. But we have undertaken to interpret the whole astronomical course *(sidd'hánta)*. Wherefore prolixity is to be shunned. CH.

[2] *Jyótpatti (jyá-utpatti)* derivation of [semi-]chords: taught in the chapter on Spherics, and to be there expounded (C. 21, § 15—21). CH.

[3] In the Chapter on the Pulverizer *(cuttacád'hyáya)* the author will treat the undermentioned topics with other heads of computation: viz. Investigation of the pulverizer *(cuttaca)*. Algorithm of symbols or colours *(varńa)*; of affirmative and negative quantities *d'hanarńa)*; of surd roots *(carańí)*. Concurrence *(sancramańa)*. Dissimilar operation *(vishama-carman)*.‡ Equation of the unknown *(avyacta-sámya)*. Equation of several unknown letters or colours *(varńa-sámya)*. Elimination of the middle term *(mad'hyamá'harańa)*. Equation involving products of unknown quantities *(bhávica)*. Affected square *(varga-pracríti)*, &c. CH.

* Nearly so. The exact square is 245⅘; or in sexagesimals 245° 26′ 40″.

† The exact square is 164 $\frac{25}{36}$; in sexagesimals 164° 41′ 40″.

‡ See Ch. 18, § 25 and *Líl*, 55—57.

CUTTACAD'HYAYA, ON ALGEBRA;

THE EIGHTEENTH CHAPTER OF THE

BRAHME-SPHUṬA-SIDD'HÁNTA,

BY BRAHMEGUPTA:

WITH NOTES SELECTED FROM THE COMMENTARY.

CHAPTER XVIII.

ALGEBRA.

SECTION I.

1. SINCE questions can scarcely be solved without the pulverizer,[1] therefore I will propound the investigation of it together with problems.

2. By the pulverizer, cipher, negative and affirmative quantities, unknown quantity, elimination of the middle term, colours [or symbols] and factum, well understood, a man becomes a teacher among the learned, and by the affected square.

3—6. Rule for investigation of the pulverizer: The divisor, which yields the greatest remainder, is divided by that which yields the least: the residue is reciprocally divided; and the quotients are severally set down one under the other. The residue [of the reciprocal division] is multiplied by an assumed number such, that the product having added to it the difference of the remainders may be exactly divisible [by the residue's divisor]. That

[1] *Cuṭṭácára, cuṭṭa, cuṭṭaca,* pulverizer. See *Lil.* § 248 and *Vij-gaṇ.* § 53.

multiplier is to be set down [underneath] and the quotient last. The penultimate is taken into the term next above it; and the product, added to the ultimate term, is the *agránta*.[1] This is divided by the divisor yielding least remainder; and the residue, multiplied by the divisor yielding greatest remainder and added to the greater remainder, is a remainder of [division by] the product of the divisors. A twofold *yuga* is a product of divisors:[2] and the elapsed portion of the *yuga* is the remainder of the two. Thus may be found the lapsed part of a *yuga* of three or more planets by the method of the pulverizer.

7. Question 1. He, who finds the cycle *(yuga)* and so forth, for two, three, four or more planets, from the respective elapsed cycles of the several planets given, knows the method of the pulverizer.

Here, for facility's sake, the revolutions, &c. of the sun and the rest are put, as follows: the sun 30; the moon 400; Mars 16; Mercury 130; Jupiter 3; Venus 50; Saturn 1; moon's apogee 4; moon's node 2; revolutions of stars 10990; solar months 360; lunar months 370; more months (lunar than solar) 10; solar days 10800; lunar days 11100; fewer days (terrestrial than lunar) 140; terrestrial days 10960.

The days of the planetary cycles of the sun and the rest are [sun] 1096; moon 137; Mars 685; Mercury and Venus 1096; Jupiter 10960; Saturn 10960; apogee 2740; node 5480.[3]

Example (a popular one is here proposed): What number, divided by six, has a remnant of five; and divided by five, a residue of four; and by four, a remainder of three; and by three, one of two?

Statement: 5 4 3 2 [Answer 59.][4]
6 5 4 3

[1] *Agránta.* The proper import of the term, as it is here used, is unexplained.

[2] This is introduced in contemplation of instances relative to planets: and so is what follows. COM.

[3]

	☉	☽	♂	☿	♃	♀	♄	☽'s Apogee.	☽'s Node.
Periodical revolutions in least terms:	3	5	1	13	3	5	1	1	1
Divisors, terrestrial days in least terms:	1096	137	685	1096	10960	1096	10960	2740	5480

[4] The divisor which yields the greater remainder, namely 6, being divided by that which yields

8. Rule for deducing elapsed time from residue of revolutions, &c. § 7. Let residue of revolutions or the like, divided by the divisor, be a remainder;

the less, viz. 5, the residue is $\frac{1}{5}$. Then, reciprocal division taking place, the quantity beneath is in the first instance to be divided by that which stands above it: and thus the quotient is 5 and the residue $\frac{0}{1}$. This is multiplied by a quantity so assumed as that the product having the difference of the remainders (namely 1) added to it, may be exactly divisible by the [residue's] own divisor 1. The quotient being solitary, the difference of remainders is in this case to be subtracted [§ 13]. The number so assumed is put 1. By that the residue 0 being multiplied, is 0; which, having subtracted from it the difference of remainders 1, makes $\dot{1}$; and this divided by the [residue's] own divisor, namely 1, yields for quotient negative unity. Statement of the first quotient and the multiplier and present quotient

5
1
$\dot{1}$

By the penultimate 1 multiplying the term next above it 5, the product is 5; which added to the ultimate $\dot{1}$, makes 4. The *agránta* thus comes out 4. Divided by the divisor yielding least remainder, viz. 5, the residue is 4: which, multiplied by the divisor yielding greatest remainder 6, produces 24; and this, added to the greater remainder 5, affords the remainder, 29, of the product of the divisors: that is to say, so much, namely 29, is the remainder of the number in question (which divided by six has a remnant of five, and divided by five a residue of four,) divided by a divisor equal to the product of the divisors, viz. 30.

Again, statement of the foregoing result with the third term

29 3
30 4

Here the divisor yielding the greater remainder, 30, being divided by that which yields the less, viz. 4, the residue is $\frac{2}{4}$. Then by reciprocal division the quotient is 2 and the residue $\frac{0}{2}$. This, multiplied by an assumed multiplier seven, produces 0; which, having subtracted the difference of remainders 26, makes $\dot{26}$; and divided by the [residue's] own divisor 2, the quotient is $\dot{13}$. Statement of the former quotient, the multiplier and the [present] quotient

2
7
$\dot{13}$

Proceeding by the rule (the penult taken into the term next above it, &c. § 5), the *agránta* comes out 1. This being divided by the divisor yielding least remainder, the residue is 1; which, multiplied by the divisor yielding greatest remainder 30, is 30, and added to the greater remainder 29, makes 59, the remainder answering to the product of the divisors, viz. 60.

Wherever abridgment of the divisors [by a common measure] is practicable, the product of divisors must be understood as equal to the product of the divisor yielding greatest remainder and quotient of the divisor yielding least, abridged [*i. e.* divided] by the common measure: and when one divisor is exactly divisible by the other, the greater remainder is the remainder required, and the divisor yielding greatest remainder is taken for product of divisors. This is to be elucidated by the intelligent mathematician, by assumption of several colours (or symbols).

Again, this number 59, of itself answers to the condition that divided by three, it shall have a residue of two.

Example of Question 1. Elapsed part of the cycles of the sun, &c. together with the divisors, as follows:

Sun.	Moon.	Mars.	Mercury.	Jupiter.	Venus.	Saturn.	Apogee ☽.	Node ☽.
1000	41	315	1000	1000	1000	1000	1000	1000
1096	137	685	1096	10960	1096	10960	2740	5480

Here the divisor yielding greatest remainder, 1096, is exactly measured by that which yields least 137: wherefore the remainder is the same, and the same [divisor] is taken for the product of divisors; 1000
1096

The sequel of the rule "A two-fold *yuga* is a product of divisors" (§ 6) is next expounded: so many days, as suffice for the commencement of exceeding months and deficient days, and the termination of the sun and moon's revolutions, to take place again on the first of *Chaitra*, light fortnight, at sunrise at *Lancá*, are days of a two-fold cycle; and this is what is termed a two-fold *yuga*. The remainder, as found, is the elapsed portion of a two-fold *yuga*. In like manner are to be understood three-fold cycles and so forth.

Again, statement of the same with the residue and divisor of Mars: 1000 315
1096 685
Here the divisor yielding greatest remainder, 1096, is divided by the divisor yielding least, namely 685; and the residue is $\frac{411}{685}$. Then the quotients resulting from reciprocal division are put one under the other
1 and the residue is $\frac{0}{137}$; which, multiplied by an arbitrary multiplier three,* makes 0; and this
1
2
lessened by the subtraction of the difference of remainders, viz. 685, and divided by its own divisor 137, yields the quotient 5. The multiplier and quotient, thus found, are put below the former quotients, one under the other: and that being done, a series is obtained 1
1
2
3
5
Proceeding as before, the *agránta* comes out 5. This being divided by the divisor yielding least remainder, 685, the residue which results is 5; which, multiplied by the divisor yielding greatest remainder and having the greater remainder added, brings out the remainder 6480. It is the elapsed portion of a three-fold *yuga*. The divisor yielding greatest remainder 1096, being multiplied by 5 the quotient of the divisor yielding least remainder abridged by the common measure 137, produces the three-fold *yuga*, 5480. But it is not fit, that the elapsed portion of a cycle should exceed the cycle: it is therefore abridged by the *yuga*; and the residue must be considered as the elapsed portion of a *yuga*. This being done, there results 1000
5480
Next statement of the same with the elapsed portion of Mercury's *yuga*: 1000 1000
5480 1096
Here either divisor at choice may be taken as the one yielding greatest remainder. Put 5480. The elapsed portion of the four-fold *yuga* is 1000. In like manner, by the operation of the pulverizer with the respective elapsed portions of *yugas* of Jupiter, Venus and Saturn, the elapsed portions of cycles come out [Jupiter] 1000, Venus 1000 and Saturn 1000; and the measure of the cycles as follows: viz five-fold *yuga* 10960; six-fold *yuga* 10960; seven-fold *yuga* 10960. In like manner the process of the pulverizer being observed with the elapsed periods of the *yugas* of the moon's apogee and node, the elapsed portion of the entire cycle for all the planets comes out 1000, and the value of such entire *yuga* 10960.

* Sic sed quære.

as also cipher divided by residue arising for one day.[1] The remainder deduced from these, being divided by residue of revolutions or the like as arising for one day,[1] is the number of [elapsed] days.

9. Question 2. He, who deduces the number of [elapsed] days from the residue of revolutions, signs, degrees, minutes, or seconds declared at choice, is acquainted with the method of the pulverizer.

Example: When the remainder of solar revolutions is eight thousand and eighty, tell the elapsed portion of the *calpas*, if thou have skill in the pulverizer.

Statement: Residue of revolutions 8080.[2] [Answer: 1000.]

The foregoing rule (§ 3) for dividing the divisor which yields the greater remainder by that which yields the less, is unrestrictive; and the process may therefore be conducted likewise by dividing the divisor which yields the less remainder by that which yields the greater.

Example: What number, divided by seventy-three, has a remnant of eight; and divided by thirteen, a remainder of three?

Statement: $\begin{matrix} 8 & 3 \\ 73 & 13 \end{matrix}$ Dividing the divisor which yields the less remainder by that which affords the greater, and the residue being reciprocally divided, the quotients are $\begin{matrix} 5 \\ 1 \\ 1 \\ 1 \\ 1 \end{matrix}$ The residue $\frac{1}{2}$.

Assumed multiplier 1. Difference of remainders 5. Here, since the process was inverted, the difference of remainders is made negative, $\dot{5}$; and, as the quotients are uneven, it again becomes affirmative, 5: consequently it is additive. Proceeding as before, the *agránta* comes out 79. This is divided by the divisor yielding the greater remainder; and the residue 6, multiplied by the divisor yielding the less remainder, 13, makes 78; and, added to the less remainder 3, brings out the quantity sought 81.

[1] By the daily increment of it.

[2] This divided by terrestrial days, and both solar revolutions and terrestrial days abridged by the common measure 10, must be put for a remainder, $\frac{808}{1096}$. Then cipher divided by residue of solar revolutions arising on one day, namely 3, is put for the [other] remainder, $\frac{0}{3}$. Proceeding by the rule (§ 3) the series is $\begin{matrix} 3^{*} \\ 270 \\ 808 \end{matrix}$ Whence, by the subsequent rule (§ 5), the remainder comes out 3000; and this, divided by residue of revolutions arising for one day, 3, gives the number of [elapsed] days 1000. In like manner, from the residue of signs and so forth, the number of [elapsed days] is to be found.

* Sic MS.

10. Rule for finding elapsed time from given residue for hours: § 8. From the result, which is derived from residues of revolutions or the like for one day, and for the proposed hours or minutes, both reduced to like denominators,[1] the number of [elapsed] days and so forth may be deduced.

Example: To what number of [elapsed] days does that amount of hours correspond, for which the residue of lunar revolutions arising is four thousand one hundred and five?

Statement: Residue of revolutions, with divisor, $\frac{4105}{8220}$. Residue as arising for a single day, with divisor, $\frac{5}{137}$. The divisor is multiplied by sixty; and thus both terms have like denominators. Cipher divided by the residue of revolutions for one day must be put as a remainder, $\frac{0}{5}$. Thus the statement is $\frac{4105}{8220}$ $\frac{0}{5}$. Statement of the same abridged by five, $\frac{821}{1644}$ $\frac{0}{1}$. Proceeding as before, the remainder is 821. This, divided by the remainder [of revolutions] for an hour, namely 1, gives the elapsed time in hours, 821; which, divided by sixty, yields quotient, days 13, hours 41. Or, with an additive in hours equal to the same divisor 1644, the elapsed time in hours is 2465 or 4109.

11—13. Rules for a constant pulverizer: § 9—11.[2] The multiplier and divisor being mutually divided, these quantities divided by the residue are [in least terms, being] irreducible by any [further common] divisor.[3] The quotients of these reciprocally divided are to be set down one under the other. The residue is multiplied by a multiplier chosen such that the product less one[4] may be exactly divisible. That multiplier is to be set down; and the

[1] That is, when the proposed residue of revolutions is calculated for elapsed time reduced to hours and minutes, then the residue of revolutions, &c. for one day must have its divisor multiplied by sixty or by three thousand six hundred; and thus the denominators are alike. COM.

[2] *St'hira-cuttaca ; drĭd'ha-cuttaca;* the steady residue, by which the given remnant of revolutions or the like is to be multiplied; and the product being divided by the divisor, the quotient is elapsed time.—COM. From *st'hira*, steady; and *drĭd'ha*, firm. *Drĭd'ha*, which the commentator makes equivalent to *st'hira* in the compound term designating this multiplier, is by BHÁSCARA employed in the sense for which BRAHMEGUPTA employs *nich'héda*, &c. See *Lil.* § 248.

[3] *Nich'héda, nirapavarta;* having no divisor; no further common measure: reduced to least terms. See *Lil.* § 248.

[4] It is so, if the quotients be even: but, if they be odd, one must be added instead of subtracted. COM. See § 13.

quotient at the end: from which the *agránta*, being found by multiplying the next superior term by the penultimate and adding the ultimate to the product, is divided by the divisor in least terms. The residue of this division is the constant pulverizer.[1]

14. Question 3. To deduce the number of days from the residue of revolutions, &c. of the sun, and the rest; tell a constant pulverizer, thou skilful mathematician who hast traversed the ocean of the pulverizer.

Here to find a constant pulverizer from a residue of revolutions of the sun, the statement of revolutions and terrestrial days is 30 / 10960 These are multiplier and divisor. Statement of them abridged by ten: 3 / 1096 The quotient of these mutually divided is 365, and remainder $\frac{1}{3}$. Multiplied by an assumed multiplier, namely 2, the product is 2; to which one is added, since the quotient is an odd number [§ 13]; and the sum divided by the divisor gives the quotient 1; and the multiplier and quotient being set under the former quotient, the series is 365 / 2 / 1 Proceeding by the rule (§ 5) the *agránta* is deduced 731: from which divided by the divisor in least terms, the residue or constant pulverizer is 731 for a residue of revolutions.

[1] When a constant pulverizer is sought, to deduce elapsed time from remainder of revolutions, then revolutions of the planet are multiplier, and terrestrial days divisor. When it is so to deduce the time from remainder of signs; twelve times the revolutions are multiplier, and terrestrial days divisor. When it is investigated to conclude the time from residue of degrees; three hundred and sixty times the revolutions are multiplier, and terrestrial days divisor. When it is so to conclude the time from residue of minutes, &c. sixty times the foregoing multiple of revolutions are multiplier, and terrestrial days everywhere divisor. The multiplier and divisor, which are thus put to find the constant pulverizer, must be reciprocally divided, and by the residue remaining the same multiplier and divisor being divided are irreducible [or in least terms]; that is, they can be no further abridged by a common measure. The same irreducible multiplier and divisor are again mutually divided until the residue in the dividend be unity. Set down the quotients one under the other. Multiply the residual unity by a multiplier taken such that the product less one, (or, if the quotients be odd, having one added) may be exactly divisible by the residue's own divisor. The multiplier and the quotient of this operation are to be set down, in order, under the former quotients. Then the *agránta* is to be computed from the bottom: by taking the penultimate into the next superior term and adding the ultimate. The *agránta* so found is divided by the irreducible divisor, and the residue is the constant pulverizer. COM.

Then the multiplier for a residue of signs is 36; and the abridged divisor 1096. From which, as before, the constant pulverizer comes out 61.

Multiplier for a residue of degrees 1080. Divisor 1096. From these, as before, the constant pulverizer is found 68.

For residue of minutes, pulverizer 129. For residue of seconds, pulverizer 9.

In like manner, the constant pulverizer for the moon and the rest must be understood, as that for the sun.

15. Rule for finding elapsed time by constant pulverizers: § 12. The given residue of revolutions, or the like, being multiplied by its pulverizer and divided by its divisor, the residue which arises is the number of [past] days; there being added a multiple of the divisor by elapsed [periods] in least terms.[1]

Example: Thou who hast traversed the ocean of the pulverizer! tell the number of elapsed days, when the remainder of degrees is four thousand and four hundred.

Statement: This remainder of degrees 4400, being abridged by the common divisor 80, as before in the investigation of the constant pulverizer, is reduced to 55: which, multiplied by the constant pulverizer 68, becomes 3740. From this divided by the divisor reduced to least terms 137, the residue which is deduced is the number of [elapsed] days 41. To find the elapsed time intended by the question, this must have added to it a multiple of the divisor by the periods gone by. In this case they are [supposed] seven, and the divisor multiplied by that, 959, being added to the number as above found 41, the number of [elapsed] days comes out 1000.

16. Rule special: § 13. So when the quotients are even. But if they be odd, what is propounded as negative, becomes affirmative; or as positive, be-

[1] *Gata-nirapavarta:* the quotient, which is obtained when the elapsed time from the beginning of the *yuga* is divided by the divisor reduced to least terms, is thus denominated. The divisor, multiplied by that, being added to the elapsed time found by the rule, the sum is the elapsed portion of the *yuga*. COM.

comes negative:[1] and the signs, negative or affirmative, of multiplicand and additive, must be reversed.[2]

17. Rule of inverse operation: § 14. Multiplier must be made divisor; and divisor, multiplier; positive, negative; and negative, positive: root [is to be put] for square; and square, for root: and first as converse for last.

18. Question 4. The residue of degrees of the sun less three, being divided by seven, and the square-root of the quotient extracted, and the root less eight being multiplied by nine, and to the product one being added, the amount is a hundred. When does this take place on a Wednesday?

Statement: 3̇—Div. 7—Root—8̇—Mult. 9—Add. 1—Giv. 100. The affirmative unity being made negative, when applied to a hundred, the result is 99. Nine, which was multiplier, becomes divisor. Dividing by that, the quotient is 11. Negative eight becomes affirmative: whence 19. The extraction of the root is converted into the raising of the square 361. The divisor seven becomes multiplier. Product 2527. The negative three becomes affirmative, and is added, 2530. This is residue of degrees; from which, the number of [elapsed] days is to be sought; until, with addition of the divisor, it come to Wednesday.

19. Question 5. He, who tells when a given residue of revolutions of the sun occurs on a Monday, or on a Thursday, or on a Wednesday, has knowledge of the pulverizer.

20. Question 6. A person, who can say when a residue of degrees or of seconds, which occurs on a Wednesday, will do so on a Monday, is conversant with the pulverizer.

21. Question 7. One, who tells when given positions of the planets,

[1] See preceding instances of the application of this first part of the rule, under Example 1st, or under Problem 3 and Rule 10.

[2] If the multiplicand were negative, it must be made positive; and the additive must be made negative: and then the pulverizer is to be sought. COM.

which occur on certain lunar days, or on days of other denomination of measure,[1] will recur on a given day of the week, is versed in the pulverizer.

22. Rule 15: The number of [elapsed] days, deduced from the given residue of revolutions or the like by means of the pulverizer, receives an addition of days of a period in least terms, repeatedly, until the intended day of the week be reached.

23. Question 8. He, who tells the number of [elapsed] days, seeing the degrees, &c. of a given [planet's] mean [place], or does so from a conjunction of two or more planets, or from their difference, is conversant with the pulverizer.

24—25. Rule 16—17: The divisor in least terms, being multiplied by the minutes, &c. in the [given] signs, &c. and divided by the minutes in a revolution, the quotient is the residue of revolutions: whence the number of [elapsed] days [may be deduced]. In like manner residues of signs, degrees, minutes and seconds, are found, and the number of [elapsed] days as before. Putting arbitrary numbers in places deficient, proceed with the rest of the process as directed.[2]

Example: Seeing past signs, degrees and minutes of Jupiter, nought, twenty-two and thirty, a person, who tells the number of [elapsed] days at that instant, is one conversant with the pulverizer.

Statement: 0
22
30 Its minutes 1350, multiplied by Jupiter's divisor in least terms 10960, and divided by the minutes of a circle; the quotient is the

[1] As solar, or siderial, &c.—COM.

[2] The place of a planet in signs, &c. being reduced to degrees, minutes, and so forth, the number of minutes is multiplied by the particular divisor in least terms, and divided by the minutes of a circle; and the quotient is the remainder of revolutions. If it be the number of degrees, &c. the quotient is then remainder of signs. If it be so of minutes, &c. the quotient is remainder of degrees. From these residues, the number of [elapsed] days is found as before. There is this difference: when (degrees being divided by the minutes of a circle) any residue arises, it is to be rejected: the quotient is taken.

residue of revolutions 685. Whence the number of [elapsed] days, as before, comes out 7535.

Example: When Jupiter and the lunar node are conjunct having passed signs, degrees and minutes, three, twenty-two and thirty; tell me the number of [elapsed] days.

Conjunction of Jupiter and the node 3 22 30 Reduced to minutes, multiplied by terrestrial days, and divided by minutes of a circle, the quotient is the sum of residues of revolutions 3425. Whence, as before, the number of [elapsed] days is deduced 685.

Example: Tell me the number of days elapsed on a day when the body of the sun, less the conjunction of Jupiter and the lunar node, is just so much.

Statement: 3 22 30. Hence, as before, the residue of degrees of the sun, less the residue of revolutions of Jupiter and the lunar node, is found 3425. Whence the number of [elapsed] days, by the rule § 7, comes out 137.

Example:[1] Signs and degrees of Jupiter have been effaced by the boy with his finger. Thirty minutes are seen: from which tell me, astrologer, the signs, degrees, and number of days, if thou have practice of the pulverizer.

Statement: 0 0 30 Here put unity in the place of signs; and in that of degrees, ten. See: 1 10 10 Hence the residue of revolutions is deduced 1233: from which the number of days, as before, comes out 411.

26. Question 9. From the residue of signs, degrees, minutes or seconds, told, or if lost assumed, he who finds the superior and intermediate terms, is a person conversant with the pulverizer.

[1] A stanza and a half.

27. Rule 18. The multiplier, by which the divisor being multiplied, and having the residue added to the product, becomes exactly divisible, is the [portion of orbit] past. The quotient is the residue. In like manner from the residue, [the place of] the planet and the number of [elapsed] days are [deduced].

Example: From the residue of seconds of the moon being eight hundred, tell the [place of the] moon and number of [elapsed] days, my friend who hast traversed the ocean of the pulverizer.

Statement: Residue of seconds 800. This, abridged by the common measure eighty, becomes 10. Making this additive; divisor in least terms, 137, dividend; and the multiplier, which serves to bring out seconds, namely sixty, the divisor; the statement is Dividend 137 Divisor 60 Additive 10. Here, by reciprocal division the the quotients are 0 2 3 1 1 Residue $\frac{1}{8}$. This is multiplied by a multiplier assumed such, that the product with one added may be exactly divisible, since the residue of seconds is additive: but, as the quotients are here uneven, one must be subtracted, [§ 13]. Such an assumed multiplier is nine; and the quotient 10. Hence, as before, the constant pulverizer is deduced. This being multiplied by the residue of seconds, namely 10, and divided by its divisor, viz. 60, the residue is the multiplier 10. So many are the seconds. The dividend being multiplied by the multiplier, and having the additive added, and being divided by sixty, the quotient is 23. This is residue of minutes. Again, make this additive, days in least terms dividend, and sixty divisor: See Dividend 137 Divisor 60 Additive 23. Hence, proceeding as before, the multiplier is 41. So many are the minutes past. The dividend being multiplied by the multiplier 41, having the additive added, and being divided by sixty, the quotient is residue of degrees, 94. Again, this is put additive, days in least terms dividend, and thirty divisor. See Dividend 137 Divisor 30 Additive 94. Hence, as before, the multiplier comes out 28. So many are the degrees. The dividend being multiplied by the multiplier, having the additive added, and being divided by thirty, the quotient is the residue of signs 131. Make this again additive, days in least terms dividend, and twelve divisor: [Dividend 137 Divisor 12 Additive 131.] The multiplier is found 5. So many are the signs past. The dividend being multiplied by the multiplier, and having the additive 131 added, and being divided by twelve, the quotient is the residue of revolutions, 68. Put days in least terms for dividend, revolutions in least terms for divisor, and residue of revolutions for additive. See: Dividend 137 Divisor 5 Additive 68. Hence, as before, the multiplier comes out 1. This is the [number of] revolutions past. The quotient is the number of [elapsed] days, 41.

From the same residue put as an assumed one, the number of [elapsed] days, in like manner, comes out 41.

Here an arbitrary multiple of 137 is additive.

28. Question 10. He, who knows the elapsed [portion of a] *yuga* from the residue of *exceeding* months told, or assumed, or from the residue of *fewer* days, or from the sum of them, is a person versed in the pulverizer.

Example: When the residue of *exceeding* months is eight hundred and eighty; and that of *fewer* days seven thousand seven hundred and twenty, and the sum of these sixteen thousand two hundred; tell, from any one of these, the elapsed [portion of the] *yuga*.

Residue of *exceeding* months 8480. Residue of *deficient* days 7720. Sum 16200.

The remainder, which is found by the rule § 7, [being divided by] residue of *exceeding* months arising for one day,[1] is the elapsed solar days of the *yuga*. Proceeding in this manner, they come out 848.

Or else let *exceeding* months be the multiplier, and solar days be the divisor, and the constant pulverizer be found by the rule § 9. Residue of *exceeding* months is to be multiplied by that. Then divide the product by the particular divisor. The residue is solar days.

In like manner, from the residue of *deficient* days, the elapsed lunar days of the *yuga* are found 293.

[1] Daily increment of the difference between lunar and solar months.

From the sum of the residues of *more* months and *fewer* days, as arising for one day,[1] and from the sum of the residues of *more* months and *fewer* days, as proposed, and reduced to lunar days, proceeding by the rule § 12, the lunar days are to be found 108.

29. Question 11. When does the square-root of three less than residue of *exceeding* months, being increased by two, and then divided and lessened by two, and squared, and augmented by nine, amount to ninety?

30. Question 12. When does the square of *deficient* days, being lessened by one, and divided by twenty, and augmented by two, and multiplied by eight, and divided by ten, and increased by two, amount to eighteen?

Here proceeding by the rule of inverse process as before taught, the residues of *more* months and *fewer* days come out 4099 and 19.

[1] Daily increment of the difference between lunar and terrestrial days.

SECTION II.

ALGORITHM.

31. Rule for addition of affirmative and negative quantities and cipher[1] § 19. The sum of two affirmative quantities is affirmative; of two negative is negative; of an affirmative and a negative is their difference; or, if they be equal, nought. The sum of cipher and negative is negative; of affirmative and nought is positive; of two ciphers is cipher.

32—33. Rule for subtraction: § 20—21. The less is to be taken from the greater, positive from positive; negative from negative. When the greater, however, is subtracted from the less, the difference is reversed. Negative, taken from cipher, becomes positive; and affirmative, becomes negative. Negative, less cipher, is negative; positive, is positive; cipher, nought. When affirmative is to be subtracted from negative, and negative from affirmative, they must be thrown together.

34. Rule for multiplication: § 22. The product of a negative quantity and an affirmative is negative; of two negative, is positive; of two affirmative, is affirmative. The product of cipher and negative, or of cipher and affirmative, is nought; of two ciphers, is cipher.

35—36. Rule for division: § 23—24. Positive, divided by positive, or negative by negative, is affirmative. Cipher, divided by cipher, is nought. Positive, divided by negative, is negative. Negative, divided by affirmative,

[1] *Shat-trinśat-paricarman.* Thirty-six operations or modes of process. See Arithm. § 1. *Víja-gaṅ.* § 3.

is negative. Positive, or negative, divided by cipher, is a fraction with that for denominator:[1] or cipher divided by negative or affirmative.[2]

[36 Concluded.] Rule for involution and evolution: § 24. The square of negative or affirmative is positive; of cipher, is cipher. The root of a square is such as was that from which it was [raised].[3]

37. Rule of concurrence and dissimilar operation: § 25. The sum, with difference added and subtracted, being divided by two, is concurrence. The difference of squares divided by [simple] difference, having difference added and subtracted and being then divided by two, is dissimilar operation.[4]

38. Rule for the construction of a rectangular figure with rational sides: § 26. Be a surd the perpendicular. Its square, divided by an assumed number, has the arbitrary quantity added and subtracted. The least is the base: and half the greater number is the flank.[5] Those [surds[6]], the product whereof is a square, are to be abridged.

39. Rule for addition and subtraction of surds: § 27. The surds being divided by a quantity assumed, and the square-roots of the quotients being extracted, the square of the sum of the roots, being divided by the assumed quantity, [is the sum,] or the square of their difference, [so divided, is the difference of the surds].

[1] *Tach-ch'héda*, having that for denominator: having, in this instance, cipher for denominator, to a finite quantity for numerator. See *Víj.-gan.* § 16.

[2] Is in like manner expressed by a fraction having a finite denominator to a cipher for numerator.

[3] The root is to be taken either negative or affirmative, as best answers for the further operations. COM.

[4] *Vishama-carman.* See *Lîl.* § 57.

[5] Let the perpendicular be put an irrational number 8; and let the assumed number be 4. Hence the figure is constructed

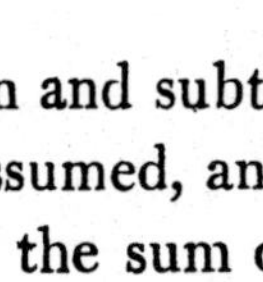

10 10
8
12

[6] Surd is understood from the preceding sentence. Those irrationals are to be abridged, the product of pairs of which is a square. COM.

Example: Tell the sum and difference of surds two and eight, and three and twenty-seven, respectively.

Statement: *c* 2 *c* 8. These surds, divided by an assumed number 2, give 1 and 4; the roots whereof 1 and 2. The squares of their sum and difference are 9 and 1; which, multiplied by the assumed number, become 18 and 2, the sum and difference of the surds.

Statement of the second Example: *c* 3 *c* 27. Proceeding as above, the sum and difference are found 48 and 12.

[39 Concluded.] Rule of multiplication: § 27. The multiplicand is put level with the [terms of the] multiplicator [placed] across, one under the other; and their products are added together.

Example:[1] The multiplicator comprises the surds two, three and eight; and the multiplicand, three with the rational number five. Tell the product quickly. Or let the multiplicator consist of the surds three and twelve less the rational number five.

Statement: Multiplicator *c* 2 *c* 3 *c* 8. Multiplicand *c* 3 *ru* 5.

Here multiplicand is placed level with the terms of the multiplicator across, one under the other:

Multiplier.	Multiplicand.	Product.
c 2	*c* 3 *ru* 5	*c* 6 *c* 50
c 3	*c* 3 *ru* 5	*c* 9 *c* 75
c 8	*c* 3 *ru* 5	*c* 24 *c* 200

Summing the products as directed by the rule, the answer comes out *ru* 3 *c* 450 *c* 75 *c* 54.

Statement of the second Example: Multiplicand *c* 3 *ru* 5. Multiplier *c* 3 *c* 12 *ru* 5.

Proceeding as before, the result of multiplication is *ru* 16 *c* 300.

40. Rule of division of surds: § 28. The dividend and divisor are mul-

[1] End of one couplet and beginning of another.

tiplied by the divisor with a selected [term] made negative;[1] and are severally summed: more than once [if occasion there be]. The dividend is then divided by the divisor reduced to a single term.

Statement of the foregoing result of multiplication as dividend, and its multiplier as divisor, for division: *ru* 3 *c* 450 *c* 75 *c* 54.
c 18* *c* 3

Put *c* 18 *c* 3̇. The dividend and divisor, multiplied by this, make *ru* 75 *c* 625. The dividend being then divided by the single surd consti-
ru 15
tuting the divisor, the quotient is *ru* 5 *c* 3.

Statement of the second Example: *ru* 1̇6 *c* 300 Here putting the surd
c 27 *c* 25
twenty-seven negative, and proceeding as before, the answer comes out *ru* 5 *c* 3.

[40 Concluded.] Rule of involution: § 28. A square is the product of two like quantities.

Example: Tell the square of the surds six, five, two and three.

Statement: *c* 6 *c* 5 *c* 2 *c* 3. Answer: *ru* 16 *c* 120 *c* 72 *c* 60 *c* 48 *c* 40 *c* 24.

41. Rule of evolution: § 29. From the square of the absolute number take surds[3] selected at choice. The square-root of the difference being added to and subtracted from the absolute number, the moieties are treated, the first as an absolute number, the second as a [radical]. surd exclusive of the rest. More than once.[4]

Example: The square as above found stated for extraction of the root: *ru* 16 *c* 120 *c* 72 *c* 60 *c* 48 *c* 40 *c* 24.

[1] That is to say, among the surd terms, which compose the surd divisor, one is selected which though affirmative is to be put negative. COM.

* Sum of *c* 2 and *c* 8.

[3] One, two or more surd terms. COM.

[4] Repeat the operation so long as there remain surd terms of the square. COM.

Subtract the three surds *c* 120 *c* 72 *c* 48 from the square of the absolute number 256. Remainder 16. Its root 4. Added to and subtracted from the rational number, and halved, it gives 10 and 6 as two surd terms. The first is treated as rational; and the second as a radical surd. Again, subtracting two surd terms 60 and 24 from the square of the rational 100, the difference is 16; of which the square-root is 4; and the moieties of the sum and difference are 7 and 3. The first is here treated as absolute; and the second as a radical surd. Again, subtracting the surd 40 from the square of the absolute 49, the remainder is 9, of which the square-root is 3; and the moieties of sum and difference are 5 and 2.

Statement of the radical surds in order as found: *c* 6 *c* 5 *c* 3 *c* 2.

42. Rule of addition and subtraction of unknown quantities, and their squares, &c. § 30. The sum and difference of like terms, whether unknown quantities, or squares, cubes, biquadrates, fifth, sixth, &c. powers,[1] are taken; but if dissimilar are severally stated.

43. Rule of multiplication, &c. § 31. The product of two like quantities is a square; of three or more, is the power of that designation.[2] The product of dissimilar quantities, the symbols being mutually multiplied, is a factum.[3] The rest is as before.

[1] *Pancha-gata*, fifth power; *shad-gata*, sixth power, &c. Literally 'arrived at the fifth, &c. [degree]'.

[2] *Tad-gata*, raised to that.

[3] *Bhávitaca*, or *bhávita*. See *Víj.-gań.* § 21.

SECTION III.

SIMPLE EQUATION.

44. RULE for a simple equation:[1] § 32. The difference of absolute numbers, inverted and divided by the difference of the unknown, is the [value of the] unknown in an equation.[2]

45. Question 13. If four times the twelfth part of one more than the remainder of degrees, being augmented by eight, be equal to the remainder of degrees with one added thereto, tell the elapsed days.

Here remainder of degrees is put *yávat-távat:* viz. *ya* 1. With one added, it is *ya* 1 *ru* 1. Its twelfth part is $\frac{ya\ 1\ ru\ 1}{12}$ This quadrupled is $\frac{ya\ 1\ ru\ 1}{3}$ Augmented by eight absolute, it is $\frac{ya\ 1\ ru\ 25}{3}$ It is equal to remainder of degrees with one added thereto. Statement of both sides tripled, $\begin{matrix} ya\ 1\ ru\ 25 \\ ya\ 3\ ru\ \ 3 \end{matrix}$ The difference of [terms of the] unknown is *ya* 2. By this the difference of absolute number, namely 22, being divided, yields the residue of degrees of the sun 11. This residue of degrees must be understood to be in least terms. The elapsed days are to be hence deduced, as before, (§ 7).

[1] The four methods of analysis (*víja-chatushťaya*) are next explained; and in the first place equation of a single colour. COM.

[2] The value of the unknown quantity, in the example, as proposed by the question, is to be put *yávat-távat ;* and, upon that, performing multiplication, division, and other operations as requisite in the instance, two sides are to be carefully made equal. The equation being framed, the rule takes effect. Subtract the [term of the] unknown in the first of those two equal sides from the unknown of the second. The remainder is termed difference of the unknown. The absolute number on the other side is to be subtracted from the absolute number on the first side: and the residue is termed difference of the absolute. The residue of the absolute, divided by the remainder of the unknown, is the value of the unknown. COM.

46. Question 14. When the residue of exceeding months, less two, being divided by three, having seven added to the quotient, and then multiplied by two, is equal to the residue of exceeding months, tell the elapsed days.

Remainder of exceeding months *ya* 1. Proceeding with this as said, there results $\frac{ya\ 2\ \ ru\ 38}{3}$. This is equal to the remainder of exceeding months *ya* 1. Statement of both sides of equation tripled $\begin{matrix} ya\ 2 & ru\ 38 \\ ya\ 3 & ru\ \ 0 \end{matrix}$ By the foregoing rule (§ 32) the answer comes out, residue of exceeding months, 38. It must be understood to be in least terms; and from it elapsed time is to be deduced as before.

47. Question 15. If the residue of deficient days, less one, being divided by six and having three added to the quotient, be equal to the residue of deficient days divided by five, tell the elapsed period.

Here the remainder of deficient days is put *ya* 1; from which, as before, results $\frac{ya\ 1\ \ ru\ 17}{6}$. It is equal to remainder of deficient days divided by five, *ya* $\frac{1}{5}$. The two sides of equation being reduced to a common denomination and the denominator dropped, the statement is $\begin{matrix} ya\ 5 & ru\ 85 \\ ya\ 6 & ru\ \ 0 \end{matrix}$ Hence, as before, the residue of deficient days is found 85; from which elapsed days are deduced as before.

SECTION IV.

QUADRATIC EQUATION.

48.[1] RULE for elimination of the middle term:[2] § 32, 33. Take absolute number from the side opposite to that from which the square and simple unknown are to be subtracted. To the absolute number multiplied by four times the [coefficient of the] square, add the square of the [coefficient of the] middle term; the square root of the same, less the [coefficient of the] middle term, being divided by twice the [coefficient of the] square, is the [value of the] middle term.[3]

49. Question 16. When does the residue of revolutions of the sun, less one, fall, on a Wednesday, equal to the square root of two less than the residue of revolutions, less one, multiplied by ten and augmented by two?

The value of residue of revolutions is to be here put square of *yávat-távat* with two added: *ya v* 1 *ru* 2 is the residue of revolutions. This less two is *ya v* 1; the square root of which is *ya* 1. Less one, it is *ya* 1 *ru* $\dot{1}$; which multiplied by ten is *ya* 10 *ru* $\dot{10}$; and augmented by two *ya* 10 *ru* 8. It is equal to the residue of revolutions *ya v* 1 *ru* 2 less one: viz. *ya v* 1 *ru* 1. Statement of both sides *ya v* 0 *ya* 10 *ru* $\dot{8}$ Equal subtraction being made
ya v 1 *ya* 0 *ru* 1

[1] Remaining half of a couplet and one whole one.

[2] *Mad'hyamáharana.* See *Víj.-gań.* Ch. 1.

[3] An equation of two sides being framed conformably to the enunciation of the instance, if there be a square or other [power] together with the unknown, then this rule takes effect. Subtract the absolute number from the side other than that from which the square and the unknown qualities are subtracted. Then equal subtraction having been so made, the numeral *(anca)* which belongs to the square of the unknown, is termed [coefficient of the] square; and that, which appertains to the unknown, is called [coefficient of the] middle term. The absolute number, which is on the second side, being multiplied by four times the square [i. e. its coefficient] and added to the square of the middle term [i. e. of its coefficient], the square-root of the sum, less the middle term [i. e. its coefficient], divided by the double of the square as it is termed [i. e. coefficient], is the middle term; that is to say, it is the value of the unknown. COM.

conformably to rule (§ 32) there arises *ru* 9 Now, from the abso-
ya v 1 *ya* 10
lute number (9), multiplied by four times the [coefficient of the] square (36), and added to (100) the square of the [coefficient of the] middle term, (making consequently 64), the square root being extracted (8), and lessened by the [coefficient of the] middle term (10), the remainder 18 divided by twice the [coefficient of the] square (2), yields the value of the middle term 9. Substituting with this in the expression put for the residue of revolutions, the answer comes out, residue of revolutions of the sun 83. Elapsed period of days deduced from this, 393, must have the denominator in least terms added so often until it fall on Wednesday.

50. Or another Rule: § 34. To the absolute number multiplied by the [coefficient of the] square, add the square of half the [coefficient of the] unknown, the square root of the sum, less half the [coefficient of the] unknown, being divided by the [coefficient of the] square, is the unknown.

In the foregoing example, equal subtraction being made from the two sides, the result was *ya v* 1 *ya* 10 Here absolute number (9) multiplied
ru 9
by (1) the [coefficient of the] square (9), and added to the square of half the [coefficient of the] middle term, namely, 25, makes 16; of which the square root 4, less half the [coefficient of the] unknown (5), is 9; and divided by the [coefficient of the] square (1) yields the value of the unknown 9. Substituting with this, the residue of revolutions comes out 83: whence elapsed days are deduced, as before, 393.

51. Question 17. When is the square of three less than the quarter of the residue of exceeding months equal to the residue of exceeding months? or the like [function] of remainder of deficient days equal to remainder of deficient days?

Remainder of exceeding months is here put *ya* 4. Its quarter less three is *ya* 1 *ru* 3̇; of which the square is *ya v* 1 *ya* 6 *ru* 9. It is equal to the remainder of exceeding months. The process being performed as before, the residue of exceeding months is found 4. Whence the elapsed period is deduced.

In like manner the remainder of deficient days likewise is 4: whence the elapsed period comes out 1031.

SECTION V.

EQUATION OF SEVERAL COLOURS.

52. RULE: § 35. Subtracting the colours other than the first from the opposite side to that from which the first is subtracted, after reducing them to a common denomination, the value of the first is derived from [the residue] divided by this [coefficient of the] first. If more than one [value], two and two must be opposed. The pulverizer is employed, if many [colours] remain.[1]

53. Question 18. He, who tells the number of [elapsed] days from the number of days added to past revolutions, or to the residue of them, or to the total of these, or from their sum, is a person versed in the pulverizer.

Example: The number of [elapsed] days together with past lunar revolutions is given equal to one hundred and thirty-nine. Tell me the number of days separately.

Here the number of [elapsed] days is put *ya* 1. Multiplied by revolutions

[1] In an example in which there are two or more unknown quantities, two or more colours, as *yávat-távat*, &c. must be put for their values: and upon those the requisite operations, conformably to the instance, being wrought, two or more sides of equation are to be carefully framed: and among them, taken two and two, equal subtraction is to be made; in this manner: the first colour being subtracted from one side, subtract the rest of the colours reduced to a like denomination, and absolute number, from the other side. The residue of another colour being divided by the residue of the first, the quotient is a value of the first colour. If many such values be obtained, they must be equated again in pairs reducing them to like denominators. But, that being done, if there be two colours in the value of another colour which is thence deduced, the coefficients *(anca)* of those two are reciprocally the values of such colours. But, if there be many colours in the value of another colour, the pulverizer must be applied to them; in this manner: excepting one colour, substitute arbitrary values for the rest, and, adding them to absolute number, form the addition. Make the coefficient of the selected colour, the dividend; and the coefficient of the colour in the denominator, the divisor. The multiplier, hence found by the method of the pulverizer, is the value of the colour in the dividend; and the quotient is the value of that in the divisor. COM.

of the moon in least terms, and divided by the divisor also reduced to the least terms, there results *ya* $\frac{5}{137}$; from which less the residue of revolutions, divided by the divisor,[1] the quotient is the [complete] revolutions; wherefore the residue of revolutions is put *ca* 1. Less that, and divided by the divisor in least terms, it yields revolutions, $\frac{ya\ 5\ \ ca\ \dot{1}}{137}$; which, added to the number of [elapsed] days, makes $\frac{ya\ 142\ \ ca\ \dot{1}}{137}$. It is equal to the sum of [complete] revolutions and number of [elapsed] days, *ru* 139. Statement of both sides of equation reduced to the same denominator, *ya* 142 *ca* $\dot{1}$ *ru* 0
ya 0 *ca* 0 *ru* 19043

Subtraction being made as prescribed by the rule (§ 35), the result is *ca* 1 *ru* 19043. Since there are several colours, the pulverizer must be
ya 142

employed. The coefficient of the colour in the dividend is dividend; that which stands with the colour in the divisor, is divisor. From these the constant pulverizer, as found by the rule (§ 9), is 141. Multiplying by this the additive 19043, divide by the divisor 142, the residue is here the multiplier sought, 127. It is the value of *cálaca.* The dividend being multiplied by the multiplier, and having the additive added, and being divided by its divisor, the quotient is the value of *yávat-távat,* 135. It is the number of [elapsed] days.

Example: When the residue of lunar revolutions, with the number of [elapsed] days, is given equal to two hundred and sixty-two, tell me the number of days.

The number of [elapsed] days is here put *ya* 1. This, multiplied by revolutions and divided by the divisor, becomes $\frac{ya\ 5}{137}$. Then *cálaca* is put for the value of quotient.[2] If the divisor multiplied by the quotient be subtracted from the number of elapsed days multiplied by the [periodical] revolutions, the residue which remains is the residue of revolutions. So doing, the result

[1] So the original: but the expression is not quite accurate, as the fraction is not again divided; but the multiple of the time by periodical revolutions, less the residue, being divided, gives the complete revolutions for quotient.

[2] Exclusive of the fractional residue.

is *ya* 5 *ca* 137. This, with the number of days, becomes *ya* 6 *ca* 137. It is equal to the sum of the number of days and residue of revolutions, 262. Statement of both sides of equation *ya* 6 *ca* 137 *ru* 0 / *ya* 0 *ca* 0 *ru* 262 The process being followed as before, the multiplier comes out 4. It is the value of *cálaca*. The quotient is the value of *yávat-távat*, 135. It is the number of [elapsed] days.

Example: If the sum of the three specified articles be equal to two hundred and sixty-six, tell me the number of [elapsed] days; or tell it from the sum of the other two.

The specified articles are [complete] revolutions, the residue of them, and the number of days. The number of days is put *ya* 1. This, multiplied by revolutions and divided by the divisor, is *ya* $\frac{5}{137}$. The quotient[1] is *ca* 1. Divisor multiplied by quotient, being subtracted from the number of [elapsed] days taken into revolutions, the remainder is residue of revolutions, *ya* 5 *ca* 137. Adding the number of days and the [past] revolutions, the total is *ya* 6 *ca* 136. This is equal to the sum of the number of [elapsed] days, the residue of revolutions and [past] revolutions, 266. Statement of the two sides of equation, *ya* 6 *ca* 136 *ru* 0 / *ya* 0 *ca* 0 *ru* 266 Hence, by equal subtraction and other process, as before, the constant pulverizer comes out 2; and the multiplier is found 4. It is the value of *cálaca;* and is the [number of] past revolutions 4. The quotient is the value of *yávat-távat*, 135. It is the number of [elapsed] days. Subtracting the sum of [past] revolutions and [elapsed] days from the sum total, the remainder is the residue of revolutions, 127.

Example: When the residue of revolutions of the moon added to the revolutions past is equal to one hundred and thirty-one, tell me the number of [elapsed] days.

Number of days *ya* 1. The residue of revolutions is found as before *ya* 5 *ca* 137. This, added to past revolutions, is *ya* 5 *ca* 136. It is equal to the sum of past revolutions and residue of revolutions, 131. The constant pulverizer comes out 4. Hence the multiplier, 4. It is the value of

[1] Exclusive of the fractional residue.

cálaca as before; and is the number of revolutions complete. The quotient is the value of *yávat-távat*, 135. It is the number of [elapsed] days.

54. Question 19. He, who tells the number of [elapsed] days from the number of days less the past revolutions, or less the residue of them, or less the sum of these, or from their difference, is a person acquainted with the pulverizer.

Example: The number of [elapsed] days, less the past lunar revolutions, is given equal to one hundred and thirty-one; tell me the number of days.

Here the value of the number of days is *ya* 1; which, being multiplied by revolutions, and lessened by residue of revolutions, put equal to *cálaca*, and divided by its divisor, becomes the number of past revolutions, $\frac{ya\ 5\ \ ca\ 1}{137}$.
The number of days, less that is $\frac{ya\ 132\ \ ca\ 1}{137}$. It is equal to the difference between the number of days and past revolutions, namely, 131. Statement of the equation reduced to a common denominator,

ya 132	*ca* 1	*ru* 0
ya 0	*ca* 0	*ru* 17947

Equal subtraction being made, as before, the constant pulverizer comes out 133. The multiplier, value of *cálaca*, 127. The quotient is the value of *yávat-távat*: it is the number of [elapsed] days, 135. The difference between the number of days and past revolutions, 131, being subtracted, the remainder is the past revolutions, 4.

Example: [Elapsed days] less the residue of the [revolutions] being eight; or less the sum of the [past revolutions and their residue] being four; or less the difference of the two being a hundred and twenty-three: tell the number of [elapsed] days.

In the first example, the value of the number of [elapsed] days is put *ya* 1. As before, the residue of revolutions, *ya* 5 *ca* 137. Taking this from the number of days, the remainder is *ya* 4 *ca* 137. It is made equal to eight; and proceeding, as before, the multiplier or value of *cálaca* is 4; and the value of *yávat-távat*, or number of [elapsed] days, 135.

In the second example, residue of revolutions, as before, *ya* 5 *ca* 137. Past revolutions *ca* 1. Their sum *ya* 5 *ca* 136. This, subtracted from the number of days, leaves *ya* 4 *ca* 136. It is equal to four. The result is value of *cálaca*, 4; and value of *yávat-távat*, the number of days, 135.

In the third example, residue of revolutions, *ya* 5 *ca* $\dot{1}37$. Past revolutions, *ca* 1. Difference *ya* 5 *ca* $\dot{1}38$. It is equal to one hundred and thirty-three. The result is, value of *cálaca*, 4: value of *yávat-távat*, the number of days, 135.

55. Question 20. He, who tells the elapsed [portion of the] cycle from the signs, or the like;[1] or the residues of them; or from past *exceeding* months; or *fewer* days; or their residues,[2] is a person conversant with the pulverizer.

Example: Forty-six, a hundred and seventy-two, a hundred and seventy-seven, and a hundred and thirty-six, are declared to be respectively the amount of the number of [elapsed] days added to past signs; or to the residue of them; or to the sum of these two; or amount of the sum of the two: tell me the number of days in the several instances.

Here the residue of lunar revolutions is, as before, *ya* 5 *ca* $\dot{1}37$. This, multiplied by twelve, becomes *ya* 60 *ca* $\dot{1}644$. Subtracting from it, the residue of signs denoted by *nílaca*, and dividing by the divisor, and adding the number of [elapsed] days, the result is $\frac{ya\ 197\quad ca\ \dot{1}644\quad ní\ \dot{1}}{137}$. It is equal to forty-six. Statement of the two sides of equation reduced to a like denomination,

$$\begin{array}{llll} ya\ 197 & ca\ \dot{1}644 & ní\ \dot{1} & ru\ 0 \\ ya\ 0 & ca\ 0 & ní\ 0 & ru\ 6302 \end{array}$$

Equal subtraction being made, the value of *yávat-távat* is $\frac{ca\ 1644\quad ní\ 1\quad ru\ 6302}{(ya)\ 197}$. Here the arbitrary value

[1] Degrees, minutes, or seconds. COM.

[2] As four problems were proposed in the preceding passage, (Question 18,) so are four to be here understood for finding the number of [elapsed] days from the number of days added to past signs; or added to the residue of them; or to the total of these [signs and residue]; or from the sum of these two. And, as four problems were proposed in the foregoing passage, (Question 19,) so are four to be inferred for finding the same from the number of days, less the past signs, and so forth. Thus the problems are eight. In like manner, from past degrees and their residue; from past minutes and their residue; from past seconds and their residue; eight problems, in each instance, are to be deduced: and as many in each case of past exceeding months, and deficient days, and the residues. COM.

of *nílaca* is assumed such, that no defect may ensue: say 131. This is residue of signs. Multiplying by it [the coefficient of] *nílaca*, and adding [the product] to the absolute number, the pulverizer is deduced, 1: it is the value of *cálaca;* that is, the past revolutions. The quotient is the value of *yávat-távat*, and is the number of [elapsed] days, 41.

In the second example, multiplying by twelve the residue of revolutions, subtracting signs multiplied by their divisor, the residue of signs is obtained *ya* 60 *ca* 1644 *ní* 137. Adding to this the number of [elapsed] days, and making the sum equal to one hundred and seventy-two, the statement of the equation is *ya* 61 *ca* 1644 *ní* 137 *ru* 0 Subtraction being made and the
ya 0 *ca* 0 *ní* 0 *ru* 172
value of *nílaca* being assumed five, the pulverizer is deduced, 1. It is the value of *cálaca.* The quotient is the value of *yávat-távat;* and is the number of [elapsed] days, 41.

In the third example, past signs are $\frac{\textit{ya}\ 60\ \ \textit{ca}\ 1644\ \ \textit{ní}\ 1}{137}$. Adding residue of signs, the sum is $\frac{\textit{ya}\ 60\ \ \textit{ca}\ 1644\ \ \textit{ní}\ 136}{137}$; to which adding the number of [elapsed] days, the result is $\frac{\textit{ya}\ 197\ \ \textit{ca}\ 1644\ \ \textit{ní}\ 136}{137}$. This is equal to a hundred and seventy-seven [to be] reduced to a common denomination. Putting 131 for the value of *nílaca,* and by means of the pulverizer, the number of days comes out 41.

In the fourth example, the sum of past signs and residue is $\frac{\textit{ya}\ 60\ \ \textit{ca}\ 1644\ \ \textit{ní}\ 136}{137}$. It is equal to a hundred and twenty-six [to be] reduced to the like denomination. With this value of *nílaca* 131, the multiplier is deduced, 2. The quotient is the number of days, 41. Or else 178; or 315. The like is to be understood also in the case of revolutions and the rest.

When the number of days less the [complete] signs is given, what is the number of days? Here, as before, the [complete] signs are $\frac{\textit{ya}\ 60\ \ \textit{ca}\ 1644\ \ \textit{ní}\ 1}{137}$. The number of days, less that, is $\frac{\textit{ya}\ 77\ \ \textit{ca}\ 1644\ \ \textit{ní}\ 1}{137}$. This is equal to a hundred and twenty-four.[1] *Nílaca* being assumed seventeen, the multiplier is deduced 4; and the quotient 135. It is the number of days.

In the second example, residue of signs *ya* 60 *ca* $\dot{1644}$ *ní* 137. Subtracting this from the number of days, the remainder is *ya* $\dot{59}$ *ca* $\dot{1644}$ *ní* 137. It is equal to a hundred and eight. Eleven[2] being put for *nílaca*, the multiplier comes out 4; and the quotient 135. This is the number of [elapsed] days.

In the third example, sum of past signs and residue $\frac{ya\ 60\quad ca\ \dot{1644}\quad ní\ 136}{137}$. The number of days, less that, is $\frac{ya\ 77\quad ca\ 1644\quad ní\ \dot{136}}{137}$. It is equal to a hundred and seven. *Nílaca* being assumed seventeen, the multiplier comes out 4; and the quotient, or number of days, 135.

In the fourth example, past signs $\frac{ya\ 60\quad ca\ 1644\quad ní\ 1}{137}$. Residue of signs, *ní* 1. Difference of these reduced to like denominators, $\frac{ya\ 60\quad ca\ \dot{1644}\quad ní\ 138}{137}$. It is equal to six. Subtraction being made on both sides, and seventeen being arbitrarily put for *nílaca*, the multiplier is found 4; and the quotient, or number of days, 135.

Next, from the sum of past degrees and number of [elapsed] days [the elapsed time is to be sought]. Here, as before, the residue of signs is *ya* 60 *ca* 1644 *ní* 137. Multiplying this by thirty, subtracting residue of degrees put equal to *pítaca*, and dividing by the divisor, the quotient is past degrees, which thus come out $\frac{ya\ 1800\quad ca\ \dot{49320}\quad ní\ 4110\quad pí\ \dot{1}}{137}$. The number of days being added, the sum is equal to 21, the assumed amount of degrees and number of days. Subtraction being made on the two sides of equation reduced to a common denominator, the result is $\frac{ca\ 49320\quad ní\ 4110\quad pí\ 1\quad ru\ 2877}{(ya)\ 1937}$. Here substituting with four for *nílaca*, and with fifty-three for *pítaca*, and adding the values so raised to the absolute number, the multiplier thence deduced is 0; and the quotient 10.

[1] This had not been previously proposed: probably from defect of the manuscript.

[2] Sic.

The like process is to be followed, [for deducing the elapsed time] from the sum of residues of degrees and number of [elapsed] days.

Next, the number of days with past seconds is [given] twenty-two: what is in this instance the number of elapsed days?[1] Here the number of days, is put *ya* 1. Whence, as before, the residue of minutes, *ya* 108000 *ca* 2959200 *ní* 246600 *pí* 8220 *lo* 137. This, multiplied by sixty, lessened by subtraction of residue of seconds equal to *haritaca*, and divided by the divisor, the quotient is seconds; which, added to the number of days, is equal to the proposed twenty-two. Subtraction being made on the two sides of equation reduced to a like denominator, the value of *yávat-távat* comes out $\frac{ca\ 177552000\ ni\ 14796000\ pi\ 493200\ lo\ 8220\ ha\ 1\ ru\ 3014}{ya\ 6480137}$. Here, substituting with four for *nílaca*, with eleven for *pítaca*, with twenty-three for *lóhitaca*, with ninety-six for *haritaca*, and adding the values so raised to the absolute number, the additive becomes 64801370. Whence, as before, the multiplier is found 0; and the quotient, or value of *yávat-távat*, 10. It is the number of [elapsed] days. Subtracting this from twenty-two, the remainder is the [past] seconds, 12.

A similar process is to be followed [for the elapsed time] from the sum of residue of seconds and number of days.

Example: If the elapsed [portion of the] cycle, added to the past *exceeding* months, be equal to three thousand, one hundred and thirty-two, tell the elapsed [portion of the] cycle.

Elapsed [part of the] cycle *ya* 1. Multiplied by the number of *exceeding* months in a *yuga* in least terms, and divided by the solar days in a *yuga* also in least terms, the result is $\frac{ya\ 1}{1800}$. From this subtracting the residue of *exceeding* months, *ca* 1, the remainder is [complete] *exceeding* months, $\frac{ya\ 1\ \ ca\ 1}{1800}$. Adding to this the elapsed [time of the] cycle, (*ya* 1,) it becomes $\frac{ya\ 1801\ \ ca\ \dot{1}}{1800}$; and is equal to 3132. Subtraction being made on the two sides of equation reduced to a like denominator, the multiplier comes out

[1] The preceding examples not being specifically proposed, like this instance, and the example of minutes and their residue being omitted, the manuscript may be concluded to be deficient.

970; and the quotient, 3130. This is the elapsed [portion of the] *yuga*. Subtracting it from three thousand, two hundred and thirty-two, the remainder is the number of past *exceeding* months, 2.

In like manner, find severally the elapsed [time of the] cycle from elapsed time added to residue of *exceeding* months, and to the sum of the past *exceeding* months and their residue, and from the sum of [complete] *exceeding* months and their residue. Four other problems are likewise to be understood for finding elapsed time from the difference between this and the complete *exceeding* months, and so forth.

Example: If the elapsed [portion of a] lunar *yuga*, added to the past [deficient] days, be equal to one thousand, nine hundred and eighty-two, tell me the elapsed lunar time.

Number of lunar days, *ya* 1. Multiplied by *fewer* days in least terms, and divided by lunar days also in least terms, the result is $\frac{ya\ 7}{555}$. Subtracting the residue of *fewer* days, for which put *cálaca* 1, the remainder is the number of *fewer* days complete, $\frac{ya\ 7\quad ca\ \dot{1}}{555}$. This, added to the number of lunar days, $\frac{ya\ 562\quad ca\ \dot{1}}{55}$, is equal to one thousand, nine hundred and eighty-two (1982). Subtraction being made on the two sides of equation reduced to like denominators, the multiplier, or value of *cálaca*, is found 386; it is the residue of *fewer* days. The quotient, or value of *yávat-távat*, is 1958. It is the number of [elapsed] lunar days.

In like manner other problems are to be understood.

56. Question 21. He, who tells the number of [elapsed] days, from the residue of minutes added to the residue of degrees of the luminary,[1] on a Wednesday[2] [or any given day], or from their difference, is a person acquainted with the pulverizer.

[1] *Bhánu*, luminary, applied especially to the sun; but here apparently intending any planet. See the following problems, and the commentator's remarks on Question 25.

[2] In this and several following instances, a day is specified; but no notice of this condition is taken in the example and its solution, until Question 23.

Example: Seeing the residue of degrees of the moon,[1] with the residue of minutes added thereto, equal to five hundred and thirty-six; or with that subtracted from it, equal to three hundred and forty-four: tell the number of days.

Here the number of days is put *ya* 1. This, multiplied by the revolutions of the sun in least terms, and divided by the divisor, is $\frac{ya\ 3}{1096}$. Subtracting from the number of days taken into the revolutions, the divisor taken into the quotient[2] represented by *cálaca*, the remainder is residue of degrees, *ya* 3 *ca* 1096. Hence, as before, the residue of degrees is found *ya* 1080 *ca* 394560 *ní* 32880 *pí* 1096. This is reserved; and multiplying it by sixty, dividing by the divisor, subtracting the divisor taken into the quotient[2] represented by *lóhitaca*, the remainder is the residue of degrees, *ya* 64800 *ca* 23673600 *ní* 1972800 *pí* 65760 *lo* 1096. Thus the sum of these residues of degrees and minutes is *ya* 65880 *ca* 24068160 *ní* 2005680 *pí* 66856 *lo* 1096. It is equal to 536. Subtraction being made, the value of *nílaca* is assumed, *ru* 1; that of *pítaca*, 10; of *lóhitaca*, 24; and multiplying the [coefficients of] those by their values, [as assumed,] and adding the products to absolute number, the amount of the absolute number becomes 2701080. Whence, as before, the multiplier is found 0; and the quotient, or number of days, 41.

The difference between the residues of degrees and of minutes is *ya* 63720 *ca* 23279040 *ní* 1939920 *pí* 64664 *lo* 1096. It is equal to 344. Subtraction being made, and putting the same values for *nílaca* and the rest, the multiplier comes out, as before, 0; and the quotient, or value of *yávat-távat*, 41. It is the number of days.

Or what occasion is there for this trouble? Putting *yávat-távat* for the residue of degrees, and multiplying by sixty, divide by the divisor. Subtracting from it the divisor taken into the quotient[2] represented by *culaca*, the remainder is residue of minutes. Then making the sum of residues of minutes and degrees equal to the proposed sum, and, equal subtraction being made, the value of *yávat-távat*, which comes out, is the residue of degrees; from which, as before, the number of [elapsed] days is to be inferred.

Or else, finding the residue of degrees, and that of minutes, as arising for

[1] So the original. But the example is wrought as an instance of the sun.

[2] Exclusive of the fractional residue.

one day, and taking their sum and their difference, the number of [elapsed] days is to be found by the constant pulverizer thence deduced.

In like manner [the modes of solution] are manifold.

57. Question 22. When is the residue of degrees of the sun, with three added, equal to the residue of minutes, on a Wednesday? or with six, seven, or eight, subtracted? Solving [the problem] within a year [the proficient is] a mathematician.

Here sun is indefinite; and the question extends therefore to any given planet. In this place an instance of the moon is exhibited. It is as follows. Value of the number of days, *ya* 1. Whence the residue of degrees of the moon, *ya* 1800 *ca* 49320 *ní* 4110 *pí* 137. So the residue of minutes is this, *ya* 108000 *ca* 2959200 *ní* 346600 *pí* 8220 *lo* 137. Here the residue of degrees, with three added, is equal to the residue of minutes. Subtraction being made on both sides, and with two put for *nílaca*, thirteen for *pítaca*, and thirty-four for *lóhitaca*, the multiplier is brought out, value of *cálaca*, 1; and the quotient, value of *yávat-távat*, 33. The value of *cálaca* is the complete revolutions; that of *nílaca*, the past signs; that of *pítaca*, the degrees; and that of *lóhitaca*, the minutes.

In like manner, making the residue of degrees less six, or that residue less seven, or the same less eight, equal to the residue of minutes, the number of [elapsed] days is to be found, as before.

58. Question 23. When is the residue of degrees of the sun equal to the [complete] degrees; or the residue of minutes, to the minutes, on a given day? Solving [this problem] within a year [the proficient is] a mathematician.

Here also sun is indefinite, and intends any planet. Therefore the residue of degrees of the moon is taken, *ya* 1800 *ca* 49320 *ní* 4110 *pí* 137. This is equal to the complete degrees, the value of which is represented by *pítaca*. Subtraction being made, and with ten put for *nílaca*, and ten for *pítaca*, the multiplier comes out 1; and the quotient 51. This is the number of [elapsed] days.

With this number of days, the residue of degrees of the moon is equal to the complete degrees.

In the very same manner, residue of minutes of the moon, *ya* 108000 *ca* 2959200 *ní* 246600 *pí* 8220 *lo* 137. This is equal to the complete minutes, the value of which is represented by *lóhitaca*. Subtraction being made on the two sides of the equation, and with nine for *nílaca*, eleven for *pítaca*, and ten for *lóhitaca*, the multiplier comes out 4; and the quotient the value of the number of days, 131. To find for the given day, the given multiple of the divisor is to be added.

Or what occasion is there for the trouble of supposing [values of] colours? Putting *yávat-távat* for the residue of degrees; and from that multiplied by thirty and divided by the divisor, subtracting the quotient[1] represented by *cálaca*, taken into its divisor, the remainder is residue of degrees. Making it equal to degrees, equal subtraction is then to be made; whence the value of *yávat-távat* is brought out. It is the residue of degrees; from which, as before, the number of [elapsed] days is to be deduced.

59. Question 24. Residue of *fewer* days, with a given quantity added or subtracted, or residue of *more* months, with the like, is equal to *fewer* days; or to *more* months. Solving [this problem] within a year [the proficient is] a mathematician.

It is as follows. Elapsed [portion of the] *yuga* 1. Multiplied by *exceeding* months in least terms, and divided by solar days also in least terms, and the quotient lessened by subtraction of *cálaca* representing the complete *exceeding* months, the result is *ya* 1 *ca* 1080. It is the residue of *exceeding* months, and is equal to *cálaca*. Whence the multiplier comes out 1, and the quotient 1081. This is the elapsed [portion of the] *yuga*.

Or equal to *cálaca* with five added. The quotient, which is the elapsed *yuga*, comes out 1086.

In the question relating to *fewer* days, the elapsed [portion of the] *yuga* is *ya* 1. This, multiplied by *deficient* days in least terms, and divided by lunar days, and lessened by subtraction of the divisor taken into *cálaca* put for the quotient, the result is *ya* 7 *ca* 555. This is equal to *cálaca*. The multiplier thus comes out 7; and the quotient, the value of *yávat-távat*, 556. It is the elapsed lunar *yuga*.

Or equal to *cálaca*, with three added; the multiplier comes out 6; and the quotient, or elapsed *yuga*, 477.

[1] Exclusive of the fractional residue.

Or equal to *cálaca*, less two, the multiplier is found 3; and the quotient, or elapsed *yuga*, 238.

60. Question 25. The sun's[1] divisor in least terms, multiplied by seventy, and lessened by the residue of degrees,[2] is exactly divisible by a myriad. Solving [this problem] within a year [the proficient is] a mathematician.

Here the sun's divisor in least terms is 1096. This, multiplied by seventy, is 76720. "Lessened by the residue of degrees:" the value of residue of degrees is put *ya* 1: less that, is *ya* $\dot{1}$ *ru* 76720. This divided by a myriad is exact. Statement $\frac{ya\ \dot{1}\ \ ru\ 76720}{10000}$. The value of the quotient is put *cálaca* 1.

Making this taken into the divisor equal to the dividend, and equal subtraction being then made, the result is $\frac{ca\ \dot{10000}\ \ ru\ 76720}{ya\ 1}$. The pulverizer comes out 1. It is the multiplier, and the value of *yávat-távat* is 76720. This is the residue of degrees: whence the number of [elapsed] days is deduced, 95.

Abridging by eighty, it is found by means of the constant pulverizer.

This is to be variously illustrated by example.

[1] Sun *(bhánu)* is here indefinite; and intends planets generally. COM.

[2] This is indefinite. Residue of revolutions, and the like, is intended. COM.

SECTION VI.

EQUATION INVOLVING A FACTUM.

61. RULE: § 36. The [product of] multiplication of the factum and absolute number, added to the product of the [coefficients of the] unknown, is divided by an arbitrarily assumed quantity. Of the arbitrary divisor and the quotient, whichever is greatest is to be added to the least [coefficient], and the least to the greatest. The two [sums] divided by the [coefficient of the] factum are reversed.

62. Question 26. From the product of signs and degrees of the sun, subtracting thrice the signs and four times the degrees, and seeing ninety [for the remainder, find the place of] the sun. Solving [this problem] within a year [the proficient is][1] a mathematician.

Signs of the sun *ya* 1. Degrees *ca* 1. Their product *ya.ca bh* 1. Subtracting from this thrice the signs and four times the degrees, the result is *ya.ca bh* 1 *ya* 3 *ca* 4. It is equal to ninety. Subtraction being made, see the result: *ya* 3 *ca* 4 *ru* 30
ya.ca bh 1
Here the multiplication of the [coefficient of the] factum and absolute number is 90. With the product [of the coeffi-

[1] In an example, in which a factum arises from the multiplication of two or of more colours, having made two sides equal, and taking the factum from one side, subtract the absolute number together with the [single] colours from the other. The equation so standing, the rule takes effect.

The multiplication of absolute number by the coefficient of the factum is termed multiplication of the factum and absolute number. That, together with the product of the two unknown, is to be divided by an arbitrary quantity. Between the arbitrary divisor and quotient, the greater is to be added to the less coefficient, and the less to the greater efficient. The two sums, divided by the coefficient of the factum, being reversed, are values of the colours. The meaning is, that, to which the coefficient of *yávat-távat* is added, is the value of *cálaca*; and that, to which the coefficient of *cálaca* is added, is the value of *yávat-távat*.

But, when a factum consisting of many colours occurs, then reserving two, and assuming arbitrary values for the rest of the colours, multiply by them the factum of the two reserved colours. Thence the rest is to be done as above directed. COM.

cients] of the unknown, namely 12, added, it becomes 102. Divided by the assumed number 17, the quotient is 6. It is "least;" and to it is added the greater coefficient 4, making 10. The "greater" is 17; to which the least coefficient 3 is added, making 20. These, divided by the [coefficient of the] factum [viz. 1], become values of *yávat-távat* and *cálaca*, 10 and 20. Signs of the sun 10; degrees 20. Its degrees [320], multiplied by the divisor in least terms, namely 1096, (making 350720,) and divided by the degrees in a revolution [360], yield as product the residue of revolutions. With unity added, that residue is 975. Whence, as before directed, the number of [elapsed] days is to be found, 325.

63—64. Rule:[1] § 37—38. With the exception of one selected colour, put arbitrary values for the rest of those, the product of which is the factum. The sum of the products of the [coefficients of] colours by those [assumed values] is absolute number. The product of the assumed values of colours and [coefficient of] the factum is coefficient[2] of the selected colour. Thus the solution is effected without an equation of the factum. What occasion then is there for it?

Here the foregoing example (Qu. 26) serves. As before, having done as directed, the two sides of equation become *ya* 3 *ca* 4 *ru* 90 / *ya . ca bh* 1 Reserving *yávat-távat*, which is selected, an arbitrary value is put for *cálaca* 20. The coefficient of *cálaca*, multiplied by that, is 80; added to absolute number, this becomes 570. Now the coefficient of the factum (1), multiplied by *cálaca*, becomes coefficient of *yávat-távat*, 20. Statement of the two sides of equation thus prepared, *ya* 3 *ru* 170 / *ya* 20 Proceeding by a former rule (§ 32), the value of *yávat-távat* comes out 10.

[1] Having thus set forth the [solution of a] factum according to the doctrine of others, the author now delivers his own method with a censure on the other. COM.

[2] *Sanc'hyá*, number: meaning coefficient usually expressed by *anca*, figure.

SECTION VII.

SQUARE AFFECTED BY COEFFICIENT.

65—66. RULE: § 39—40. A root [is set down] two-fold: and [another, deduced] from the assumed square multiplied by the multiplier, and increased or diminished by a quantity assumed. The product of the first [pair], taken into the multiplier, with the product of the last [pair] added, is a "last" root.[1] The sum of the products of oblique multiplication is a "first" root. The additive is the product of the like additive or subtractive quantities. The roots [so found], divided by the [original] additive or subtractive quantity, are [roots answering] for additive unity.[2]

[1] The terms familiarly used for the practice of solution of problems under this head are here explained: viz.

Canisht'ha or *ádya (pada* or *múla)* least or first root; that quantity, of which the square multiplied by the given multiplicator and having the given addend added, or subtrahend subtracted, is capable of affording an exact square root.

Jyésht'ha or *antya (pada* or *múla)* greatest or last root: the square-root which is extracted from the quantity so operated upon.

Pracrĭti, the multiplier [the coefficient of the first square].

Cshépa, cshipticá, chipti, additive, or addend: the quantity to be added to the square of the least root multiplied by the multiplicator, to render it capable of yielding an exact square-root.

Sód'haca, subtractive, or subtrahend: the quantity to be subtracted for the like purpose.

Udvartaca, the quantity assumed for the purpose of the operation.

Apavarta, abridger, common measure: the divisor, which is assumed for both or either of the quantities.

Vajra-bad'ha, forked or oblique [that is, cross] multiplication. See *Víj.-gań.* § 77. COM.

[2] The root of any square quantity is to be set down twice; that is, being repeated, the second is to be put under the first. These two are "least" roots. Then multiplying by the multiplicator the square of the "least" root, consider what quantity, added or subtracted, will render it capable of yielding an exact root. The quantity, of which the addition effects that, is "additive." That, of which the subtraction effects it, is "subtractive." So doing, the root, which is afforded, is "greatest" root. This also is to be set down in two places, in front of the "least" roots. Being so arranged, the product of the two least roots multiplied by the multiplicator, with the product of the two greatest, is a "greatest" root. That is to say, it is so by composition. The product of multiplication crosswise, or obliquely, like forked or crossing lightning, is product of oblique multiplication. That is, the least and greatest roots are twice multiplied cornerwise. The

67. Question 27. Making the square of the residue of signs and minutes on a Wednesday, multiplied by ninety-two and eighty-three respectively, with one added to the product, [afford, in each instance] an exact square, [a person solving this problem] within a year [is] a mathematician.

Here the assumed square is put 1. Its root is "least" root. Set down twice, L 1 / L 1. Again the same square, multiplied by ninety-two, and having eight added [to make it yield a square-root], amounts to 100: the root of which is "greatest," 10. Statement of them in order L 1 G 10 A 8 / L 1 G 10 A 8. Then, proceeding by the rule (the product of the first taken into the multiplier, &c. § 39—40) the "least" and "greatest" roots for additive sixty-four are found L 20 G 192 A 64. By the concluding part of the rule (§ 40) the "least" and "greatest" roots for additive unity come out L $\frac{5}{2}$ G 24 A 1. Again, from this, by the combination of like ones, other least and greatest roots are brought out L 120 G 1151 A 1. Here the "least" root is residue of degrees. Whence, as before, the number of [elapsed] days is deduced, 65.

In the second example, the square assumed is 1. Its square-root is "least" root, 1. From the assumed square, multiplied by eighty-three, and lessened by subtraction of two, the square-root extracted is "greatest," 9. Proceeding by the rule (§ 39), the "greatest" root comes out, 164; and by the sequel of it (§ 40), the "least" root, 18; and these, divided by the subtractive, namely 2, become roots for additive unity, L 9 G 82 A 1. The "least" is residue of minutes: whence, as before, the number of [elapsed] days is found, 22.

68. Rule: § 41.[1] Putting severally the roots for additive unity under roots for the given additive or subtractive, "last" and "first" roots [thence deduced by composition] serve for the given additive or subtractive.[2]

sum of these two products is a "least" root by composition. But the additive by composition amounts to the product of the two like additives or subtractives. Then the least and greatest roots, so derived from composition, being divided by the number of the [original] additive, or by that of the subtractive, are roots serving when unity is the additive. COM.

[1] When the additive is many [i. e. more than unity]. COM.

[2] Under least and greatest roots, which serve for the given additive or subtractive, are to be placed least and greatest roots serving for additive unity; then the roots, which are found by the foregoing rule (§ 39—40), are roots which also serve for the same given additive or subtractive. COM.

An example will be given further on.[1]

69. Rule: § 42. When the additive is four, the square of the last root, less three, being halved and multiplied by the last, is a last root; and the square of the last root, less one, being divided by two and multiplied by the first, is a first root, [for additive unity].[2]

70. Question 28. Making the square of the residue of revolutions or the like, multiplied by three and having nine hundred added, an exact square; or having eight hundred subtracted; [a person solving this problem] within a year [is] a mathematician.

Here the assumed square is put 4. Its root is a least root 2. Its square 4, multiplied by the coefficient 3, is 12; and with 4 added, 16. Its square-root is a greatest root, 4. L 2, G 4. From this are to be found roots for additive unity. Here the last root is 4: its square 16; less three, 13; halved, $\frac{13}{2}$; multiplied by last root, 26; this is a last root for additive unity. Again, square of the last root, 16; less one, 15; divided by two, $\frac{15}{2}$; multiplied by the first root, gives a first root, 15. The meaning of the rest of the question is shown [farther on].[3]

71. Rule:[4] § 43. When four is subtractive, the square of the last root is twice set down, having three added in one instance and one in the other:

[1] See Qu. 32.

[2] Under the rule "A root is set down twofold," &c. (§ 39), if four be the additive, then the [original] last root being squared, and lessened by three, the half of the remainder, multiplied by the last root, is a last root, answering, however, to additive unity. Again, the square of the [original] last root, lessened by one, divided by two, and multiplied by least root, is a least root, the additive being unity. COM.

[3] See Qu. 32.

[4] Rule to find roots answering for additive unity, from roots which serve when four is subtractive. COM.

Of least and greatest roots serving when four is subtractive, the square of "last" root is twice set down, having three added in the one instance, and one added in the other. The moiety of the product of those reserved quantities is also to be twice set down, having one subtracted in the one instance, and as it stood in the other. That which is diminished by one, is next "multiplied by the former less one;" that is to say, multiplied by the square of "last" root [having three added] less one. So doing, the result is "greatest" root for unity additive. The moiety of the product, which was set down as it stood, being multiplied by the product of "least" and "greatest" roots, is "least" root for additive unity. COM.

half the product of these sums is set apart, and the same less one. This, multiplied by the former less one, is "last" root. The other, multiplied by the product of the roots, is "first" root answering to that "last."

72. Question 29. The square of residue of *exceeding* months, multiplied by thirteen, and having three hundred added, or the cube of three subtracted, affords an exact square. A person solving [this problem] is a mathematician.

Here the assumed square is put 1. Its root is 1: it is a least root. The square of this multiplied by thirteen, and lessened by four, is 9. Root of the remainder 3: a greatest root is thus found. From these least and greatest roots, L 1 G 3, roots are to be found for additive unity. In this case, the last root is 3: its square twice, 9 and 9; with three and one added, 12 and 10.[1] Again, half the product of those reserved quantities, 60: multiplied by the product of the least and greatest roots, namely, 3, makes 180. The least root is thus found. The purport of the rest of the question is shown further on.[2]

73. Rule:[3] § 44. If a square be the multiplier, the additive [or subtractive] divided by any [assumed] number, and having it added and subtracted, and being then [in both instances] halved; the first is a "last" root; and the last, divided by the square-root of the multiplier, is a "first."[4]

74. Question 30. The square of a residue of revolutions, or the like, multiplied by four, and having sixty-five added, or having sixty subtracted, is a square. Solving [this problem] within a year [the proficient is] a mathematician.

[1] Something is here wanting in the MS. $\left(\frac{12\times 10-1}{2}\right)\times(3^2+3-1)=59\times 11=649=(180^2\times 13)+1$.

[2] See below.

[3] To find roots, when the coefficient is an exact square. COM.

[4] When a square number is the multiplier, the additive must be divided by any number arbitrarily put. The quotient must then have the same assumed number added in one place and subtracted in another. Having thus formed two terms, halve them both. The first of these moieties is "greatest" root. The second, divided by the square-root of the multiplier, is "least" root. COM.

In one case, however, the first of the moieties, divided by the square-root of the coefficient, is "least" root, and the second is "greatest" root; as is remarked under the following example.

Statement: Multiplier 4. Additive 65. Here the additive divided by an arbitrarily assumed number, 5, is 13. This, increased and lessened by the assumed number, becomes 18 and 8. The half of the first of these is "greatest" root, 9. The moiety of the second, divided by the root of the multiplier, 2, gives the "least" root, 2.

Statement of the second example: Multiplier 4. Additive 60. "Additive" in the rule (§ 44) is indefinite and intends subtractive also. Here let the assumed number be 2. By this, the subtractive, namely, 60, being divided, makes 30. This, with the assumed number added and subtracted, gives 32 and 28: the moieties of which are 16 and 14. The text expresses "the first is a last root," (§ 44): but that is a part only of the rule. The second then is "greatest" root, 14. The first, divided by the square-root of the multiplier, is the "least" root, 8.

75. Rule: § 45. If the multiplier be [exactly] divided by a square, the first root is [to be] divided by the square-root of the divisor.[1]

76. Question 31. The square of a residue of *deficient* days, being multiplied by twelve, and having a hundred added, or having three subtracted, is a square. Solving [this problem] within a year [the proficient is] a mathematician.

Here multiplier 12, divided by the square 4, yields 3. Hence, least and greatest roots answering to additive a hundred are deduced, 10 and 20. The least root, 10, being divided by the square-root of that square, gives the "least" root for the multiplier twelve, viz. 5. Thence, as before, the elapsed [portion of] *yuga* is 793. The "greatest" root is the same, 20.

In the second example, the least and greatest roots, as found for a multiplier divided by the square number, are 2 and 3; for subtractive three. The first, divided by the square-root of that square, is the "least" root, 1; the "greatest" is the same 3; for the multiplier twelve. Here the "least" root is the residue of *deficient* days.

[1] If the multiplier can be abridged by a square, then reducing to its least term, let roots be found as before. But the first root so found being divided by the square-root of the abridging divisor, is "least" root. The "greatest" root remains the same.

But, if the coefficient be multiplied by a square quantity, it of course follows, that the first root, multiplied by the square-root of that square, is the "least" root. COM.

Rule: [45 completed.] If the additive be exactly divisible by a square, the roots must be multiplied by the square-root of the divisor.[1]

77. Question 32. The square of residue of revolutions, or the like, multiplied by three, and having nine hundred added, or eight hundred subtracted, is a square. Solving [this problem] within a year [the proficient is] a mathematician.

Statement: Multiplier 3. Additive 900. Here the additive is divided by the square number 900; and the quotient is 1: whence the "least" and "greatest" roots are deduced, 1 and 2. These, multiplied by the square-root of the abridging divisor, namely, 30, become the "least" and "greatest" roots for the additive nine hundred, 30 and 60.

Statement for the second example: Multiplier 3. Subtractive 800. Here the subtractive, being divided by four hundred, becomes 2. Whence "least" and "greatest" roots are deduced, 1 and 1. These, multiplied by twenty, are 20 and 20, "least" and "greatest" roots serving for subtractive eight hundred.

In any instance, where the additive is exactly divisible by a square, the least and greatest roots, which are thence deduced, being multiplied by the square-root of the abridging divisor, become roots adapted to that additive or subtractive. And further, by composition of the roots so found for the given additive, with roots serving for additive unity, other roots are derived for the same additive. For instance, L 30 G 60 A 900 By the preceding rules
L 1 G 2 A 1
(§ 39—40 and 41) other "least" and "greatest" roots are here found, 120 and 210. So in all similar cases.

78. Rule: To find a quantity such, that, being severally multiplied by two multipliers, and having unity added in each instance, both sums may afford square roots: § 46. The sum of the multiplier. being multiplied by

[1] If the additive can be abridged by any square, devising least and greatest roots as before, for the abridged additive, both being then multiplied by the square-root of the abridging divisor of the additive, become adapted to their proper additive.

Of course, if the additive be raised by multiplication by any square multiplier, the least and greatest roots, which are thence deduced, must be divided by the square-root of the additive's multiplier, and thus become least and greatest roots adapted to their proper additive. COM.

eight, and divided by the square of their difference, is the quantity [sought]. The two multipliers, tripled and added to the opposite, and divided by the difference, are the roots.[1]

79. Question 33. The residue of seconds of the moon, severally multiplied by seventeen, and by thirteen, and having one added, [becomes in both instances] a square. Solving [this problem] within a year [the proficient is] a mathematician.

Here the multipliers are 17 and 13. Their sum 30: multiplied by eight, 240. Difference of the multipliers 4: its square 16. Quotient of the division, 15: it is the number [sought]; and it is the residue of the seconds of the moon.

To find the roots: multipliers 17, 13; multiplied by three, 51, 39: added to the reciprocals, 56, 64. Divided by the difference of the multipliers 4, the roots come out 14, 16.

80. Rule: §47. A square, with another square added and subtracted, being multiplied by the quotient of the sum of that sum and difference divided by the square of half their difference, produces numbers, of which both the sum and difference are squares; as also the product with one added to it.[2]

81. Question 34. The residue of minutes of the sun on Wednesday, having the residue of seconds on Thursday added and subtracted, yields in both instances an exact square; and so does the product with one added. A person solving [this problem] within a year is a mathematician.

Here let an assumed square be 16; with another square, as 4, added and

[1] The proposed multipliers are to be added together: and the sum, being multiplied by eight, and divided by the square of the difference of these multipliers, is the quantity [sought]. How are the roots found? The author proceeds to reply: multiplying the multipliers severally by three, add to the two products the opposite multiplier respectively. Then dividing by the difference of the multipliers, the quotients are the roots. COM.

[2] Some square of an arbitrary number is to be set down; and the square of another arbitrary number is to be added in one place and subtracted in another. The sum of these two quantities is divided by the square of half their difference: the quotient is their multiplier. Multiplied by it, they are the numbers sought: of which if the sum be taken, it is a square; if the difference, it also is square; if the product with unity added, this again is square. COM.

subtracted, 20 and 12 : sum of these 32. Divided by the square of half the difference of these quantities, namely 16, the quotient is their multiplier, 2. Multiplied by it, the two quantities come out 40, 24. The first is residue of minutes of the sun, 40. Hence, as before, the number of [elapsed] days is deduced, 3385. The second is residue of seconds of the sun, 24: whence the number of days, 27. Adding five times the divisor, 5480, the number of [elapsed] days on Thursday comes out 5509. So, by virtue of suppositions, manifold answers may be obtained.

82. Rule: To find a quantity, such, that having two given numbers added, or else subtracted, the results may be exact squares: § 48. The difference of the numbers, by addition or subtraction of which the quantity becomes a square, is divided by an arbitrary number and has it added or subtracted: the square of half the result, having the greater number added or subtracted, is the quantity which answers in the case of addition or subtraction.[1]

83. Question 35. Making the residue of minutes of the sun on a Wednesday, with the addition of twelve and of sixty-three, and with the subtraction of sixty and of eight, an exact square; [the proficient solving the problem] within a year is a mathematician.

Two questions are here proposed. The numbers, which are to be added to the quantity, are separated, 12, 63. Their difference, 51; divided by an arbitrary number, as 3, gives 17; with the same added, since addition is in question, the sum is 20: its moiety, 10; the square of which is 100. The greater of the two additive quantities is 63. Subtracting this, the result is

[1] Of the two quantities, the addition of which makes the quantity in question an exact square, or the subtraction of which does so, the difference is in every case to be taken. This step is common to both methods. Dividing the difference then by an arbitrary number; the quotient must have added to it the same arbitrary number; if addition were given by the question: but, if subtraction were so, the same quotient must have the arbitrary number subtracted. Then the quantity resulting in either case is to be halved; and the half, to be squared. [From which subtracting the greater number, the remainder is the quantity which answers *] if the condition were addition: but, if it were subtraction, the square of the moiety appertaining to the case being added to the greater number, the sum is the quantity sought. COM.

* The original is deficient: but may be thus supplied from comparison of the text, and of the example as wrought.

the quantity sought, 37. With either twelve or sixty-three added, it is an exact square.

In the example of subtraction; the two numbers which are to be subtracted to make a square, are 60, 8. Their difference, divided by an arbitrary number, namely two, yields 26: less the arbitrary number, leaves 24: its moiety 12; the square of which is 144. Here the greater of the two subtractive quantities is 60. This added to the square is 204. It is the quantity, which lessened by sixty, affords a square root; or by eight. It is the residue of minutes of the sun. Hence, as before, the number of [elapsed] days on Wednesday, is to be deduced.

In like manner, by virtue of suppositions, manifold answers may be obtained.

84. Rule: §49. The sum of the numbers, the addition and subtraction of which makes the quantity a square, being divided by an arbitrarily assumed number, has that assumed number taken from the quotient: the square of half the remainder, with the subtractive number added to it, is the quantity [sought].[1]

85. Question 36. Making the residue of seconds of the sun on Wednesday, with ninety-three added, or with sixty-seven subtracted, an exact square, [a proficient solving this problem] within a year [is] a mathematician.

Here the subtractive number is 67; the additive number, 93: their sum, 160: divided by an assumed number 4, makes 40: less the assumed number, leaves 36; the half of which is 18: its square, 324: added to the subtractive quantity 67, the quantity is found 391. It is the residue of seconds of the sun.

86. Question 37. Making the residue of seconds of the sun on Thursday lessened and then multiplied by five, an exact square, or by ten, [the proficient solving this problem] within a year [is] a mathematician.

[1] If a pair of quantities equal or unequal be given such, that a quantity, which lessened by the first, is an exact square, is also a square when increased by the second; then the two proposed quantities are to be added together; and their sum is to be divided by some arbitrary number. From the quotient subtracting the same arbitrary number, the half of the remainder is taken: and the square of that moiety, added to the number the subtraction of which renders the quantity in question a square, is in every case the quantity sought. COM.

This comprises two examples. In the first, let residue of seconds be *ya* 1. This less five is *ya* 1 *ru* $\dot{5}$; and then multiplied by five, *ya* 5 *ru* $\dot{25}$. It is a square. Put it equal to the square of the arbitrary number ten; and from this equation a value of *yávat-távat* is obtained, 25. Or by equating it with the square of five, the value of *yávat-távat* comes out 10. This is a residue of seconds of the sun.

In the second example, the value of residue of seconds is put *yávat-távat*, *ya* 1. This less ten, and multiplied by ten, becomes *ya* 10 *ru* $\dot{100}$. It is a square. Put it equal to the square of ten arbitrarily assumed. By this equation the value of *yávat-távat* is brought out 20. It is residue of seconds of the sun.

By virtue of suppositions the answers are manifold.

87. Question 38. Making the residue of revolutions or the like of a given object, lessened by ninety-two, and multiplied by eighty-three, and with unity added, an exact square, [the proficient solving this problem] within a year [is] a mathematician.

Since the residues of revolutions or the like are many, put *yávat-távat* for the value of such residue; *ya* 1. This less 92 is *ya* 1 *ru* $\dot{92}$. Multiplied by eighty-three, it is *ya* 83 *ru* $\dot{7636}$. With one added, it becomes *ya* 83 *ru* 7635. It is a square. Put it equal to the square of unity as an assumed number. By this equation the value of *yávat távat* comes out 92. Thus the residue of revolutions of some planet is found.

This also, by means of suppositions, admits manifold solutions.

SECTION VIII.

PROBLEMS.

88. QUESTION 39. From residue of seconds of the moon, finding residue of minutes of the sun, or residue of degrees, or the mean place of the proposed planet, [a proficient solving this problem] within a year [is] a mathematician.

89. Question 40. From residue of seconds of Jupiter, find Mars; or from residue of minutes of the moon, the sun; or from residue of revolutions of the moon; [a proficient solving this problem] within a year [is] a mathematician.

90. Rule: § 50. The number of [elapsed] days deduced from the given residue for the given planet, is added to a multiple of the divisor in days by the elapsed [periods] in least terms. From that the residue for another planet, or its place, [may be found].[1]

Example: When the residue of degrees of the moon is equal to seven thousand five hundred and twenty, tell me the mean [place of the] sun, if thou be conversant with the pulverizer.

Residue of degrees 7520. Abridged by the common divisor of revolutions of the moon and terrestrial days, namely 80, it is 94. For the residue of

[1] From the residue of revolutions or the like, as given, relative to the proposed planet, the number of [elapsed] days is to be found, as before. It must be converted into the number of elapsed days of the *yuga*. How? The rule proceeds to answer. The number of days, which comes out by the pulverizer, must necessarily be short by the divisor in days: for it is the elapsed portion of the present *yuga* of the planet. Therefore, whatever number of its *yugas* may be past, by that number as the elapsed periods in least numbers, multiply the divisor in days, and add the product; the sum is the complete number of elapsed days from the beginning of the *yugas*. From that elapsed time, by the [periodical] revolutions of the other planet and terrestrial days, find the residue of revolutions and so forth. Or the mean [place of the] given planet may be deduced. COM.

degrees of the moon, finding the constant pulverizer, the multiplier is 101. Whence the number of [elapsed] days 41. Here the number for the elapsed periods of the moon in least terms is 7. The days in the divisor 137, multiplied by that, are 959: which, added to the number of elapsed days by the pulverizer, 41, makes the elapsed [portion of the] *yuga* 1000. Hence the residue of revolutions of the sun, 8080; from which the sun's place is to be deduced, as before.

It is to be in like manner understood in all [similar] examples.

91. Rule to find the time or number of days, after which the same residues of revolutions or the like of two planets, or of more, which occur on a given day, will recur: § 51.

The divisors in least terms are inverted.[1] The result being added to the number of days, the residues [occur] again in that [time]. In the same manner for three or more [planets]. Proceed as before for the given days.

Example: At the foregoing number of days, 41, the residues of revolutions of the sun and moon come out 123 and 68. When do the same residues of revolutions occur again? To find this, the divisors of sun and moon in least terms are taken 1096, and 137. Their greatest common measure is 137. Abridged by this, the quotients are 8, and 1: by which the divisors in least terms being multiplied become 1096 and 1096. With this, being the amount of the equal denominator, added to the number of days 41, the sum is 1137; at which the residues are again the same. With this addition, the number of elapsed days may be many ways [augmented].

In the same manner for three or more. The divisor for two planets is 1096; that for a third, Mars, is 685. Their greatest common divisor 137;

[1] Under the residues of revolutions relative to the two planets, as deduced from the number of [elapsed] days on the given day, their respective divisors in least terms stand. Some common measure of them is to be assumed by which they may be reduced to the lowest terms. Being divided by that common measure, the quotients serve to multiply reciprocally those same divisors of the planets in least terms. This being done the denominators are equal. Then add that result to the number of [elapsed] days; and the same residues of revolutions or the like are deduced from the sum. It is so for two planets; and the method is precisely the same for three or more. Thus the equal denominator arising from the divisors of three given planets is considered as one divisor; and the third divisor of a planet, as the second. From these, as before, an equal denominator is to be deduced. The same must be understood in regard to four or more. COM.

with which finding the quotients [8 and 5] and proceeding as before, the result [5480] added to the number of days, 41 makes 5521, when the residues are similar.

It must in like manner be understood for four or more planets.

Other examples are now propounded.

92. Question 41. From residue of fewer days, making out the number of [elapsed] days, and the mean or the true [places of] sun and moon, and the lunar day and the planet, [a proficient solving these problems] within a year is a mathematician.

Here are five questions. The solution of them is delivered in four couplets.

93—96. Rule: § 52—55. Residue of *fewer* days for a single day, being multiplied by some quantity and lessened by unity or by revolutions of sun or of moon, is exhausted being divided by terrestrial days; or lessened by unity [and divided] by lunar. The rule is as follows:[1]

[1] These problems are relative to the planetary revolutions as taught in the author's own astronomical course. He has, therefore, here specified the constant pulverizers adapted to them. To propound which this first rule is delivered. Its meaning then is this:—The residue of *fewer* days (terrestrial than lunar) for a single day, abridged by its own divisor, being multiplied by some number, and having then unity subtracted, is exactly divisible by terrestrial days reduced to least terms by the divisor of *fewer* days. This rule, for finding a multiplier and divisor to deduce the number of [elapsed] days from the residue of fewer days, is as here follows in the second and succeeding couplets (§ 53—55); including the finding of revolutions. Residue of *fewer* days, as proposed in the problem, and reduced to least terms, being multiplied by 108455 and divided by 3506481, the remainder of the division is the number of [elapsed] days. From that number, the mean places of sun and moon are to be found: and then to be converted into the apparent planets: whence the lunar days; and from these the planet sought is to be deduced. Thus the solution of all the problems is effected. Nevertheless [a direct method of] finding the residue of the sun's revolutions is taught by the third couplet (§ 54). The proposed residue of *fewer* days in least terms, being multiplied by 3249624 and divided by 3506481, the remainder of the division is residue of sun's revolutions: whence the sun, as before. In like manner the finding of the number of [elapsed] days being effected, the deducing of lunar days was also accomplished: the author shows how to find the same by another [and direct] method, in the fourth couplet (§ 56). The residue of *fewer* days for the given time being in least terms, and multiplied by 110179 and divided by 3562220, the remainder of the division is lunar days: under which is the fraction of such days. For the hours, minutes, &c. under days, which are quotient of residue of *fewer* days divided by terrestrial days, according

The [proposed] residue of fewer days [in least terms] being multiplied by a hundred and eight thousand, four hundred and fifty-five, and divided by three millions, five hundred and six thousand, four hundred and eighty-one, the remainder is the number of elapsed days.

The proposed residue of fewer days [in least terms] being multiplied by three millions, two hundred and forty-nine thousand, six hundred and twenty-four, and divided by three millions, five hundred and six thousand, four hundred and eighty-one, the remainder which results is the residue of solar revolutions.

The [proposed] residue of fewer days [in least terms] being multiplied by one hundred and ten thousand, one hundred and seventy-nine, and divided by three millions, five hundred and sixty-two thousand, two hundred and twenty, the remainder is the lunar days.

97. Question 42.[1] Knowing the sum of the residues of both degrees and minutes, and their difference, say what are the residues?

Rule : § 56.[2] The sum set down twice and having the difference added and subtracted and being in both instances halved, the moieties are the residues.

98. Rule : § 57. If the difference of their squares and their [simple] difference be given : the difference of the squares being divided by the [simple] difference and having the same added and subtracted and being divided by

to the rule delivered in the chapter on the solution of problems respecting mean motions,* are the fractional part of lanar days.

As in finding the number of [elapsed] days from residue of *fewer* days, the residue of fewer days for a single day was put for dividend, unity for the subtractive quantity, and terrestrial days for divisor; so, in finding residue of solar revolutions, the revolutions of the sun are the subtractive quantity; and, in finding residue of lunar revolutions, the revolutions of the moon are the subtractive quantity; and, in finding lunar days, unity is so : but here days of the moon are the divisor. Such is the difference. The author has specified the constant pulverizers. In like manner, when a proposed planet is sought in any example from residue of *fewer* days, the subtractive quantity is to be put equal to the revolution of such planet, and the constant pulverizer is to be thence brought out. COM.

[1] Example of the rule of concurrence.—COM. See § 25.

[2] Second half of the couplet. Its first half proposes a problem.

*Ch. 13 § 22.

two, the quotients are the residues; whence the number of elapsed days [may be found].

99. Rule: § 58. From twice the sum of the squares subtract the square of the [simple] sum of the residues. The sum of residues, having the square-root of the remainder added and subtracted, and being divided by two, yields the residues severally.

For instance, the sum of the squares of the residues is 9365; and the sum of those residues is 117. The sum of the squares doubled is 18730; from which subtracting 13689 the square of 117, the remainder is 5041; of which the square-root is 71. This being added to and subtracted from the sum of residues and then halved, the residues are found 94 and 23.

100. Rule: § 59. From the square of the difference of residues added to four times the product of residues, extract the square-root, which added to, and subtracted from, the difference of residues, and halved, yields the residues severally.

Example: Product of residues, 2162, multiplied by (4) square of two, 8648, added to square (5041) of the difference of residues 71, makes 13689: of which the root is 117. This, having the difference of residues added and subtracted, becomes 188 and 46; which halved are 94 and 23: and the residues are found.

101. These questions are stated merely for gratification. The proficient may devise a thousand others; or may resolve, by the rules taught, problems proposed by others.

102. As the sun obscures the stars, so does the proficient eclipse the glory of other astronomers in an assembly of people, by the recital of algebraic problems, and still more by their solution.

103. These questions recited under each rule, with the rules and their examples, amount to an hundred and three couplets: and this Chapter on the Pulverizer is the eighteenth.

Rules, (*sútra,*) sixty-one and a half couplets : problems, (*pranśa,*) forty-one and a half.

Interpretation of the Algebraic Pulverizer in the *Brahma-sidd'hánta* composed by BRAHMEGUPTA.

FINIS.

London: Printed by C. Roworth, Bell-yard, Temple-bar.